Exploratory Data Analytics for Healthcare

Innovations in Big Data and Machine Learning

Series Editors:
Rashmi Agrawal and Neha Gupta

This series will include reference books and handbooks that will provide the conceptual and advanced reference materials that cover building and promoting the field of Big Data and Machine Learning which will include theoretical foundations, algorithms and models, evaluation and experiments, applications and systems, case studies, and applied analytics in specific domains or on specific issues.

Artificial Intelligence and Internet of Things
Applications in Smart Healthcare
Edited by Lalit Mohan Goyal, Tanzila Saba, Amjad Rehman, and Souad Larabi

Reinventing Manufacturing and Business Processes through Artificial Intelligence
Edited by Geeta Rana, Alex Khang, Ravindra Sharma, Alok Kumar Goel, and Ashok Kumar Dubey

Convergence of Blockchain, AI, and IoT
Concepts and Challenges
Edited by R. Indrakumari, R.Lakshmana Kumar, Balamurugan Balusamy, and Vijanth Sagayan Asirvadam

Exploratory Data Analytics for Healthcare
Edited by R. Lakshmana Kumar, R. Indrakumari, B. Balamurugan, and Achyut Shankar

Information Security Handbook
Edited by Abhishek Kumar, Anavatti G. Sreenatha, Ashutosh Kumar Dubey, and Pramod Singh Rathore

For more information on this series, please visit: https://www.routledge.com/Innovations-in-Big-Data-and-Machine-Learning/book-series/CRCIBDML

Exploratory Data Analytics for Healthcare

Edited by
R. Lakshmana Kumar, R. Indrakumari,
B. Balamurugan, and Achyut Shankar

CRC Press
Taylor & Francis Group
Boca Raton London New York

CRC Press is an imprint of the
Taylor & Francis Group, an **informa** business

First edition published 2022
by CRC Press
6000 Broken Sound Parkway NW, Suite 300, Boca Raton, FL 33487-2742

and by CRC Press
4 Park Square, Milton Park, Abingdon, Oxon, OX14 4RN

CRC Press is an imprint of Taylor & Francis Group, LLC

Library of Congress Cataloging-in-Publication Data
Names: Kumar, R. Lakshmana, editor. | Indrakumari, R., editor. |
Balamurugan, B., editor. | Shankar, Achyut, editor.
Title: Exploratory data analytics for healthcare / edited by R. Lakshmana
Kumar, R. Indrakumari, B. Balamurugan, Achyut Shankar.
Description: First edition. | Boca Raton, FL : CRC Press, 2022. |
Series: Innovations in big data and machine learning | Includes index.
Identifiers: LCCN 2021042071 (print) | LCCN 2021042072 (ebook) |
ISBN 9780367506919 (hbk) | ISBN 9780367506926 (pbk) | ISBN 9781003050827 (ebk)
Subjects: LCSH: Medical informatics—Research. |
Medicine—Data processing—Case studies.
Classification: LCC R858 .E9895 2022 (print) | LCC R858 (ebook) |
DDC 610.285—dc23
LC record available at https://lccn.loc.gov/2021042071
LC ebook record available at https://lccn.loc.gov/2021042072

ISBN: 978-0-367-50691-9 (hbk)
ISBN: 978-0-367-50692-6 (pbk)
ISBN: 978-1-003-05082-7 (ebk)

DOI: 10.1201/9781003050827

Typeset in Times
by codeMantra

Contents

Preface

> It is not the strongest of the species that survives, nor the most intelligent, but the one most responsive to change.
>
> – *Charles Darwin*

Exploratory data analysis (EDA) introduces the significant process of making fundamental investigations on data to identify patterns, detect anomalies, examine the hypothesis, and validate assumptions with the guidance of summary statistics and graphical illustrations. These methods are typically applied before formal modeling commences and can help inform more complicated statistical models. Experimental techniques are also essential for eliminating or sharpening potential hypotheses about the world that the data can address. Exploratory data analysis is a significant step after feature engineering and acquiring data, and it should be done before any modeling. The purpose of EDA is to use summary statistics and visualizations to understand data better, and it also helps to find clues about the tendencies of the data and its quality, and to formulate assumptions and the hypothesis of our analysis.

This book explores the concepts and techniques of EDA in healthcare and explicitly discriminates the combination in understanding various techniques of EDA in healthcare. This study will stimulate the wisdom of the IT professionals, researchers, and academicians towards the next revolutionary technologies. The principal aspiration is to assist both practitioners and researchers in sharing their views and recent research in healthcare through EDA among academicians and industry people. The contributors are expected to present their technical evaluation and comparison with existing technologies. The theoretical explanation and experimental case studies related to real-time scenarios are expected. The wide variety of topics it presents offers readers multiple perspectives on a variety of disciplines.

Chapter 1 by Kalpana Hazarika, A. Ambikapathy, R. Shobana, and Amit Agrawal details the scope and challenges of visual analytics, principles, challenges, and applications pertaining to data analytics in healthcare applications. They have touched on the viability of humongous and heterogeneous data and evidence-based decisions to improve healthcare services. Tools and techniques used are also detailed.

Chapter 2 by Areeba Kazim, Achyut Shankar, and Muskan Jindal discusses the statistical methods and applications for the healthcare industry.

Chapter 3 by G. Shyamala and A. Ilavendhan discusses the groundwork of machine learning-based data classification techniques used in healthcare. They have detailed various supervised and unsupervised machine learning algorithms used in healthcare for various applications.

Chapter 4 by M. Arvindhan, D. Rajeshkumar, and Anupam Lakhan Pal has further taken this book's insights by exploring deep learning techniques for computer-aided medical diagnosis. They have pondered the changes that have led to the penetration of deep learning techniques in healthcare, major deep-learning models used, and their applications.

Chapter 5 by N. Pooranam, M. Diwakaran, and T. Vignesh stresses that automatic classification plays a major role in the medical field. They have discussed various case studies of healthcare (COVID-19, SARS, and tumor) about machine learning classifiers.

Chapter 6 by S. Suganthi and T. Poongodi discusses interactive visualization for understanding and analyzing medical data. They give a brief introduction to interactive visualization and its requirement in healthcare industry.

Chapter 7 by R. Indrakumari, Priyanka Shukla, and Akanksha Sehgal discusses the heart disease prediction using exploratory data analysis. They have used Tableau, a business intelligence tool, for the prediction of infectious diseases and to develop a recommendation system. It also helps in identifying and handling the outlier data.

Chapter 8 by Geraldine Bessie Amali, Gaurav Ramtri, Anukriti Kacker, and Siddharth Menon covers the deep learning framework using AlexNet for early detection of pancreatic cancer.

Chapter 9 by Gagandeep Kaur, Sachin Minocha, and Satish Saini discusses the applications of the Map-Reduce Programming Framework to clinical big data analysis and its current landscape and future trends.

Chapter 10 by Kayal Padmanandam discusses applied intelligence and impact of machine learning in healthcare data analytics. The impact of healthcare systems is discussed by utilizing the efficient exchange of health records, speedy and accurate medical diagnostics, clinical decision-making, different medical research, drug development research, and many more in the future at a reduced cost.

Chapter 11 by Kunal Pant, Nikhil Sati, Divyansh Agarwal, and Deepa Dangwal undermines the machine learning algorithms for healthcare data analytics. They focus on the basic concepts of machine learning algorithms.

Chapter 12 by R. Shobana, A. Ambikapathy, Kalpana Hazarika, and Amit Kumar Gupta introduces the field of machine learning applications in areas of healthcare. They also show that machine learning has given healthcare a new tool to work with and identified a new path to focus on early predictions of many deadly diseases.

Chapter 13 by Roshan Lal and Sandhya Tarar introduces clever and powerful coronary illness forecast techniques utilizing information mining. They have explained that the information mining has the potential to create an information-rich condition that can help to essentially improve the nature of clinical choices.

Lakshmana Kumar Ramasamy
Indrakumari Ranganathan
Balamurugan Balusamy

Editors

Dr. Lakshmana Kumar Ramasamy is the head of the Center of Excellence for Artificial Intelligence and Machine Learning at Hindusthan College of Engineering and Technology, Tamil Nadu, India. He is currently pursuing his post-doctoral fellowship at Thu Dau Mot University, Vietnam. He is also an offshore director for Research & Development (AI) in a Canadian-based company (ASIQC) in the Vancouver region of British Columbia, Canada. He represents the Technical Group Committee – National Cyber Defence Research Centre (NCDRC), Government of India. He is the founding member of IEEE SIG of Big Data for Cyber Security and Privacy, IEEE. He serves as a core member in the Editorial Advisory Board of Artificial Intelligence group in Cambridge Scholars Publishing, UK. He is a member of IEEE. He was invited as a keynote speaker for AVIS 2020 (Asia Artificial Intelligence Virtual Summit 2020) which is Asia's first biggest Virtual Summit on Artificial Intelligence held in Malaysia in June 2020. He is a global chapter lead for MLCS (Machine Learning for Cyber Security). He researches on and has expertise in artificial intelligence and blockchain technologies.

Ms. Indrakumari Ranganathan is an assistant professor in the School of Computing Science and Engineering, Galgotias University, NCR Delhi, India. She has received an M.Tech. in Computer and Information Technology from Manonmaniam Sundaranar University, Tirunelveli, Tamil Nadu, India. Her main thrust areas are big data, Internet of Things (IoT), data mining, data warehousing, and its visualization tools like Tableau and Qlikview. She has several top-notch conferences on her resume and has published over 45 quality journal papers, conference proceedings, and book chapters combined.

Dr. Balamurugan Balusamy has served up to the position of associate professor in his stint of 14 years of experience with VIT University, Vellore. He has completed his Bachelor's, Master's, and Ph.D. degrees from top premier institutions. His passion is teaching, and he adapts different design thinking principles while delivering his lectures. He has written around 30 books on various technologies and visited 15 plus countries for his technical discourses. He has several top-notch conferences on his resume and has published over 150 quality journal papers, conference proceedings, and book chapters combined. He serves on the advisory committee for several start-ups and forums, and does consultancy work for the industry of Industrial IoT. He has given over 175 talks at various events and symposiums. He is currently working as a professor at Galgotias University and teaches and does research on blockchain and IoT.

Dr. Achyut Shankar is currently working as an assistant professor at Amity University, Noida, India since March 2018. He obtained his Bachelor's degree in Computer Science and Engineering from Dr. MGR University, Chennai, India. Then he obtained his Master's degree in Computer Science and Engineering from SRM University, Chennai, and obtained a Ph.D. in Computer Science and Engineering majoring in wireless sensor network from VIT University, Vellore, India. He has published more than 20 research papers in reputed international conferences and journals. He is a member of ACM and has received a research award for excellence in research for the years 2016 and 2017. He had organized many special sessions with Scopus Indexed International Conferences worldwide, the proceedings of which were published by Springer, IEEE, Elsevier, etc. He is serving as a reviewer of *IEEE Transactions on Intelligent Transportation Systems* and *IEEE Sensors Journal*. His areas of interest include wireless sensor networks, cloud computing, blockchain, and IoT.

Contributors

Divyansh Agrawal
School of Computing Science and Engineering
Galgotias University
Greater Noida, Uttar Pradesh, India

Amit Agrawal
ABES Engineering College
Ghaziabad, Uttar Pradesh, India

Geraldine Bessie Amali
School of Computer Science and Engineering
Vellore Institute of Technology
Vellore, Tamil Nadu, India

A. Ambikapathy
Galgotias College of Engineering and Technology
Greater Noida, Uttar Pradesh, India

M. Arvindhan
School of Computing Science and Engineering
Galgotias University
Greater Noida, Uttar Pradesh, India

Deepa Dangwal
Uttranchal University
Uttarakhand, India

Uttranchal M. Diwakaran
Department of Information Technology
Sri Krishna College of Engineering and Technology
Coimbatore, Tamil Nadu, India

Amit Kumar Gupta
ABES College of Engineering
Ghaziabad, Uttar Pradesh, India

Kalpana Hazarika
Galgotias College of Engineering and Technology
Greater Noida, Uttar Pradesh, India

Ilavendhan A.
Department of CSE
School of Computing, Vel Tech Rangarajan Dr. Sagunthala R&D Institute of Science & Technology
Chennai, Tamil Nadu, India

R. Indrakumari
School of Computing Science & Engineering
Galgotias University
Greater Noida, Uttar Pradesh, India

Muskan Jindal
Department of CSE
Amity School of Engineering & Technology, Amity University
Noida, Uttar Pradesh, India

Anukriti Kacker
School of BioSciences and Technology
Vellore Institute of Technology
Vellore, Tamil Nadu, India

Gagandeep Kaur
Department of Electronic Communication and Engineering
RIMT University
Mandi Gobindgarh, Punjab, India

Areeba Kazim
Department of CSE
Amity School of Engineering & Technology, Amity University
Noida, Uttar Pradesh, India

Anupam Lakhan Pal
School of Computing Science and Engineering
Galgotias University
Greater Noida, Uttar Pradesh, India

Roshan Lal
Department of Information and Communication Technology
Gautam Buddha University
Greater Noida, Uttar Pradesh, India

Siddharth Menon
School of Electronics Engineering
Vellore Institute of Technology
Vellore, Tamil Nadu, India

Sachin Minocha
School of Computing Science and Engineering
Galgotias University
Greater Noida, Uttar Pradesh, India

Kayal Padmanandam
Department of Information Technology
BVRIT Hyderabad College of Engineering for Women
Hyderabad, Telangana, India

Kunal Pant
School of Computing Science and Engineering
Galgotias University
Greater Noida, Uttar Pradesh, India

T. Poongodi
School of Computing Science and Engineering
Galgotias University
Greater Noida, Uttar Pradesh, India

N. Pooranam
Department of Computer Science and Engineering
Sri Krishna College of Engineering and Technology
Coimbatore, Tamil Nadu, India

D. Rajeshkumar
School of Computing Science and Engineering
Galgotias University
Greater Noida, Uttar Pradesh, India

Gaurav Ramtri
School of Information Technology and Engineering
Vellore Institute of Technology
Vellore, Tamil Nadu, India

Satish Saini
Department of Electronic Communication and Engineering
RIMT University
Mandi Gobindgarh, Punjab, India

Nikhil Sati
School of Computing Science and Engineering
Galgotias University
Greater Noida, Uttar Pradesh, India

Akanksha Sehgal
School of Computing Science & Engineering
Galgotias University
Greater Noida, Uttar Pradesh, India

Achyut Shankar
Department of CSE
Amity School of Engineering & Technology, Amity University
Noida, Uttar Pradesh, India

Shobana R.
Galgotias College of Engineering and Technology
Greater Noida, Uttar Pradesh, India

Priyanka Shukla
School of Computing Science & Engineering
Galgotias University
Greater Noida, Uttar Pradesh, India

G. Shyamala
Department of Biomedical Engineering
Mahendra College of Engineering
Salem, Tamil Nadu, India

S. Suganthi
Department of Computer Science
Cauvery College for Women
Tiruchirappalli, Tamil Nadu, India

Sandhya Tarar
Department of Information and Communication Technology
Gautam Buddha University
Greater Noida, Uttar Pradesh, India

T. Vignesh
Department of Mechatronics Engineering
Sri Krishna College of Engineering and Technology
Coimbatore, Tamil Nadu, India

1 Visual Analytics
Scopes and Challenges

Kalpana Hazarika, A. Ambikapathy, and Shobana R.
Galgotias College of Engineering & Technology

Amit Agrawal
ABES Engineering College

CONTENTS

1.1 INTRODUCTION: CONCEPT OF VISUAL ANALYTICS

The volume of information exchange is growing exponentially in almost all sectors, and even in healthcare. This is a sector that deals with human beings, where the cost of a life is not replenished. It is a domain that certainly cannot be excluded and needs more accuracy in its data interpretation for accurate prediction of future or immediate course of actions. Hence, it has to work efficiently for its exact prediction and conclusion to generate the next step of actions during normal or critical situations.

Visual analytics (VA) is the solution to these complex data and gives a visual interactive information clustering-relevant statistics as per end-users' needs. Every sector of healthcare needs a different set of information for its analysis and

DOI: 10.1201/9781003050827-1

interpretation corresponding to its objectives. This demands nonhomogeneous data combination for analytical or visual analysis to represent the statistical model of the original data. VA merges computer-aided evaluation methods with communicative visualizations for better understanding and decision-making based on logical inferencing with the high volume of complex datasets [1]. In simple words, this is a methodological study supported by communicative visual interfaces. It integrates cognitive intellect with synthesizing efficacy of the enhanced methods to gain hidden understanding into multifaceted problems for decision-making. Hence, the key objectives of VA are to design operational tools and procedures to combine meaningful data and develop comprehension from huge, vigorous, confusing, and sometimes contradictory information. Also, its due responsibility is to deliver judicious, strong, and logical assessment from the discovered insights and communicate assessment effectively for action [1].

1.1.1 Process of Visual Analytics

As discussed, VA combines intuitive and graphical evaluation techniques with a close humanoid interaction to discover facts from unstructured information. Here, the automatic data evaluation performs a systematic process of scrutinizing, cleaning, transmuting, and demonstrating data for further analysis, whereas graphical evaluation techniques will have a communicative panel on the monitor for experts to choose data sources and display patterns for visual display, simple data interpretation, and accessibility to competent authorities [1].

But identifying the best-automated algorithm for healthcare data analysis is another challenge. Another important requirement for all related examinations is developing a strongly integrated solution for the successful integration of computerized evaluation procedures with suitable imagining and communication techniques. These analyses require a huge data set; hence, yet another prerequisite of visual or automatic analysis lies in integrating heterogeneous data sources. Automated Analysis through data mining algorithms plays an important role in data cleaning, normalization, grouping, and integration. The primary goal of knowledge discovery in healthcare is to pull out meaningful facts from large databases. The knowledge discovery in databases is nothing but a programmed, exploratory analysis and modeling of vast data storehouses.

The VA control panel has various engines involving visual graphics to represent the analytical results. The results are displayed on the screen after computational algorithms work. VA discovers actionable commercial facts for business intelligence support. To do so, it takes advantage of deep learning technologies to detect uncovered tendencies and characteristics of large data.

1.1.2 Need and Benefits of VA in Healthcare

For a decade, VA has played an instrumental role in healthcare data analysis to give a big momentum to the overall performances of the sector. The governance of health systems is now open to adopting new tools and technologies. It has also recognized that for the survival of healthcare institutions, data analytics is an integrated part of the sector.

The factors for which the acceptance pace of analytics is accelerating in healthcare are stated under the following:

- Using Electronic Health Record (EHR), a complicated set of structured and unstructured data that VA can use to quickly demonstrate the results in a sensible way for further course of action. The data may include a patient's medical history maintained over time, including demographic data, development notes, complications, and treatments to provide need-based services.
- If technology needs future improvement in the healthcare system, then the EHR data used by doctors should be empowered by the emerging power of analytics and machine learning. Using modern analytical techniques, superior information can be provided to doctors for patient care. Meaningful inferences and predictions can be pulled out from it, like the probability of a patient's successive risk to vital organ failure or catastrophic failure based on blood pressure reading, significant signs of illness, test results, genetic background, and recent clinical research data [2].
- Cloud computing infrastructure allows dynamic data processing with massive data. The extensive application of remote server accessibility in healthcare endorses policymakers to efficiently optimize operational management cost of treatment and offers individualistic care plans for better outcomes. It holds the potential to produce vast data compared to warehouses a decade ago [3].

1.2 VA TECHNOLOGIES AND TOOLS

These tools and technologies are playing a key role in the sustainable progressive development of small and big enterprises. It helps in exploring and visualizing the current status of any business operations for long-term strategic planning and policymaking. By identifying important patterns and deriving key insights, it helps to take actionable decisions for accomplishing set goals. To perform these analyses, VA plays a significant role in a large volume of data handling from various sources and processing to give a need-based visual presentation [1].

1.2.1 General Features of Visual Analytics Tools

1.2.1.1 Data Visualization

It encompasses scientific visualization – a multidimensional scientific data representation to support clinical experts to identify the pattern from biological data, information visualization – the study of communicative pictorial representations of descriptive data to improve human intellectual, and visual analytics – which combines them both with an emphasis on analytical reasoning through an interactive visual interface [4].

The professional data visualization engineers must have the following technology expertise for developing interactive visual interface tools:

- **Basic mathematics**: understanding of linear equations, geometric procedures, and trigonometric function.
- **Graphics**: knowledge on canvas, vector-based graphics, Web Graphics Library, computer graphics, and network theory for creating or developing rich 2D and 3D interactive graphics.
- **Engineering algorithms**: arithmetical, basic, and general layout algorithms.
- **Data analysis**: scrutinizing, cleaning, transmuting, and demonstrating data for inherent knowledge discovery, drawing conclusions, and reviewing operational strategies.
- **Design aesthetics**: designing ethics, visual and color judgement, collaboration, thoughts, etc.
- **Visual basis**: visual coding (visual programing languages such as R, Scala, and Python Programming Languages), visual analysis, and graphical interaction.
- **Visualization solutions**: timely decisions through an in-depth analysis of raw data that helps in visualization of common business scenarios.

1.2.1.2 Dashboards

These are control panels that show the major implementation of data visualization. It generates a dynamic pictorial graphical presentation of specific data for uninterrupted updates to keep us informed. These panels are incredible in providing a quick view on a need-based analysis if one needs to see it at a glance. By using dynamic panels, the status can be observed as and when required. This feature allows the decision-makers to immediately address the issue with a suitable solution. Apart from inventory control, it is used for run-time KPI monitoring, steering other responsibilities, and ensuring effective utilization of personnel and resources. The major benefits are summarised as quick and easy access, forming of better decisions, meeting growing business demands, improving an inefficient reporting system, and displaying reliable data.

However, healthcare professional dashboards are as follows:

- **Clinical dashboards**: This dashboard integrates correlated data into a single point of display or "dashboard" to optimize worker efficiency, accelerate medical judgements, rationalize workflow, and minimize human error in clinical practice. Here, clinical data are regularly recovered, recorded, examined, and shared to intended users for better care of patients. This also helps in tracking public health eruptions due to any flu epidemic to plan employment needs effectively by displaying the current loadable capacity of patients for emergency rooms, bed occupancy rates, nursing unit tallies, etc. [5].

- **Hospital dashboards**: It assists hospitals to keep a close and vigilant eye on significant functions of all departments for workflow and patient care quality improvements. The main objective of this dashboard is to follow patient satisfaction, optimum doctor staff-to-patient ratio, optimize the cost of care, and take care of required logistics. For this, it maintains a KPI structure that conducts surveys on the following parameters: Patient Satisfaction, Operational Workload, Operation Costs, Patient Volume, and Patient Experience. This includes surveys like patients admission, average waiting time, doctors' treatment plan, confidence in treatment, the average cost per patient and satisfaction, cost incurred by the department for patients' care, patients per doctor, number of doctors, etc. [6].
- **Patient dashboards**: This healthcare dashboard displays key information relating to patients' overall health. With this health summary dashboard, viewers can observe the following information: patients' vital signs, i.e., heart rate, temperature, and blood pressure, patients sleeping duration and the quality of sleep, daily step goals. It also helps to follow the aggregate duration of patient stay and laboratory waiting hours. Further, it may help physicians decide where to focus their treatment and preventive care efforts [7].
- **Physician dashboards**: This assists medical professionals to follow patient satisfaction, patient count, and new patient registration updates as well as track the metrics like the volume of cases treated, probable mortality, and reporting of adverse events, along with quality and process like significant parameters. To provide on-time treatment, appropriate appointment setup, lesser waiting time, increase patient safety, and reduce patient dissatisfaction, it is important to analyse healthcare plans for physician's allocation. By improving the ratio of patients per physician to 100:1, healthcare institutions can deliver better care by streamlining physician allocation through this dashboard.
- **Quality and risk management dashboards**: It supports in keeping a close eye on complex compliance issues, managing patient welfare information, hospitals recommended treatment count, and comparing hospital standards with in-house performance [8].

1.2.1.3 Integration with Multiple Data Sources

The act of merging data from several sources into a unified set is called data integration. During this process, data is cleaned and transformed to enable accurate analysis. Through data integration, dissimilar data sources are reviewed together to provide invaluable business intelligence to support better business decision-making. The common sources of the medical database are Healthcare Data Warehouse, NoSQL, Cloud Services, etc. Healthcare data is collected to improve the cost and quality of healthcare and to generate meaningful insights for patients and researchers [9]. In this case, the visual analysis tool plays a vital role in communicating with all of the different data sources and extracting the desired information through visual presentation.

1.2.1.4 Collaboration

This feature allows multiple healthcare professionals to view and work with the same analysed data set, which helps to select evaluation methodology to work in a group. It facilitates sharing one's analysis with others, enables to give inputs on their dissimilar concepts, and helps holds conferences in the VA platform for better understanding of the case and summarisation. Collaboration leads to the following benefits in the world of healthcare [10].

- **Enhances confidence in data quality, reliability, and balance**: collaboration increases the sample size; sharing helps in enhancing data quality. Inferencing the pattern of a particular case from a large data sample gives a more accurate prediction than using a small data sample. If a large scale of data is indicating the same issue, naturally, it enhances the dependability on data to achieve more accuracy in analysis.
- **Enables better healthcare products**: Accurate analysis from massive data gives better product development. It assists developers to generate need-based healthcare products. Also, statistics helps in ensuring the availability of drugs and logistic supports in the market.
- **Permits superior monitoring of healthcare trends**: Professionals can more effectively monitor patients with vital sign depending on healthcare trends and inferencing experience. Researchers and scholars are extensively exploring to use cutting-edge technology in healthcare (sensors, wireless communication, non-invasive wearables with Internet of Things) for innovative global healthcare development and promoting remote monitoring. This helps in keeping vigilance on the emotional and physical health status of patients by identifying and transferring data (like ECG rate, electrocardiogram, temperature, breath count, chest echoes, and patient blood pressure) to address clinical support through a single healthcare assistive platform to reduce dependencies of patients [11].
- **Preserves transparency and trust**: With the advancement of digital technologies, the collaboration of data is playing a dominant role in driving the global economy all around. As discussed, there are several data sources available for collaboration. But improper and unethical use of these data may lead to unsafe and unwanted results. Moreover, vague, incomplete, and non-trusted data sets may also lead to some wrong conclusions and solutions that may become a hindrance in the way to create a more comprehensive, pioneering, robust society. Data sharing, linking across organizational frontiers, generates many other consequences like safety, security, authenticity, reliability, etc. Collaboration reveals new perceptions and identifications that often were not projected. So, the WEFTDI has organized an extensive examination of influencing factors for promoting evolution in data collaboration among government and self-driven organizations. The major suggested concern of the whole community (i.e. business intelligence, government, researchers,

scholars, and leaders) is to generate a realistic agenda for pioneering innovation and protecting them against expected dangers. Some tools used for strengthening trusts are maintaining guidelines for data responsibility, data ethical assessment to maintain essential privacy, data privacy, ethics, data protection compliance systems, humanitarian society's data security, data exchange agreement format, etc. [12]. Maintaining transparency among service providers is paramount for building everlasting faith between the medical system and customers.

1.2.2 Visual Analytics Tools

Healthcare visual analytical tools must have the potential of fast data filtering and be integrative, multifactorial, generalizable, resilient, and chronological, i.e., time-based representation feature. A few most promising healthcare visual analytical tools are as follows:

- **Board**

  Board gives a wide range of instrument selection, data assessments, plots, graphs, thingamajigs, data desks, and other data informative stuff for previsualization and data demonstration. This provides communicating dashboards and cutting-edge graphical features and communicating plots that give users the chance to obtain multidimensional understandings of their data. It includes bi-dimensional color-coded phenomenon studies, treemaps, progressive bar graph analysis, three-dimensional analysis, and radar. Sufficient visualization options are provided to users to make sensible data analysis. It provides cutting-edge analytical abilities for empowering users to build instinctive and communicating dashboards, analyses, or reports. For precise forecasting and consequence analysis, it involves Board Beam, a computer-aided analytical demonstrating methodology. It facilitates foreseeing the consequences with the impact of external factors too. The additional features of business analysis algorithms are conventional min/max, average, and standard deviations, like statistical analysis. It is useful to estimate and follow up KPIs for frequency and quality, to prolong performance dip, or to enhance growing rate [13].
- **Tableau**

  It is a Big Data analytics tool that provides resilient information imagining competencies, empowering operators to generate smart corporate reports through its dynamic feature. It permits operators to represent information in diverse well-defined charts, dynamic presentations to imagine fluctuations among numerous groups, or else at a different interval. It also helps in supporting local communication and permits operators to control spatial data directly in the Tableau for influential exploration.

  The important application of this tool in healthcare is to understand profitability by specialties keeping efficiency and productivity moving in the right direction. It helps in visualizing healthcare reporting metrics across hospitals, programs, and regions by sharing data in different ways with key employees to manage the business more effectively [14].

- **TIBCO Spotfire**

  It provides more strong information imaging capacity with excellent graphical representation for data-driven decision-making. It gives easily interpretable intuitions for identifying important tendencies and characteristics. Operators can generate directed knowledge to communicate difficult ideas by incorporating visuals, explanations, and directives to address their discoveries. It suggests the finest conceivable illustrations of the data and provides local communication. It helps in creating and viewing dynamic graphs forming all-inclusive links between data and location, assists in data interpretation, produces intuitions, and predicts the possible outcomes. It supports healthcare insurers, plans, and pharmacy benefit management systems as a whole.

- **SAS Visual Analytics**

  It offers a resilient, instantaneous way of communicating pictorial explanation to help in creating an attractive diagnostic previsualization platform. It permits plotting important activities before analysing data and allows collaboration and previsualization through MS Office incorporation.

  It has features of a self-generating plan, allowing operators to begin assessments from an ideal fact and comfortably pick the finest graphical information presentation. It assists local assimilation with GIS mapping software or free editable mapping skills. It is also empowered with communicating moving pictures and graphs proposals.

- **Microsoft Power BI**

  It provides an adaptable prior-built panel that unites important metrics to create a sole view showing dynamic informs on all devices. It supports generating tallies that include cutting-edge cleaning, steered directions, communicative exploration, and imagining. Moreover, it provides a different table and moving picture to represent data in an adoptive and more descriptive mode. Its visualization technology supports data analysis with necessary communicating reports and local studies. Power View Add-in functionality facilitates to display and share understandings with sharing partners. It supports local assimilation with Structured Query Language local engine, free mapping, and GIS mapping softwares. Its application in the healthcare system is observed for developing business intelligence to improve the quality, performance, and productivity of the system.

1.3 SCOPE OF VA IN DIFFERENT SECTORS OF MEDICAL SCIENCE

- **VA in Healthcare Industry**

  Data analytics is the prerequisite of VA, used to examine huge data for deriving inferences computed from the analytical results using some standard qualitative and quantitative techniques for improving production efficiency, commercial expansion, and other services of the organization. Most organizations are using this for efficient decision-making and overall performance improvement. Data analytics are primarily used to cross-verify the current working strategies and philosophies to achieve the intended goal. Based on industry needs and requirements, specific data analytic technics are adopted for its adequate and optimum utilization. Data analytics is a process of investigating, cleaning, transforming, and demonstrating data to discover expedient unseen or uncover information. This helps in drawing the summon on studies to frame the new decisions. Data analysis encompasses vivid domain-specific features to make it applicable in all sectors like Science, Politics, Business and Commercials, and Healthcare. Big Data can bring a remarkable change in healthcare's data analytics for good decision-making to improve its product and services. This has the power to shrink budgets of treatment, help prevent preventable diseases and deaths, foresee a rising occurrence, and improve general life quality. Although data analytics play a critical role in the health industry, it still faces some challenges like large data collection from different sources in different formats, maintaining privacy, and protecting patient personal data [15–17]. Researchers and developers have a strong claim that to assist in Quality Improvement (QI) and Clinical Research (CR), Big Data and VA play a vital role too. QI deals with parameters like safety, effectiveness, access, and patient-centeredness, whereas CR is using unidentified health data and practical clinical judgements to limit and prevent various health disorders. It also includes experimental drug trials, investigative techniques, and therapeutic approaches, before sending it to the market and before it is prescribed to the patients [18]. Twelve major applications of Big Data in healthcare are as follows: to forecast the expected daily patient's income for staff management, for safe sharing of EHRs, to generate alert to pay urgent attention, for preventing opioid abuse in the US, for increasing patient awareness for their health monitoring, for well-versed strategic planning, for extensive research for cancer treatment, for prognostic analytics, in shrinking fraud and increasing data security, for practicing telemedicine, for assimilating medical images for a larger diagnosis, and to prevent unnecessary patients visits.

- **Clinical Data Prediction**

  Clinical data is complex and messy. Exploratory VA tools fill a much-needed gap. However, exploratory tools alone do not address their predictive desires. There is a strong role for visualization in predictive tasks. Initially, for clinical trial a SAS Visual Analytics tool has been developed for all users that had some basic features like facilitating all users to access the home page, designated information view rights, following actions of end-users like a type of data search, generating documents, inspecting reports, delivering management activities, gathering data, supervision of various systems, encouraging to use searching, and adding motivation for users collaboration. Generally, it is an exceptional business intelligence that offers evaluations at RAM's speed, and it is enhanced with LASR Server. Due to its exceptional graphic design, communicative, and self-learning environment, users get the opportunity to empower analytic skills in their clinical research. They search for new information, analyse it and generate visualizations to expose hidden characteristics, and helps to share these in reports through the internet and smart devices [19]. But with the increasing complexity, tools and technologies have also taken their place. For a few years, consistent effort has been made by researchers and developers to bring a high degree of accuracy in clinical data analysis for providing better clinical care to the patients by advancing the technologies. There is a constant development going on for better visualization and prediction by using deep learning methodologies like Trained Neural Network Models, Machine Learning Models, Explainable AI (XAI) Models, etc. [20]. Although developers have acclaimed that these tools are playing a vital role in clinical decision-making, still there are some limitations left in their evaluations and validations.

- **Diagnostic**

  The gauge of improvement in healthcare predominantly depends on the early and accurate diagnosis of disease. It plays a vital role in almost all sections, i.e., primary, secondary, and tertiary healthcare, and includes both lifestyle diseases and child and mother care. Till about a decade earlier, there was scarce use of visual image recovery mechanisms in clinical practice. At that time, in the field of radiology, content-based image recovery mechanism was considered a very efficient technology. Gradually, the tools, technologies, and algorithms were derived after the rapid advancement in the fields of data warehouses, statistical information collection from images using machine learning practices, electrical signal mining like sound, images, and measurements, artificial intelligence-trained visual world interpretation by computer, etc. Also, the content-based access mechanisms led to the need for correlation among health practitioners and computer scientists [21]. Extensive research has also indicated that to support diagnostic analysis, a combined effort of Visual Analytics Tools developers, pathologists, and radiologists is most essential in VA tools development. It is claimed that medical imaging plays an important role in this type of analysis, and hence this must be supported by advanced scientific reasoning based on radiologist's reviews. The key characteristics of diagnostic image review include cognitive task overview, knowledge characteristics and usage, exploratory components, and differences within radiologist subgroups for comparative analysis incorporation. The characterization presented is intended to contribute to the future. VA research in diagnostic imaging is based on the selection of relevant research questions to pursue and guide domain-specific adaption of developed methods [22].

- **Decision-making and Policymaking**

  In the present-day scenario, it is observed that organizations are beholding the opportunity to elevate business processes with nifty and rapid decisions. Healthcare industries are also subjected to this. In bringing this elevation, decision management is playing a key role to improve and optimize business processes. To support it, tools and technical assistance are of utmost need to deal with huge unstructured data and facilitate many automated decisions. It clearly states that policymakers and strategic planners need technology collaboration incorporating prognostic analytics, Artificial Intelligence (AI), business guidelines and process management, and multifaceted event handling. It is being realized that VA plays a significant role in decision management. First, it helps business experts to identify the problematic areas to optimize and derive automated decisions. Second, it improves the cognitive interfacing ability of experts and facilitates tools for monitoring gradual development in applied areas. Even a few advanced VA tools used in business facilitates managers to simulate and observe the impact of decision management in forecasting event analysis. Operational performance visualization can be achieved by adding BI reporting, analytics functionality, public relationship administration, customer services, ERP, the flow of goods and services, etc. Experts get introduced to the applications and enforce them to follow the program optimization approach by going through organizational protocols. Clinical researchers use VA tools to perform longitudinal research where multiple databases with real-world clinical data need to be integrated for a better understanding of critical diseases. Administrators use it to understand organizational performance and productivity, service gaps, effective outcomes, and customer satisfaction monitoring [23]. Other than business development decision-making, VA plays an important role in treatment-oriented critical decision-making too. Till now, it is well understood that health-related data of a patient is unstructured in nature as it is generated, gathered from a variety of resources to include vital information like symptoms, disease, lab test results, various diagnostic results, complications, treatment schedules, and prescriptions suggested by multiple specialists. Decision-making in proceeding with future treatments may be even more complex; critical decision-making may not rely on single patient health information. The decision may demand the experience of similar patients. VA is an emerging technology that shows a vital commitment in dealing with associated challenges in massive data handling. It also helps in taking other administrative decisions like workflow, doctor and staff management, hospital scheduling, emergency bed scheduling during the flu epidemic, economic decisions, drugs and treatment-oriented logistics management, etc. [24,25]. It also helps in Peak-Preserving Prediction analysis for overall hospital management [17].

- **Medicine**

  To empower a new generation of personalized evidence-based medicine, the human intellectual via interactive visual interfaces with computational power is being utilized widely. If the hypothesis is to be believed, an evidence-centric healthcare ecosystem can drive healthcare transformation, yielding lower costs and improved outcomes [26].

- **VA in Drug–Drug Interaction**

  More often, it is heard that death is caused by Adverse Reaction of Drugs (ADRs). It may be caused due to some unknown/unaccounted for effect of prescription drugs caused in addition to the desired medication. These side effects may vary based on an individual's physical condition, age, body weight, sex, and background. Also, it may be due to taking multiple drugs simultaneously. So, it is urgently required to identify the groups of drugs that cause this catastrophic effect. Testing all drugs virtually is not feasible. To address the same, FDA plays an important role in analysing Multiple Drug Adverse Reactions (MDARs) after the drugs reach the market. The organization has launched a common reporting platform in the name of Adverse Event Reporting System (FAERS) to receive real feedback from health professionals and patients regarding their experience that may indicate an ADR. Safety evaluators and FDA analyse these received data to identify MDARs caused due to the combination of drugs referred to or consumed by the stockholders. For a high volume of FAERS data analysis, interventions of analysing tools and techniques have become now the most required concern. Looking at cutting-edge technology development, a few recent developments are discussed here. To meet the present demands, several developing teams are putting their enormous effort to make ease in this sector. Addressing the same issue, Xiao Qin and Tabassum Kakar developed the Multi-Drug Adverse Reaction Analytics Strategy (MARAS) system and also designed a visual paradigm, which they called Drug–Drug Interactions via Visual Analysis (DIVA). These systems use connotation rule mining to identify potential MDARs from FAERS reports and provide an interactive visualization of the two drug rules mined by MARAS. Further Brian Andrew, Huy, Brian in association with Xiao and Tabassum unified the MARAS and DIVA systems into a single application, the Multi-drug Interaction Analytics Platform (MIAP) recently. In addition to its basic features, MIAP has included the prioritization of potential MDARs and subsequent analysis. It minimizes the task of Safety Evaluators to prioritize potential drug combinations by sorting the severity of the associated adverse reactions for further investigation and retrieve all reports related to each drug combination before applying domain knowledge to determine how to proceed with a clinical investigation or potential drug regulation [27]. Researchers such as Lamya, Berthelota, Favreb, Ugona, Duclosa, and Venota have also developed a new drug comparative website where the comparison is done with similar existing drugs based on drug properties, cost, age, gender, new drug clinical trials results, contraindications, drug interactions, adversarial effects, exceptions, dosage count, management, and official document links. Designing this website with emerging visual presentation techniques was not an easy task for the developers. The difficulties they encountered when comparing drug properties in the context of safety were contradictory results, connections, and adversarial effects. It is simplified by visualization techniques development with two approaches. First, they pick up the tables used by physicians and experts. To enhance visibility and understandability, certain modifications were introduced by incorporating symbols and icons, highlighting exclusive properties, and table content with dynamically filtering. Second, to represent the comparisons of contraindications or adverse effects with visual representation, they adopted Vann Diagram, Bar Charts, and Visualization of Concept in Medicine (VCM) icons. But due to its technical limitations, finally, the authors have developed a new website using metaphysics and visual techniques. It is supported on Java, Python, and HTML pages with Cascading Style Sheet. It is accessed by a dedicated metaphysics programming tool and managed with PyMed Termino [28].

- **Quality Improvement and Clinical Research**

  Researchers and Developers have a strong claim that assists the same Big Data and VA playing a vital role. The QI focuses on parameters like safety, effectiveness, access, and patient-centeredness, whereas CR focuses on novel health data and experimental clinical trials. More practical, broad, and cost-effective clinical research can be carried out using random records. It also includes clinical drugs trial, investigative techniques trials, and other medication strategies trials, before providing service to the market [37].

- **Mental Healthcare**

  If the literature is to be believed, changing lifestyle, increasing workload, peer competitions, the gap in competencies, and shrinking job opportunities create a tremendous uncertainty at work front. Keeping pace with fast-changing technology competencies is another challenge for young generations. Moreover, increasing academic and extracurricular work pressure on children makes their life more pathetic. The adverse effect is engulfing all around. Such a life schedule finally left most of the people with no time to spend some quality time with family members, to discuss the difficulties, pay attention to their older people and children. Even, hardly there is any chance to cherish oneself. Such a lifestyle is increasing mental depression and disability exponentially over the last few decades. We need to adopt advanced technology to support counseling for identifying solutions with evidence-centered decision-making.

VA can contribute a lot with its exclusive features discussed earlier. Novel visualization technologies with AI techniques in association with domain experts can help in achieving evidence-informed knowledge discovery from the database. It allows domain experts to explore both local and global data for in-depth analysis. It helps in framing more realistic health plans and policies. "Chrono-View" and "Brain Expression Spatio-Temporal (BEST)" are some recently developed VA tools that have the potential to view dysregulated gene expression patterns for determining the similarity and anomaly of temporal patterns. It helps to connect disease-specific genetic studies and mechanism studies, gives shreds of evidence for key gene and molecular system identification, and helps in analysing particular brain regions and age stages [29–31].

1.4 CHALLENGES TO FACE

- **Challenges in biomedical image analysis**: Multiple difficulties developed in image evaluations are object recognition, pattern extraction, image separation, and listing. If these difficulties are resolved, the development of worthy analytic capacities can help as valuable input to data analytics of healthcare.
- **Challenges in heritable data analysis**: It analyses the genetic information in a chromosome of the living organ. A wide understanding of the associations among various genetic signs, transformations, and ailments/disorders may contribute as an important input in developing gene treatments to cure these. Converting genetic findings into a personalized medicine is an important assignment with huge unexplained difficulties. For example, the genetic sceneries in complicated diseases like cancers are vast and complex and have a huge degree of diversity among peoples.
- **Challenges in medical transcript mining**: It is extremely challenging to automatically analyse and convert an unstructured medical transcript to a structured format. It is difficult because of the nonhomogeneous nature, varied formats, and changing situations across diverse consultants and patients. However, Natural Language Processing and Entity Extraction contribute significantly in inferencing meaningful information from big volume of medical transcripts by routinely encoding methods. But the process encounters difficulties when needing to extract inherent information from grammatically incorrect, brief, and abbreviated phrases and transcriptions [15].
- **Challenges in evaluating and validating**: Challenges associated with the performance of Clinical Predictive Models for drawing sense-making understandings after carrying out investigations of the analysis of cases by researchers and scholars. Comparison of public investigation solutions for better and appropriate test design and visualization, etc., is also a challenge [32–34].

- **Challenges in VA**: The fastest-growing health-associated pieces of evidence develop criticality to structure effective modes of exploring bulk data by man–machine communication and diagrammatic interfaces. It should have the capacity to explore and spott useful characteristics in multimodal medical information and narrate this for a better understanding of diseases and medical workflow management. The multidimensional, noisy, nonhomogeneous, global patterns of medical data create a hindrance in processing the data and getting an accurate understanding from it. Although in healthcare, open-access data from multiple resources bring transparency and trust in data collaboration for obtaining the most effective clinical solution to revolutionize the healthcare sector and to serve better humanity, domain-specific methods, software, and server-specific VA create limitations in greater collaboration at large. So, the current era demands some more advanced technology development that can allow working in global and common dashboard support systems irrespective of tools, technology, and branding differences.
- **Conceptual challenges**: Involvement of multiple Stakeholders brings the following bridging gaps like competence gap, analysis gap, iteration gap, etc. for VA decision support, and this introduced another challenging point, which will be addressed soon.
- **Technical challenges**: Associated issues are investigating and observing decision procedures, exploring and examining manuscript collections – doing extensive analysis and comparing backers views and arguments, going through a wide range of analysis and comparing experiential performance pointers, studying, analysing, and comparing the effects of resolutions, analysing for finding optimum results, etc. need immediate attention to minimize its limitations.
- **Other associated challenges are**: providing medical professionals sensible data in correct layout, extending toward visual tools networking, facilitating team decision-making, characterizing and understanding similarity, presenting risk and uncertainty, evaluating VA, advancing data visualization technology forming combined values, and generating methodical supple VA procedures that endorse comprehensive analysis when finding any new insights. These dependable processes may help in iterating analysis with the new database, classification of complicated arrangements of VA tasks so that we can reuse the management system review tracks to new data set and share it with concerned stakeholders [16, 35–37].

1.5 CONCLUSION

Although VA is gaining recognition in the health sector at different levels for service optimization and business development, it still could not be able to utilize the technology effectively for tertiary healthcare issues and treatment. Technology itself needs more development in all terms for addressing these issues too. Also, it needs to be more cost-effective, compatible, simple, and self-explanatory to make it more resilient.

REFERENCES

1. Keim DA et al., 2008. Visual Analytics: Definition, Process, and Challenges. First publication in: *Information Visualization*, A. Kerren (eds.). Berlin: Springer, 154–175.
2. Shailaja K, Seetharamulu B, Jabbar MA, 2018. Machine Learning in Healthcare: A Review. *Second International Conference on Electronics, Communication, and Aerospace Technology (ICECA)*, DOI: 10.1109/ICECA.2018.8474918.
3. Cloud Standards Customer Council, 2017. Impact of Cloud Computing on Healthcare Version 2.0 [Original Report].
4. Keim DA, Mansmann F, Schneidewind J, Thomas J, Ziegler H, 2008. Visual Analytics: Scope and Challenges. First publication. in: *Visual Data Mining*, SJ Simoff, MH Böhlen, A Mazeika (eds.). Lecture Notes in Computer Science, No. 440. Berlin: Springer, 76–90.
5. Egan M, 2006. Clinical Dashboards Impact on Workflow, Care Quality, and Patient Safety. *Critical Care Nursing Quarterly*, Vol. 29, No. 4: 354–361.
6. Pestana M et al., 2018. A Productivity Dashboard for Hospitals: An Empirical Study. http://www.researchgate.net/publication/326506889.
7. Dowding, D et al., 2015. Dashboards for Improving Patient Care: Review of the Literature. *International Journal of Medical Informatics*, Vol. 84, No. 2: 87–100. ISSN 1386-5056.
8. U.S. Department of Commerce & National Institute of Standards and Technology, 2012. Guide for Conducting Risk Assessments. NIST Special Publication 800-30 Revision 1.
9. Epplera MJ, Aeschimann M, 2009. A Systematic Framework for Risk Visualization in Risk Management and Communication. [original work] *Risk Management*, Vol. 11, No. 2: 67–89.
10. Van de Wetering R, Versendaal J, 2018. How a Flexible Collaboration Infrastructure Impacts Healthcare Information Exchange, DOI: 10.18690/978-961-286-170-4.30.
11. Vitabile S et al., 2019. Medical Data Processing and Analysis for Remote Health and Activities Monitoring. In: *High-Performance Modelling and Simulation for Big Data Applications*, J Kołodziej, H González-Vélez (eds.). Lecture Notes in Computer Science. Cham: Springer, Vol. 11400, 186–220.
12. World Economic Forum, 2019. Data Collaboration for the Common Good: Enabling Trust and Innovation through Public-Private Partnerships, [Original work] an Insight Report Produced in Collaboration with McKinsey & Company.
13. Institute of Healthcare Department, 2018. Visual Management Board, [Original work], Tool Originally Developed through a Partnership with the HRET and the AHRQ Safety Program For Ambulatory Surgery.
14. Ko I, Chang H, 2017. Interactive Visualization of Healthcare Data Using Tableau. *Healthcare Informatics Research*, Vol. 23, No. 4: 349–354.
15. Sarada J, Lakshm NV, 2017. An Introduction to Data Analytics in Healthcare Industry. *International Journal of Advanced Scientific Technologies, Engineering and Management Sciences*, Vol. 3, No. 1: 23–27.
16. Shneiderman B, Bradford W, 2013. Improving Healthcare with Interactive Visualization. Published by the IEEE Computer Society: http://ComputingNow.computer.org.

17. Hao MC et al., 2011. A Visual Analytics Approach for Peak-Preserving Prediction of Large Seasonal Time Series. First published in: *Computer Graphics Forum*, Vol. 30, No. 3: 691–700.
18. Kamal N et al., 2014. Big Data and Visual Analytics in Health and Medicine: From Pipe Dream to Reality. *Journal of Health & Medical Informatics*, Vol. 5, No. 5, DOI: 10.4172/2157-7420.1000e125.
19. Becker M, 2013. Visualizing Clinical Trial Data. *SAS® Visual Analytics. Proceedings of the SAS Global Forum 2013 Conference*, SAS Institute, Cary, NC.
20. Li Y et al., 2020. A Visual Analytics System for Multi-Model Comparison on Clinical Data Predictions. *Visual Informatics*, DOI: 10.1016/j.visinf.2020.04.005.
21. Müller H et al., 2005. Benefits of Content-based Visual Data Access in Radiology1, *Radio Graphics*, Vol. 25, No. 3: 849–858.
22. Lundström C, Persson A, 2011. Characterizing Visual Analytics in Diagnostic Imaging. *International Workshop on Visual Analytics (2011).* Centre for Medical Image Science and Visualization (CMIV), Linköping University, Sectra-Imtec AB, Linköping, Sweden.
23. Stodder D, 2015. Visual Analytics for Making Smarter Decisions Faster Applying Self-Service Business Intelligence Technologies to Data-Driven Objectives. TDWI Best Practices Report, TDWI Research, cosponsored by SAS, 2015. https://tdwi.org.
24. Ruppert T, 2017. Visual Analytics to Support Evidence-Based Decision Making. Dissertation submitted to Darmstadt, Technische Universität.
25. Sudha G, Sangaiah S, 2018. A Survey on Contribution of Visual Analytics in Health Care Domain. *Computing Communications and Data Engineering Series*, Vol. 1, No 1: 1–6.
26. Perer A, 2012. *Visual Analytics for Evidence-Based Medicine.* Hawthorne, NY: Healthcare Analytics Research Group, IBM T.J. Watson Research Centre.
27. McCarthy B, Schade A, Tran H, Zylich B, 2018. Multiple-Drug Interaction Analytics Platform: Developing a Minimum Viable Web Application for Visualizing Multi-Drug Interactions. Project thesis, submitted to WPI.
28. Lamy JB, Berthelot H, Favre M, Ugon A, Duclos C, Venot A, 2017. Using Visual Analytics for Presenting Comparative Information on New Drugs. *Journal of Biomedical Informatics*, DOI: 10.1016/j.jbi.2017.04.019.
29. Misue K et al., 2013. Visualizing the Overview of Temporal Patterns of Patients' Activities. *Proceedings: Workshop on Visual Analytics in Healthcare (VAHC 2013)*, Washington, DC.
30. Guo L et al., 2019. BEST: A Web Server for Brain Expression Spatio-Temporal Pattern Analysis. *BMC Bioinformatics*, DOI: 10.1186/s12859-019-3222-6.
31. Chunga Y et al., 2020. Role of Visual Analytics in Supporting Mental Healthcare Systems Research and Policy: A Systematic Scoping Review. *International Journal of Information Management*, Vol. 50: 17–27.
32. Kang Y, Gorg C, Stasko J. Evaluating Visual Analytics Systems for Investigative Analysis: Deriving Design Principles from a Case Study. [White Paper] School of Interactive Computing & GVU Center, Georgia Institute of Technology.
33. Scholtz J, Plaisant C, Whiting M, Grinstein G. 2013. Evaluation of Visual Analytics Environments: The Road to the Visual Analytics Science and Technology Challenge Evaluation Methodology, VAST Challenge Repository.
34. Battle L, et al. 2018. Evaluating Visual Data Analysis Systems: A Discussion Report. *HILDA'18.* Association for Computing Machinery, Houston, TX. ACM. ISBN 978-1-4503-5827-09/18/06.

35. Kumar V et al., 2014. Exploring Clinical Care Processes Using Visual and Data Analytics: Challenges and Opportunities [White paper]. Tennenbaum Institute, Atlanta, GA, Corpus ID: 12293684.
36. Aigner W, Federico P, Gschwandtner T, Miksch S, Rind A, 2012. Challenges of Time-oriented Data in Visual Analytics for Healthcare. *Conference: IEEE Vis Week Workshop on Visual Analytics in Healthcare (VAHC).* http://www.mobiguide-project.eu/.
37. Gotz D, Borland D, 2016. Data-Driven Healthcare: Challenges and Opportunities for Interactive Visualization. *IEEE Computer Graphics and Applications*, Vol. 36: 90–96.
38. Brian, M, Andrew, S, Huy, T, Brian, Z, 2018. Multiple-Drug Interaction Analytics Platform Developing a Minimum Viable Web Application for Visualizing Multi-Drug Interactions. Degree of Bachelor of Science. Worcester Polytechnic Institute.

2 Statistical Methods and Applications

A Comprehensive Reference for the Healthcare Industry

Areeba Kazim, Achyut Shankar, and Muskan Jindal
Amity University

CONTENTS

DOI: 10.1201/9781003050827-2

2.1 INTRODUCTION

Prediction of the spread or outbreak of an epidemic or predicting climate conditions has always been a "data-oriented task" [1,2], where the majority of outcomes depend on arduous efforts made for analyzing data, i.e., the value of Exploratory Data Analysis (EDA) increases as, before applying any kind of machine learning or AI2 algorithm, one requires erudite knowledge of data, thereby requiring efficient EDA [3]. But the presence of outlier and absconded data has always been an issue to be handled by EDA algorithms; such sets of data cannot be analyzed and thus are left out the majority of the time, which may not hamper results in many cases but when predicting epidemic spread it creates erroneous results. EDA has been used in the field of healthcare in varied areas, and thereby it has gigantic scope. For example, many studies use Exploratory Spatial Data Analysis (ESDA) to detect the relationship between risk factors and mortality rate of neural diseases [4] or visualize EDA results to develop links between mortality rate and birth rate considering geographical factors [5]. Other established areas of applications in the healthcare domain include handling "lattice data" using ESDA [6] and using heat-map or item map to evaluate already existing statistical software and models [7]. EDA has been a major aid and a robust concept in epidemic data analysis and the Spatial EDA analyses how season and climate change affects the spread of viruses [8]. Few epidemic researchers utilizes EDA[9] for various pandemics like Ebola, poliomyelitis to name a few using geographical demographics, climatic changes, etc. as attributes. Some use rainfall cycle or severe changes in temperature and hygiene as deciding factors, supporting their results by Spatial EDA for epidemics like dengue [10]. Consequently, EDA has been the preferred approach for research enthusiasts, especially in the health care domain for the reason that it provides a simpler and robust method to analyze and derive significant conclusions from raw and unsorted data with the least amount of time and effort. The requirement for quick, online algorithms to break down high information rate estimations is an indispensable component, all things considered, in all settings. Given the ever-increasing number of data sources, the increasing complexity of uses, and the exceptional task at hand designs, irregularity detection systems should have low-unpredictability and work constantly. In numerous cutting-edge applications, the information shows up in a gushing manner. Thus, the hidden presumption of old-style strategies that the information is an example from a basic distribution is not true, and Gaussian and nonparametric-based strategies are deficient. Spilling information can be viewed as an ever-changing superposition of distributions. But the issue arises when the presence of outlier and absconded datasets increases so much so that it may affect final results and analysis like in sensitive cases of prediction of pandemic spread or outbreak. In such matters of great consciousness, this study aims to provide a specific well-defined methodology to identify outlier datasets using Machine Learning Technique of Clustering [11] and handle or categorize outlier data by (1) converting dataset into the hierarchal structure using concept hierarchy, and then (2) performing backtracking to categorize outlier into the structure of hierarchy. This problem is explained in detail in the next section.

2.2 CONCEPTUAL FRAMEWORK OF THE METHODOLOGY

This section of the study describes details about the theoretical base and origin of the concepts and methodology, which includes answers to some rudimentary questions like [12]

1. What are outliers and how they handled in the available literature?
2. What is clustering and how it is implemented using various algorithms and techniques?
3. Elaborated explanation on how this study uses clustering to fulfill the aim of handling outliers.
4. What is concept hierarchy and how it is implemented in EDA?
5. How this research will outline the categorization of outliers using concept hierarchy?

Let's begin with the first query.

2.2.1 Outlier Definition

An outlier is an observation that resides as an anomaly and as compared to discordant abilities in an irregular example in a bundle of data points. It's not wrong to state that the mentioned description submits it over to the examiner (or to the respective technique) to choose what will be viewed as irregular or anomaly in the given dataset. Before strange insights can be identified and excluded, it is exigent to label mundane data points.

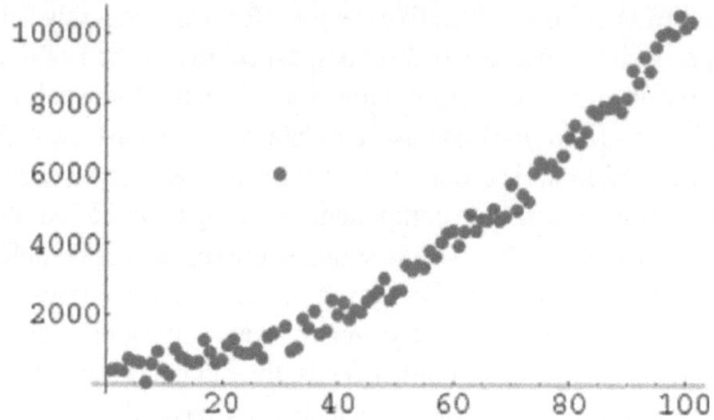

FIGURE 2.1 Graphical representation of identification of an outlier using standard clustering technique using a line graph.

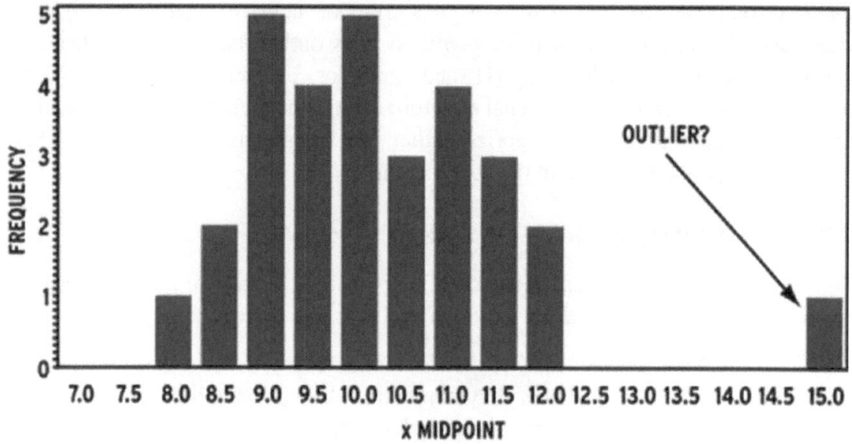

FIGURE 2.2 Graphical representation of identification of an outlier using standard clustering technique using a bar graph.

In simple words, an outlier is an observation point in an incomplete dataset that is absconded or different and distant from other observations; thus, they are often called "left-out datasets" or "outliers," thereby this aberrant issue from the usual is taken as an anomaly.

Some graphical examples are mentioned in Figures 2.1 and 2.2 for a pictorial understanding of the definition.

2.2.2 Overview of Techniques and Motivation for Outlier Detection and Handling

Outlier existence in datasets has always been a roadblock towards coherent analysis in both EDA and Data Mining as it prevents analysis of complete and integrated data all at once, thereby creating discrepancies in absolute results and misunderstanding in model development and discordant analysis of the model afterward.

Having lucid knowledge of Outlier Detecting methods is a fruitful venture to invest in as it can be pertained in various fields like clinical research trials, testing various combinations for vaccine development, market analysis and prediction, administration tasks, software algorithms, performance analysis of athletes, singers, actors, or employees in multinational corps and numerous other data mining tasks. An outlier may occur due to several reasons like there may be extreme differences in given recorded data, or occurrences of any human error while recording the data and there may be the existence of some unanticipated complexities between dependent and independent relations and variables of the dataset. Any aberrant factor can result in anomalies that can, in turn, result in presence of outliers in the raw data. Thus, outlier detection and handling is a nonpareil part of data pre-processing and must be performed with great responsibility before starting with data analysis or model development. There are multiple methodologies for outlier detection; taxonomy of outlier detection methods from a statistical perspective are multivariate and univariate outliers, while the contrary perspective of taxonomy is parametric and non-parametric methods. Parametric techniques include a presumption of some hidden circulation, for example, typical appropriation though there is no such prerequisite with non-parametric methodology. Also, you could do a univariate investigation by considering a solitary variable at once or multivariate examination where you would concentrate on more than one variable simultaneously to recognize exceptions. A data point consisting of an anomaly like an extreme value on just one variable is a multivariate outlier, that is, it consists of an amalgamation of unusual scores in at least two variables, while univariate consists of extreme data points on one variable.

Some of the sagacious and popular methods of outlier analysis are listed in the following.

2.2.2.1 *Z*-Score or Extreme Value Analysis (Parametric)

It is a successful technique if you can portray the qualities in the element space with a Gaussian distribution (Parametric). The execution is exceptionally simple utilizing pandas and scipy.stats9libraries.

2.2.2.2 Probabilistic and Statistical Modeling (Parametric)

It is just helpful to use in a low-dimensional element space, in a little too medium measured dataset. It is not suggested when disseminations cannot be thought to be parametric.

2.2.2.3 Linear Regression Models (PCA, LMS)

It is a hugely successful strategy when the appropriation of qualities in the component space cannot be accepted. It functions admirably if the component space for looking through exceptions is multidimensional i.e., at least three measurements).

Sci-kit learn's execution is anything but difficult to utilize and the documentation is eminent. Visualizing the outcomes is simple, and the strategy itself is natural.

2.2.2.4 Proximity-Based Models (Non-Parametric)

The proximity of an outlier data point diverges ominously from that of the majority of the rest in the given datum.

This brings us to the end of the discussion on outliers, and we will move forward to the second query.

2.3 DEFINITION OF CLUSTERING

It is the assignment of untying the masses and datum that pertains to "n" number of congregations with the ultimate goal of providing data that is aggregated in similar congregations and that are gradually like other data focused on a similar congregation and, unlike the statistics, concentrated in diverse congregations. It is principally a medley of data points centered on comparability and disparity inherent to it. For example, the information focused in the chart beneath that has been clustered together can be characterized into one single gathering. One could recognize the clusters, and we can distinguish that there are three groups in Figure 2.3.

Clusters may or may not be spherical in shape; it purely depends on datasets and their similarities, as in Figure 2.4.

The importance of clustering can be questioned from the aspects of both machine learning and data mining. Clustering is predominantly eminent because it formulates the existing congregation in the unlabeled information. There are no specific standards for cladding or "good clustering". Contingent upon the consumer, the benchmarks they may utilize cater to their unique need or requirement of the hour. For example, one could concentrate on learning agents for consistent gatherings (data decline), in uncovering "natural clusters" and depict their ambiguous characteristics ("natural data types"), in discovering valuable and appropriate alliances ("helpful" data classes), or in discovering abnormal information

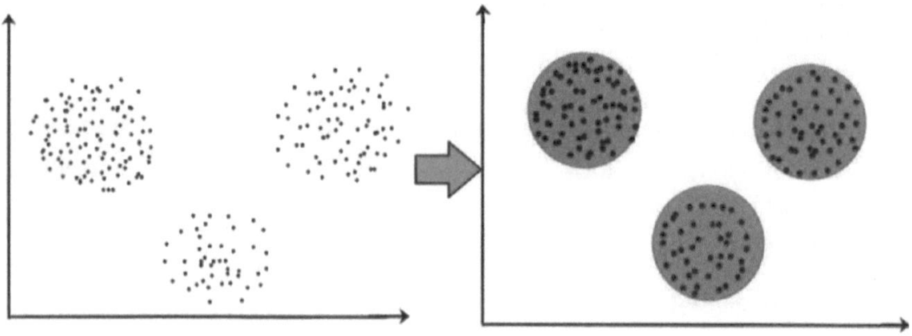

FIGURE 2.3 Visual representation of identification of clusters in an unsupervised format.

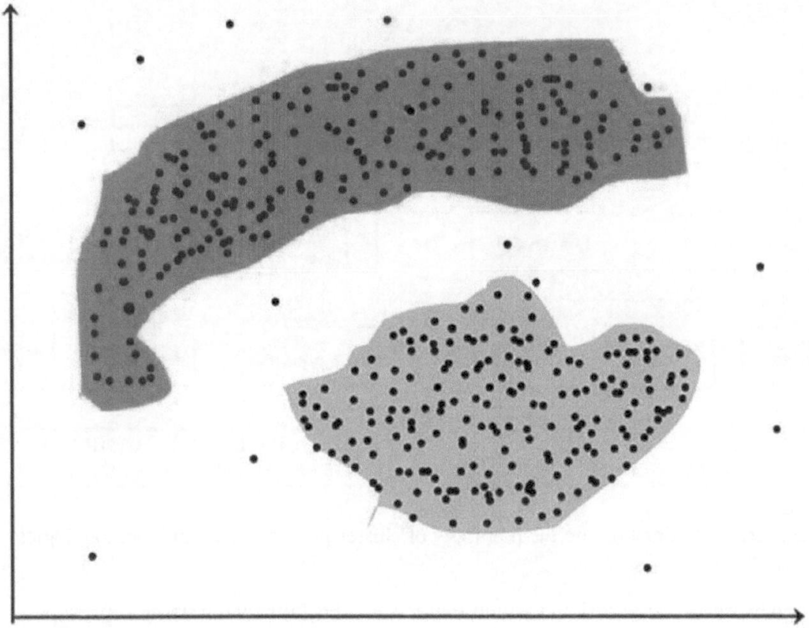

FIGURE 2.4 Visual representation of outliers, clusters in a semi-supervised format.

objects ("outlier detection"). This calculation must make a few conjectures to establish the comparability of focuses, and every suspicion make extraordinary and correspondingly legitimate clusters.

Numerous "distance techniques" and methodologies are used for the determination of the clusters with great congeniality for outlier detection as well, which will be elaborated upon in the next section. Clustering methodologies are described with detailed explanations in Figure 2.5.

2.3.1 Density-Based Methods

These stratagems work under the assumptions that the clusters have a dense area, i.e., having some closeness or sparse district of the proximity. These techniques have great exactness and capacity to blend two clusters. Some algorithms that come under this category are "DBSCAN" (density-based spatial clustering of applications with noise") and "OPTICS" (ordering points to identify clustering structure) to name a few.

1. **Hierarchical-based methods**: Here, the clusters are framed into a tree-like construction reliant on the hierarchy. Fresh clusters are shaped utilizing the recently framed one. It is separated into given classification, as mentioned below:
   Agglomerative (bottom-up approach)
   Divisive (top-down approach)

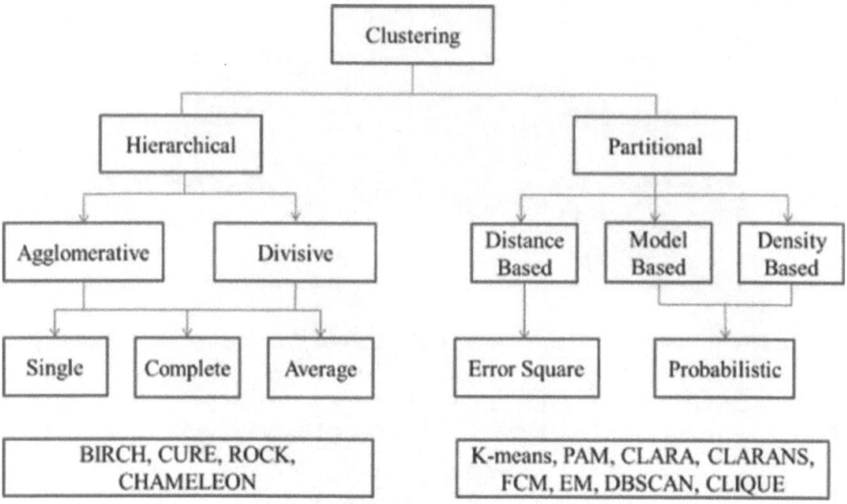

FIGURE 2.5 Understanding the hierarchy of clustering techniques and methodologies.

Algorithms that follow the same are "CURE" (clustering using representatives), "BIRCH" (balanced iterative reducing clustering and using hierarchies) [13], etc.

2. **Partitioning methods**: This approach bifurcates or divides the data points to let us say "k" clusters, and each division makes up to one cluster. It can be applied when one wants to improve the "objective criterion similarity function," likewise in the case where distance is a main stricture example in the algorithm "K-means" [14], "CLARANS" (Clustering Large Applications based upon Randomized Search) [15], etc.
3. **Grid-based methods**: Here the spatial plan or popularly known as the Euclidian plan is articulated into a limited amount of compartments that form a "grid-like construction". The clustering operations performed on these identified grids and autonomous number of data objects. Example algorithms are "STING" (Statistical Information Grid) [16], wave cluster [17], and "CLIQUE" (Clustering in Quest) [18].

Apart from the above methods of performing clustering, there are various algorithms for performing clustering, the taxonomy of these methods are discussed in the following.

2.3.2 Distributed Based Algorithms

It is a clustering prototype where fitting of the information occurs on the likelihood that how it might have a place with a similar appropriation. The gathering done might be typical or Gaussian. Gaussian dispersion is progressively conspicuous such that there is a constant population of appropriations and the forthcoming

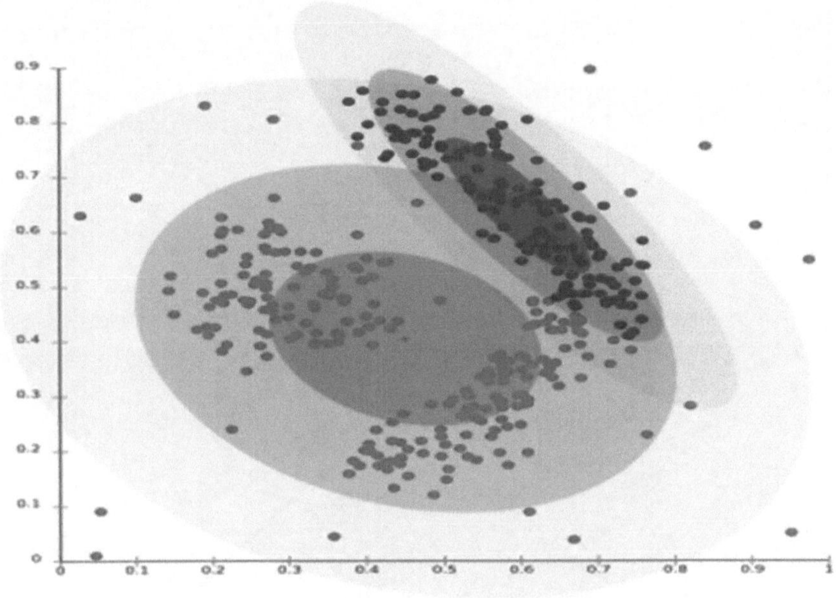

FIGURE 2.6 Visual representation of distributed clustering-based algorithms.

information is "fitted" into it with the end goal that the conveyance of information may get expanded as shown in Figure 2.6.

2.3.3 Centroid-Based Algorithms

This is fundamentally one of the iterative clustering algorithms in which the clusters are shaped by the closeness of information focuses to the centroid of groups. Here, the group community for example centroid is framed to such an extent that the separation of information focuses is least with the inside. This problem is essentially one of the NP-Hard problems [19] and hence arrangements are normally approximated over various preliminaries.

K-means [20] algorithm is one of the popular examples of this algorithm displayed in Figure. 2.7.

2.3.4 Connectivity-Based Algorithms

The center thought of connectivity-based model is like centroid-based model which is essentially characterizing groups based on the closeness of information focuses. Here we take a shot at an idea that the information focuses which are nearer have comparative conduct when contrasted with information focuses that are more remote. It's anything but a solitary apportioning of the informational index, rather it gives a broad order of groups that converge with one another at specific separations. Here the decision of separation work is abstract. These

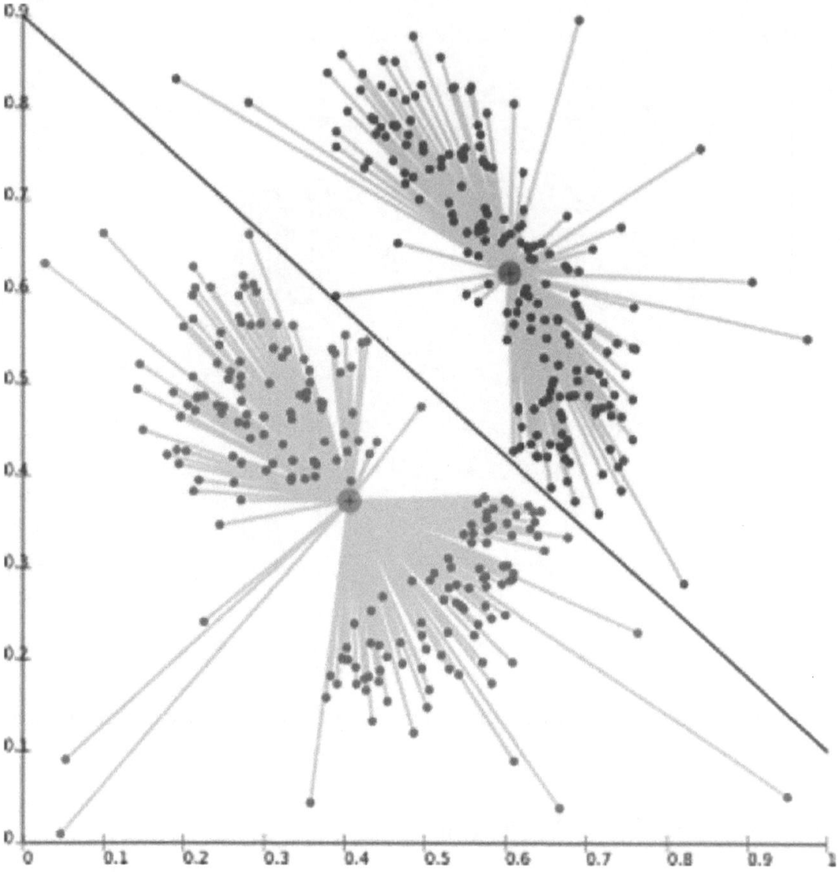

FIGURE 2.7 Visual representation of centroid clustering-based algorithms. K-means algorithm.

models are anything but difficult to decipher however it needs versatility as shown in Figure 2.8.

2.3.5 Density-Based Algorithms

In this clustering model, there will be a probing of data space for areas of the diverse density of data points in the data space. It segregates various density regions based on different densities present in the data space.

For example, DBSCAN and OPTICS are displayed in Figure 2.9.

2.3.6 Subspace Algorithms

Subspace clustering is an unsupervised learning issue that targets gathering information focuses into different groups so information point at a single bunch

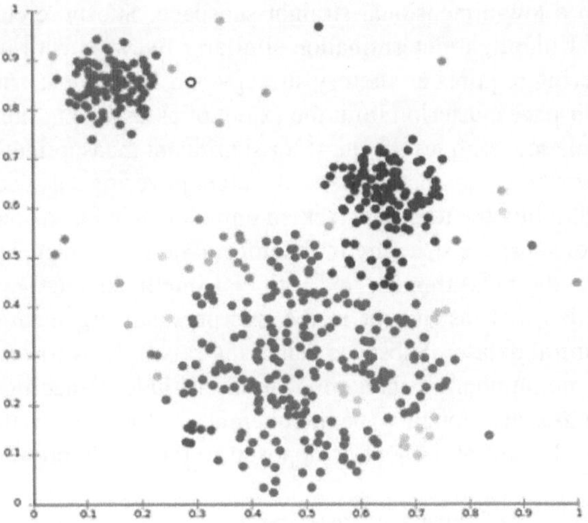

FIGURE 2.8 A visual sneak-peak into connectivity-based clustering algorithm.

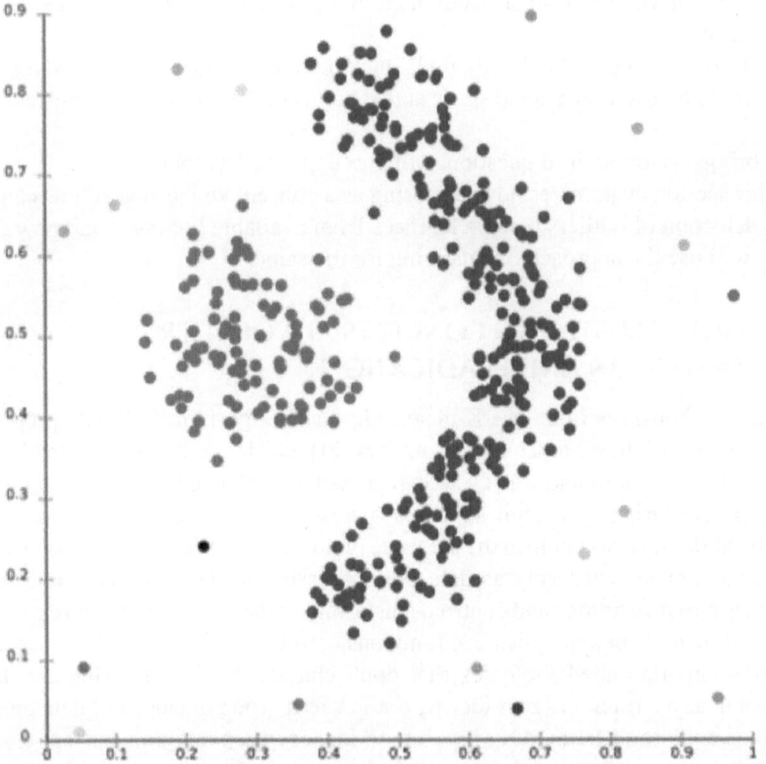

FIGURE 2.9 A visual sneak-peak into density-based clustering algorithm. DBSCAN and OPTICS.

lies roughly on a low-dimensional straight subspace. Subspace clustering is an augmentation of highlight determination similarly likewise with include choice subspace clustering requires an inquiry strategy and assessment criteria however what's more subspace clustering limit the extent of assessment criteria. Subspace clustering calculation confines the quest for significant measurement and permits them to discover the group that exists in different covering subspaces. Subspace clustering was initially the reason to tackled unmistakable PC vision issue having an association of subspace structure in the information yet it increases expanding consideration in the measurement and artificial intelligence (AI) people group. Individuals utilize this instrument in the interpersonal organization, film suggestion, and natural dataset. Subspace clustering raises the worry of information security the same number of such applications include managing touchy data. Information focuses are thought to be in-coherent as it just ensures the differential security of any element of a client as opposed to the whole profile client of the database.

Here are two parts of subspace clustering dependent on their hunt technique:

- Top to bottom methodology locate an underlying clustering in a complete arrangement of measurement and assess the meta-space of each group.
- Bottom to top methodology finds thick or dense or highly populated districts in less dimensional space at that point consolidate to shape clusters.

This brings us to the third question, which is explained as follows:

This section anatomizes how clustering as a concept and its algorithms can be used detection of outliers using references from available literature and how this study will use the approach of clustering for the same.

2.4 HOW CLUSTERING CONCEPTS AID OUTLIER DETECTION AND ERADICATION

Research enthusiasts often use K-means clustering algorithm post data preparation to identify interior and external outliers [21] which are then eliminated by an expert of the domain and this cycle is repeated and checked after each iteration until the clustering algorithm approves of results that display no outliers post which validation tests confirm the presence of no outliers. While another approach that amalgamates clustering as data pre-processing to detect outliers is to use distance-based functions and centroid clustering methods to find out extreme outliers and then again apply distance functions on them for their elimination, there are contemporary methodologies that don't classify outliers are singular data points but as a cluster, i.e., considering outliers as a group or cluster of data points and then applying LDBSCAN algorithm to identify such outlying clusters. Some novel and new methods of defining and eliminating clusters include identifying cluster-based local outliers by analyzing local attributes and density distribution which are proposed to be rather efficient and time-saving as compared to the already available techniques of the same. Some of the clustering approaches work on outlier detection by redefining the problem via considering spatial distribution instead of feature distribution and then using graphing techniques to identify and eradicate the spatial data points that are outliers. Other sagacious practices somehow determine the degree of abnormality or in other words how much of the data point outlying from the domain's dataset and this determines whether it's classified as an outlier or not post which K-Nearest Neighbor is applied on selected clusters.

This study uses the clustering algorithm K-Nearest Neighbor visualized with spatial distribution on the graph to show pre and post results for identification and eradication of outlier data which is then saved in another dataset which is will be thrown more light upon in the later sections of this chapter.

2.5 DEFINITION OF CONCEPT HIERARCHY

Concept hierarchies can be utilized to diminish the data by gathering and supplanting low-level concepts with more elevated-level concepts. In the multidimensional model, data are sorted out into various measurements, and each measurement contains different degrees of reflection characterized by concept hierarchies. This association gives clients the adaptability to see data from alternate points of view. Data mining provides good results on some smaller datasets and less useful results on others. As a result of these advantages, discretization methods and concept hierarchies are normally applied before data mining, as opposed to during mining.

In Figure 2.10 is a modest instance of concept hierarchy.

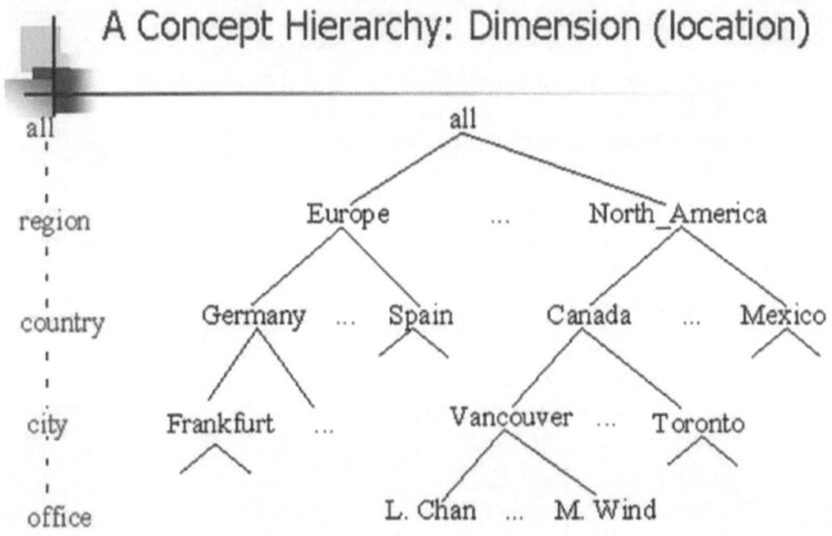

FIGURE 2.10 Comprehending the concept hierarchy in a visual format.

There different ways to implement it on numerical data and some popular methods are displayed below.

2.5.1 Binning

Here, primarily sorting of the information occurring and parceling it into (equal-depth) bins, consequently there are operations done to "smooth by bin means", "smooth by bin median", "smooth by bin boundaries" and so forth as indicated in Figure 2.11.

2.5.2 Histogram Analysis

The histogram is a well-known information decrease method and is implemented by separating information into buckets and store normal (whole) for each can be built ideally in one measurement utilizing dynamic programming. Identified with quantization issues a display of the same is shown in Figure 2.12.

2.5.3 Clustering Analysis

Partition dataset into clusters, such that a clustering analysis is performed. Hierarchical clustering can be stored in multidimensional index tree structures as indicated in Figure 2.13.

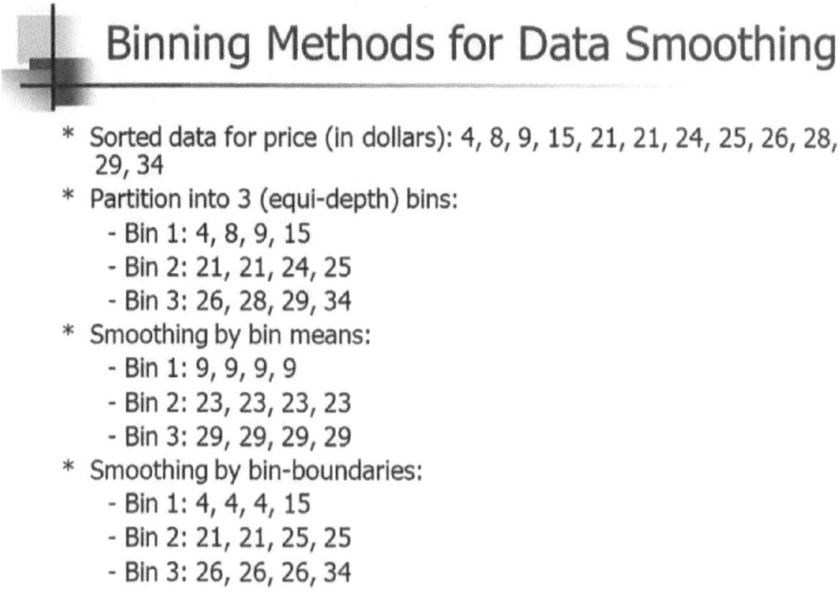

FIGURE 2.11 An example of various binning methods for data smoothing.

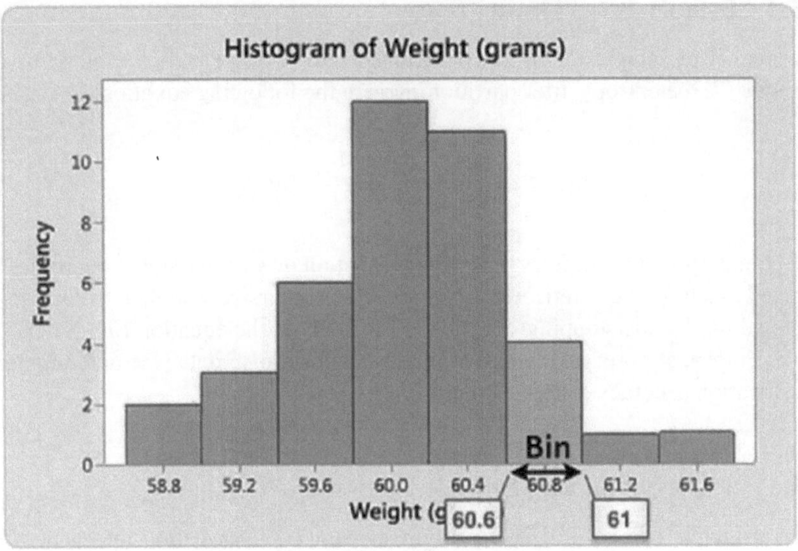

FIGURE 2.12 Performing binning via histograms.

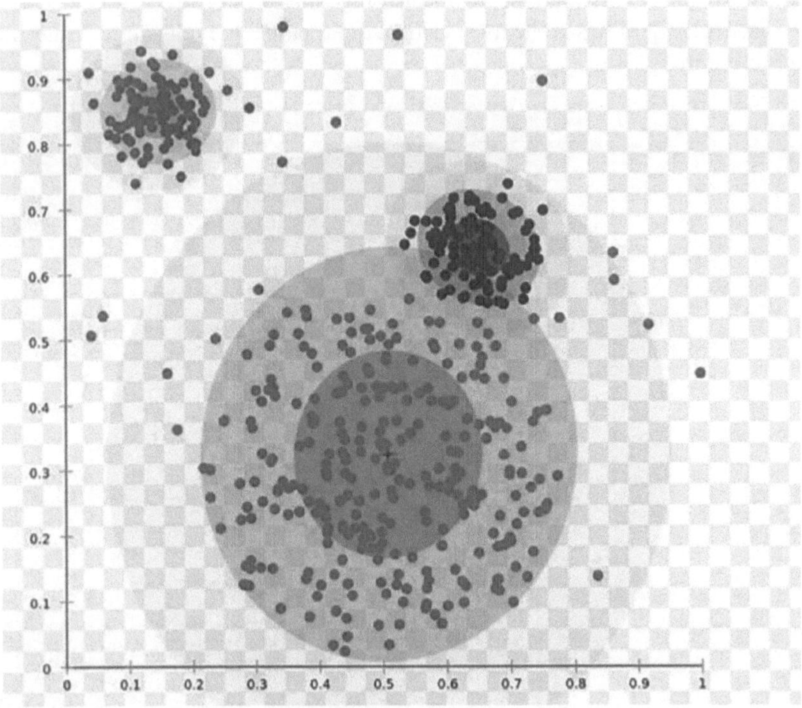

FIGURE 2.13 Index tree structures based on clustering in a visual format.

2.5.4 Entropy-Based Discretization

Given a set of samples S, if S is partitioned into two intervals S_1 and S_2 using boundary T, the entropy after partitioning is in the following equation:

$$E(S,T)=\frac{|S_1|}{|s|}Ent(S_1)+\frac{|S_2|}{|s|}Ent(S_2)$$

The boundary that minimizes the entropy function over all possible boundaries is selected as a binary discretization. The process is recursively applied to partitions obtained until some stopping criterion is fulfilled, by the equation $Ent\ (S)-E\ (T,\ S)>\delta$. Empirical observations prove that it may diminish data size and augment classification accuracy (refer to Figure 2.14).

2.5.5 Segmentation via "Natural Partitioning"

3-4-5 principle can be utilized to fragment numeric information into generally uniform, "characteristic" interims. On the off chance that an interim spreads 3, 6, 7, or 9 unmistakable qualities at the most noteworthy digit, parcel the range into 3 equal-width interims. On the off chance that it covers 2, 4, or 8 particular qualities at the most noteworthy digit, segment the range into 4 interims. If it covers 1, 5, or 10 particular qualities at the most critical digit, parcel the range into 5 interims (refer to Figure 2.15).

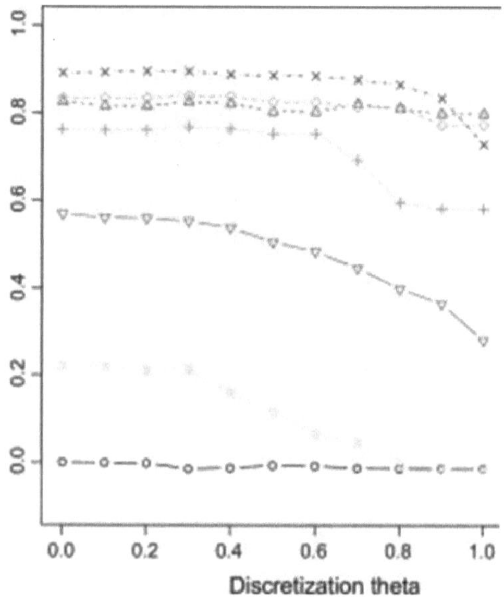

FIGURE 2.14 Binary discretization-based clustering in a visual format.

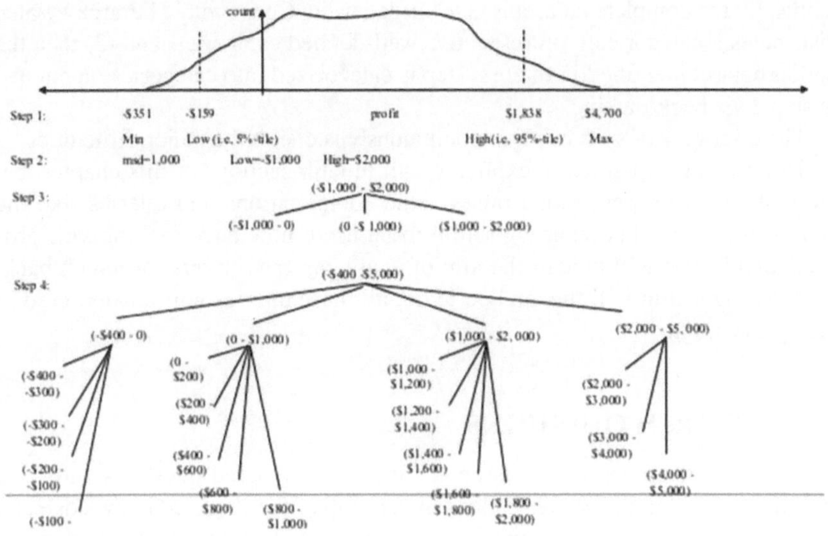

FIGURE 2.15 Understanding 3-4-5 rules via visual representation.

2.6 PURPOSE OF THE STUDY

The absolute and sole drive of this study is to provide a transformed approach towards improving the process of EDA by proposing a novel perspective to define, detect and analyze outlier data, unlike most the pre-existing studies which propose to detect and remove outlier data, it defines and develops a well-defined model to not only detect but analyze and fit outliers instead of removing or modifying these datasets. This sets the proposed methodology apart as by analyzing instead of deleting or manipulating datasets, it sanctuaries the authenticity and legitimacy of collected data, thereby also increasing the correctness of the results or any kind of algorithm or conclusion derived from it.

The requirement for quick, online algorithms to break down high information rate estimations is an indispensable component, out of all things considered and relevant settings. Given the ever-expanding number of information sources combined with expanding intricacy of uses, and the outstanding task at hand designs, irregularity identification techniques ought to be of low unpredictability and must work continuously. In numerous cutting-edge applications, the information shows up in a gushing manner. Thusly, the hidden presumption of old-style strategies to handle complex data is an example from a basic distribution that is obsolete now, and Gaussian and nonparametric-based strategies are deficient. Spilling information can be viewed as an ever-changing superposition of distributions. Thus, there is a bearing and exigent need of proposing a colossus approach to rally the process of EDA not only for outlier's sake but as a whole.

It aims to do so by using (1) data mining techniques like Clustering to identify outlier data points graphically and create a new dataset of the identified data points, (2) the complete dataset is categorized using Conceptual Hierarchy which distributes it categorically in respective well-defined categories, and (3) then the outlier dataset identified in the first step is categorized into categories mentioned in step 2 via backtracking.

The discovery of such changes continuously is one of the major difficulties.

This will be explained in explicit detail in later sections of this chapter, but the hypothesis proposed also raises some complications and queries like the details about the clustering algorithm to be used, how backtracking will proceed, and how it will benefit the aim of analyzing the outliers, or which backtracking algorithm will the applied to do, all these queries will be answered in the next section.

2.7 RESEARCH QUESTIONS

This section illustrates and calls upon various unanswered questions and open ends that are left in the above sections in the heap of pre-required knowledge and assumptions, all such queries and questions answered lucidly with detailed explanation for the reader, here is the list.

What is the difference between outlier and minority datasets and how is it related to imbalanced classification?

An imbalanced classification issue is a case of a classification issue where the dissemination of models over the realized classes is one-sided or skewed. The dispersion can change from a slight predisposition to an extreme irregularity where there is one model in the minority class for hundreds, thousands, or a great many models in the greater part class or classes.

Imbalanced classifications represent a test for predictive modeling as the greater part of the AI calculations utilized for classification were planned around the presumption of an equivalent number of models for each class. This outcome in models that have a poor predictive presentation, explicitly for the minority class. This is a problem because typically, the minority class is more important, and therefore the problem is more sensitive to classification errors for the minority class than the majority class.

2.8 WHAT ARE THE CAUSES AND CHALLENGES OF IMBALANCED CLASSIFICATION?

Various challenges occur in EDA while handling imbalanced [22] data which occur mainly due to two crucial reasons: (1) insufficient data to build a model, in case the minority class has only a few examples (similar to dealing with small samples/small datasets), and (2) many "special cases" in the minority class, so that in the class itself, some kind of sub-clustering occurs, which might lead again to insufficient examples for correctly identifying such a sub-cluster.

Thus, other less general names may be used to describe these types of classification problems, such as:

- "Rare event" prediction
- "Extreme event" prediction
- "Severe class" imbalance

Errors may have been mentioned when gathering and recording the data. One kind of mistake may have been applying inappropriate class names to numerous models. On the other hand, the procedures or frameworks from which models were gathered may have been harmed or debilitated to cause the imbalance. Regularly in situations where the imbalance is brought about by an inspecting predisposition or estimation blunder, the imbalance can be remedied by improved testing strategies, as well as adjusting the estimation mistake. This is because the preparation dataset can't reasonable portrayal of the issue space that is being tended to.

2.9 WHAT ARE THE ISSUES THAT OCCUR WHILE ANALYZING ACTIVE DATASETS OF EPIDEMICS?

Epidemic datasets are not just datasets with certain persistent attributes but they are affected by vivid factors which change continuously and are never constant, some of the factors are the determination of infection rate that changes every minute, rate spread, household, country, state, and per square feet specific distribution of cases, factors that affect the spread of the epidemic, distribution of population in terms of population density along with distribution of cases of deaths and survival age and region-wise. All these data are collected, and then after rigorous pre-processing of data by cleaning and handling missing or outlier data, finally the dataset is ready to be processed.

Major problematic issues occur in data pre-processing as a lot of data collected at such sensitive times are often not authentic and complete, and thus requires outlier detection and handling with scaling of each data point.

2.10 WHAT IS ACTIVE LEARNING?

Active learning can be understood as a sort of "design description methodology" like transfer learning, which can likewise be utilized to use limited quantities of marked data. Having the option to appropriately use active learning will give you an integral asset that can be utilized when marked data are deficient. The main assumption in active learning is that if a learning calculation can pick the data it needs to gain from, it can perform superior to conventional strategies with generously less data for preparing.

Here, there exist generally three situations in which the student will question the names of occasions. The three most prominent circumstances that have been taken into course in the writing are:

- **Membership query synthesis**: Here the novice generates/constructs an instance (from some underlying pre-existing or already present distribution).

- **Stream-based selective sampling**: Here, the novice makes the hypothesis that acquiring an unlabeled instance is of no major cost to the process. Under this presumption, selecting each unlabeled instance individually and consequently persuading the learner to understand and see whether it intends to question the label of the instance or scrap it deliberating its formativeness.
- **Pool-based sampling**: In this scenario, one considers that there exists is a colossus pool of unlabeled and unorganized datum, since deliberating the "stream-based" selective sampling. Instances can be then strained from the pool according to some formative measure or attribute. This method is then constructed onto all presently available instances in the pool (or some subset of the pool if it is too arduous to handle all at once), and then the most prominent instances providing salient information are selected. Pool-based sampling is the most commonly adopted scenario among the currently existing active learning algorithms.

2.11 WHAT ARE THE CHALLENGES FACED IN DATA MINING?

- **Security and social challenges**: Decision-making techniques are implemented by the use of information distribution; thus, it becomes pivotal to enlist and enable security provisions to prevent any kind of information leakage to authorities with malicious intent. Sequestered or personal data related to individual profiles which are restricted to a particular authorization, i.e., petulant data are congregated by analyzing and pre-processing client profiles. Illicit entrée access to the classified and petulant idea of information can be a noteworthy incident of concern as it is considered as an act of perfidy by the client who owns the data.
- **User interface**: The knowledge discovery process in data mining using various algorithms and soft wares is helpful only if it is stimulating enough or relatively reasonable by the client. Beginning with insightful visualization and detailed analysis of provided information, excavation or KDD (Knowledge Discovery process) results can be expedited and comprehend effectively while fathoming their volatile demands and requisites. To receive a more perspicacious visualization of data, "n" number of procedures are performed in the field on tons of datasets and to manipulate mined knowledge for obtaining business intelligence to aid decision-making.
  (i) Abstraction-based mining procedure: The process of "data mining" demands to show cooperation because it provides the freedom of choice to clients to focus on pattern verdict, understanding, and optimizing requisites for data mining based on outputs received post-procedure.
  (ii) Integrating of contextual information: Previously obtained datum may have been used to display and visualize freshly discovered patterns to give some direction to the exploration and discovering the procedure and to express discovered patterns in a more meaningful way.

- **Mining methodology mhallenges**: "Data Mining" or Knowledge Discovery may have great efficacy, but if not provided with clean and noise-free input data or "pre-processed" data, the results are not likely to be accurate as available data has many discrepancies like "redundancy", "noise" and missing terms which elude the actual results. Thus "data pre-processing" is one of the most primordial yet challenging tasks in the mining procedures.
- **Multifarious datasets**: Standardized and approved data is often heterogeneous in nature, and it could be a simpler or complex form of multimedia data like heterogeneous datasets of pictures, audio, multimedia datasets, "transient data", "spatial data", time contingent data, and characteristic language content to name a few. It becomes cumbersome to assay and analyze vivid sort of data and extract the necessary and useful datum.
  (i) Multifaceted data types: A databank of this kind often possess multifaceted data points and figures like "graphical data", "spatial data", "non-continuous streams", and "temporal data" along with data streams or data from different data sources in varied categories. Mining of this kind of data cannot be practically be done on one device or by implementing one kind of methodology.
  (ii) Mining from a wide range of sources: The data is congregated via vivid bases courtesy of multiple databases and datasets available. The data source is often natured to be of different categories and forms. Thus depending on the nature of the databank required, the collected information points are kept in formats like "structured", "semi-structured" or "unstructured" data, "stream data"(continuous or non-continuous), "free data", etc.

2.12 DEFINITION OF TERMS, KEYWORDS, ALGORITHMS, AND METHODOLOGIES

This section of the chapter defines all the technical terms floated in the previous and upcoming sections in simple terms so as for the better understanding of the reader who may or may not form the technical domain.

1. **Machine learning**: It is a problem solving and applied presentation of AI that provides a complete and explicit framework the capacity to consequently take in and improve for a fact without being unequivocally customized. Machine learning centers around the advancement of PC programs that can get to information and use it learn for themselves.
2. **Ebola**: It is a primordial yet destructive contagion resulting in health issues including fever, expedited body throbs, along with looseness of the bowels, and here and there, seeping inside and outside the body. As the infection spreads through the body, it harms the invulnerable framework and organs. At last, it causes levels of blood-thickening cells to drop, prompting lethal scenarios.

3. **Poliomyelitis**: Also popularly known as polio, short for poliomyelitis, or lossless motion, is an enervating illness caused by the presence of the "poliovirus". In almost 0.5% of the total cases recorded, there are muscle shortcomings bringing about powerlessness of movement.
4. **Dengue**: It is a mosquito-borne viral infection. There are four serotypes of the malady, or in other words, is possible to be come infected "four" times. Prominently, most of the germs produce very mild illness, dengue can cause an acute flu-like illness which is difficult to detect from symptoms alone and requires a blood test. Though, sporadically if not treated properly leads to lethal health conditions.
5. **Gaussian distribution**: The Gaussian distribution is a persistent capacity that approximates the specific binomial distribution of occasions. The mean worth is "$a=np$", where n is the number of occasions and p the likelihood of any whole number estimation of x (this articulation continues from the binomial distribution).
6. **scipy.stats**: It is a statistical function very widely used (scipy.stats), and this unit comprehends a myriad of probability distributions as well as a growing library of statistical functions which has numerous applications in multiple fields.
7. **Multidimensional database**: A multidimensional database, also abbreviated as MDB, is a type of database that is specifically optimized for data warehouse and online analytical processing (OLAP) applications. An OLAP application that accesses data from a multidimensional database is known as a MOLAP (multidimensional OLAP) application.
8. **Sci-kit learn**: It is one the few most prominently used libraries for machine learning in Python in various algorithm applications. The ski-learn library has proven its efficacy over the several years of applications as it comprises tools for machine learning and statistical modeling including classification, regression, clustering, and dimensionality reduction.
9. **Visualizing of database**: "Data visualization" is simply the graphical presentation of information and acquired data. By using visual elements like pie charts, bar graphs, histograms, and maps, data visualization tools provide an accessible way to see and understand trends, outliers, and patterns in data to comprehend a more protean image of the data provided.
10. **DBSCAN (density-based spatial clustering of applications with noise)**: Prominently, abbreviated as "DBSCAN," clustering can identify outliers; that is, observations that won't belong to any cluster. Since DBSCAN clustering identifies the number of clusters as well, it is very useful with unsupervised learning of the data when we don't know how many clusters could be there in the data.
11. **OPTICS (ordering points to identify clustering structure)**: Here, the "ordering" points to attempt to identify the clustering structure (OPTICS) is an algorithm for finding density-based clusters in spatial

data. Additionally, a special distance is stored for each point that represents the density that must be accepted for a cluster so that both points belong to the same cluster.

12. **CURE (clustering using representatives)**: Simple clustering observations are graded together based on similarity, hoping to make accurate conclusions based on the clusters. Back to DBSCAN. DBSCAN is a clustering method that is used in machine learning to separate clusters of high density from clusters of low density.
13. **BIRCH (balanced iterative reducing clustering and using hierarchies)**: Is an "unsupervised data mining algorithm" used to perform "hierarchical clustering" over comparatively large datasets. In most cases, BIRCH only requires a single scan of a database.
14. **K-means**: K-means clustering is primordial & comparatively simplistic in nature. After determining the number of clusters and assuming it as "k", the last step is to fathom the center or centroid of these clusters.
15. **CLARANS (clustering large applications based upon randomized search)**: Is compelling and is the best practice for spatial information mining. CLARANS applies a system to look in a specific chart. Here, k is the predefined incentive to pick the k medoids; accordingly, the hubs in the chart are a lot of predefined informational collections.
16. **STING (statistical information grid)**: It for the most part uses the "density-based" and/or "grid-based" techniques for mining bunches in a huge multidimensional space wherein groups are viewed as denser locales than their environmental factors.
17. **Wave cluster**: This is one of the grid-based bunching calculations, which is frequently utilized for multidimensional huge scope information that contains an enormous number of disconnected exceptions.
18. **CLIQUE (clustering in quest)**: This algorithm, CLIQUE, actually is an abbreviation of Clustering QUEST is an IBM data mining system. It was developed by a group of researchers at IBM. It is known for its density-based, grid-based subspace clustering capabilities.
19. **Distance-based functions**: A distance function has extraordinary viability in giving separation between the components of a set. On the off chance that the separation is zero, at that point components are proportionate or else they are unique about one another though an edge separation is frequently considered to separate the two. A separation work is essentially a scientific recipe utilized by separation measurements. The separation capacity can vary across various separation measurements.
20. **LDBSCAN**: Popularly abbreviated as DBSCAN (density-based spatial clustering of applications with noise) is a common unsupervised learning method utilitarian in the task of model construction and various other machine learning algorithms.
21. **Skewed**: "Skewness" alludes to contortion or asymmetry in a symmetrical chime bend, or normal distribution, in a lot of data. On the off chance that the bend is moved to one side or one side, it is said to be skewed. A

normal distribution has a skew of zero, while a lognormal distribution, for example, would display some level of right-skew.

22. **Predictive modeling**: Predictive modeling is a procedure that utilizes data and stats calculated from data to anticipate outcomes with the aid of vivid data models. These models can be used to anticipate anything from sports outcomes and TV ratings to technological advances and corporate earnings, and are widely used as business and economic analysis models in the industry to predict sales, earning, loss, etc.
23. **Cyber security**: It refers to the infrastructure and organization of machinery, developments, and performances deliberated to guard networks, devices, programs, and data from attack, damage, or unauthorized access of any external or internal party.
24. **Point outliers**: Outliers differing from mundane datasets by a very small value, corresponding to a specific data point from the dataset while analyzing huge clusters of data.
25. **Home location register (HLR)**: It is the database maintained by the mobile service provider for the permanent subscriber and contains all the important info about call history, location, duration, the basic profile of the subscriber, etc.

2.13 APPLICATIONS AND FUTURE SCOPE OF THE STUDY

EDA is the most eminent aspect of data science, acting as a "filibuster" between business solutions and various concepts and methodologies of data science. Its lauds have been well elaborated upon in previous sections along with a detailed study on its various techniques, and this section concerns the "real-world applications and uses" of EDA and outlier detection. There are numerous fields like cyber security, network enhancement, data mining applications, and business intelligence where it is used, and their detailed explanations are as follows.

2.13.1 Credit Card Fraud Detection

- Fraud detecting regarding the issue of credit card
- Fraud identification for use of credit card

Information access and action records are divided across changeable classes, for example, the client identification subtleties like individual data identified with client profile [23]. The fraud detection and any criminal behavior are normally reflected in value-based records (recognized by point outliers). Accessibility of named records is certainly not a reprobate assignment since credit organizations have this total information accessible in expand subtleties. Besides, the information falls into unmistakable profiles based on the credit card client and individualistic movement. Consequently, profiling and clustering-based procedures are normally utilized in this space, by the use of EDA and outlier detection, both insurance fraud detection and credit card detection. This issue can be addressed by using three techniques, namely, (1) neural networks [24,25], (2) rule-based clustering [26], and (3) active learning aided with clustering [27].

This application is very frequently used by both police and government authorities, and is even mundane for credit companies to use as it helps them keeping track of the customer activities to detect patterns for business intelligence or for providing special benefits for regular customers on basis of credit rating, i.e., is calculated based on credit card activity analysis. Credit card rating is also very commonly used by banks for sanctioning loans to customers. Credit card monitoring is a very systematic and organized task of banks; data of every activity, transaction, and expenditure is noted and maintained in a database; and these details can easily be accessed by the government authorities in case of fraud detection.

2.13.2 Mobile Fraud Detection

In case of stolen mobiles & tracking the activities of certain phone stations, EDA, and robust outlier detection algorithms are used to understand, analyze activities for mobile monitoring, scanning the data in case of misuse of the account. All the activity data with elaborate details are stored in databases of network service providers, and such databases are called HLR and can be easily extracted from there. Calling activity is usually represented with call records. Each call record is a vector of continuous (e.g., call duration) and discrete (e.g., calling city) features which can be detected easily by clustering or visualization of the data on graphs or by using data visualization techniques and function available in predefined libraries. However, there is no inherent primitive representation in this domain because data may be organized and complete but is not properly scaled, thus requiring nascent developments to get a lucid picture of the activities on an account. Calls are aggregated by criteria like time, duration of the call, location of the caller, location of receiver, messages & media. The outliers correspond to a high volume of calls or calls made to unlikely destinations or call durations with likely longer or shorter time, and this depends upon what one is looking for.

Some techniques and algorithms to perform the needful are as follows.

Statistical Profiling Using Histograms [28,29]: data stream analysis like data streams in the form of a number, texts, messages, characters, locations, and duration is done by the application of KDD techniques like regression, classification, and time series analysis. With a specific emphasis on granularity, multiple alarms, and timely alarms, a framework is developed depending on the kind of activity analysis desired and attributes dedicated for the purpose. Some other methodologies used to do so are parametric statistical modeling [30], neural networks [31], and rule-based systems [32].

2.13.3 Insurance Claim Fraud Detection

An important problem in the property-casualty insurance industry is claims fraud, e.g., automobile insurance fraud and property insurance fraud to name a few.

Individuals and conspiratorial rings of claimants and providers manipulate the claim processing system for unauthorized and illegal claims. The data in this domain for fraud detection comes from the documents submitted by the claimants. The techniques extract different features (both categorical and continuous) from these documents. Typically, claim adjusters and investigators assess these fraud claims. These manually investigated cases are used as labeled instances by supervised and semi-supervised techniques for insurance fraud detection. Insurance claim fraud detection is quite often handled as a generic activity monitoring problem [17,33].

2.13.4 Insider Trading Detection

Insider trading is a marvel found in stock markets, where individuals make unlawful benefits by acting on (or spilling) inside data before the data is made open. Insider data can be of various structures by and large alluding to any data which would influence the stock costs in a specific industry. It could be information about a pending merger/obtaining, a psychological militant assault influencing a specific industry, or a pending enactment influencing a specific industry. The misrepresentation must be identified in an online way and as right on time as could be expected under the circumstances, to forestall individuals/associations from making illicit benefits. The data has temporal associations since the data is collected continuously. The temporal and streaming nature has also been exploited in certain techniques [34,35].

2.13.5 Medical Public Health Outlier

The data normally comprises patient records which may have a few distinct sorts of highlights such as patient age, blood gathering, and weight. The data may have a transient as well as spatial aspect to it. The data can have outliers because of a few reasons such as an unusual patient condition or instrumentation blunders or recording mistakes. A large portion of the present outlier detection methods in this domain target detecting odd records (point outliers). Regularly the marked data has a place with the sound patients, and consequently, a large portion of the methods embrace a semi-supervised approach. Another type of data handled by outlier detection strategies in this domain is time series data, such as electrocardiograms (ECG) and electroencephalograms (EEG). Collective outlier detection strategies have been applied to recognize outliers in such data. Several systems have additionally focused on detecting disease episodes in a particular region. Subsequently, outlier detection is a basic problem in this domain and requires a high level of exactness. The most challenging aspect of the outlier detection problem in this domain is that the expense of classifying an outlier as typical can be high [36–38].

2.13.6 Industrial Damage Detection

Industrial units suffer damage because of continuous utilization and typical mileage. Such damages should be distinguished ahead of schedule to forestall further heightening and misfortunes. The data in this domain is generally sensor data recorded using various sensors and collected for analysis [39–41].

2.13.7 Image Processing

Outlier detection here expects to distinguish changes in an image after some time (motion detection) or in regions that seem abnormal on the static image. This domain includes satellite imagery, digit recognition, spectroscopy, mammographic image, and video surveillance. The outliers are caused by motion or insertion of foreign objects or instrumentation errors. The information has spatial just as temporal qualities. Every datum point has a couple of continuous properties, for example, color, delicacy, and surface. The intriguing outliers are either irregular focuses or regions in the images (point and contextual outliers). One of the key difficulties right now is the enormous size of the info. The test is more prominent when managing video information which is continuous in nature.

Some of the areas where it is used are satellite imagery, digit recognition mammographic image analysis, and spectroscopy video surveillance [42–46].

2.13.8 Outlier Detection in Text Data

Outlier detection strategies right currently on a very basic level perceive novel focuses or events or reports in an arrangement of records or reports. The outliers are caused by another charming event or a sporadic point. The data were right now ordinarily high dimensional and scanty. The data also has a transitory point of view since the records are congregated over some time. A typical and viable test for outlier detection techniques to deal with the huge assortments in records having a spot with one class or points. Some algorithms to do the needful are support vector machines, statistical profiling using bar graphs, pie charts, mixture of models, box plots, whisker plots, histograms, and neural networks [47–50].

2.13.9 Sensor Networks

Sensor systems have recently become a critical subject of investigation from a data assessment perspective since the data congregated from differing remote sensors has a couple of uncommon characteristics. Exemptions in such cases can either recommend at any rate one imperfect sensor (sensor culpability detection applications), or the sensors are distinguishing events ("intrusion detection" applications) that are intriguing for inspectors. A lone sensor framework may incorporate a mix of sensors that get together different sorts of data, for instance, twofold, mix, discrete, nonstop, sound, video, and sight and sound in particular. The data is made in a spilling mode, and the accumulated data as often as possible contains racket and missing characteristics due to requirements constrained by the association condition and correspondence channel. This speaks to a lot of novel troubles. The ceaselessly gushing data calls for exemption detection strategies to work in an online technique. The genuine resource objectives call for lightweight detection strategies. The data accumulated in a scattered way requires a dispersed data mining approach to manage to separate the data. In conclusion, the nearness of commotion in sensor data makes anomaly detection additionally testing, since it needs now to recognize intriguing exceptions and the undesirable qualities (noisy/missing datum). Some algorithms to do the needful are Bayesian networks, parametric statistical modeling, nearest neighbor-based techniques, and spectral techniques [51–54].

2.13.10 Miscellaneous Domains

Other domains of applications include traffic monitoring, speech recognition, originality recognition in robot behavior, clickbait protection, detecting issues in web applications and software in the industry, detecting outliers in biological data for research purposes, detecting outliers in census data for prediction, and detecting or matching relation and pattern among criminal activities [55–61].

Various research works like the hybrid method for mining rules based on enhanced Apriori algorithm with sequential minimal optimization in the healthcare industry [62] and handling failures in semantic web service composition through replacement policy in healthcare domain are happening nowadays [63].

2.14 CONCLUSIONS

The number of novel pharmaceutical items and technologies submitted to and authorized by regulatory organizations has decreased considerably in recent years. The medical product development process is no longer able to keep up with rising technologies, science, and innovations, and the objective is to develop new scientific and technological tools, and to make product development procedures more efficient and effective. Statistical methods in healthcare focus on the use of statistical approaches to analyze prospective alternatives, maximize performance, and establish the usefulness of those that merit further investigation. In this chapter, the statistical approaches used in health care planning, delivery, and monitoring, as well as specific statistical elements of pharmaceutical and medical device research are elaborately discussed.

REFERENCES

1. Ndanguza, D., Tchuenche, J. M., & Haario, H. (2013). Statistical data analysis of the 1995 Ebola outbreak in the Democratic Republic of Congo. *AfrikaMatematika*, 24(1), 55–68.
2. Billings, L., & Schwartz, I. B. (2002). Exciting chaos with noise: unexpected dynamics in epidemic outbreaks. *Journal of Mathematical Biology*, 44(1), 31–48.
3. Komorowski, M., Marshall, D. C., Salciccioli, J. D., & Crutain, Y. (2016). Exploratory data analysis. In MIT Critical Data (Ed.). *Secondary Analysis of Electronic Health Records* (pp. 185–203). Springer, Cham.
4. Wu, J., Wang, J., Meng, B., Chen, G., Pang, L., Song, X., Zhang, K., Zhang, T., & Zheng, X. (2004). Exploratory spatial data analysis for the identification of risk factors to birth defects. *BMC Public Health*, 4(1), 23.

5. Haug, D., MacEachren, A. M., Boscoe, F., Brown, D., Marrara, M., Polsky, C., & Beedasy, J. (1997, October). Implementing exploratory spatial data analysis methods for multivariate health statistics. In *Proceedings of GIS/LIS* (Vol. 97, pp. 28–30).
6. Wise, S., Haining, R., & Ma, J. (1997). Regionalisation tools for the exploratory spatial analysis of health data. In Fischer, M. M., & Getis, A. (Eds.). *Recent Developments in Spatial Analysis* (pp. 83–100). Springer, Berlin, Heidelberg.
7. Stelmack, J., Szlyk, J. P., Stelmack, T., Babcock-Parziale, J., Demers-Turco, P., Williams, R. T., & Massof, R. W. (2004). Use of Rasch person-item map in exploratory data analysis: a clinical perspective. *Journal of Rehabilitation Research & Development*, 41(2), 233–241.
8. Stone, L., Olinky, R., & Huppert, A. (2007). Seasonal dynamics of recurrent epidemics. *Nature*, 446(7135), 533–536.
9. Trevelyan, B., Smallman-Raynor, M., & Cliff, A. D. (2005). The spatial structure of epidemic emergence: geographical aspects of poliomyelitis in north-eastern USA, July–October 1916. *Journal of the Royal Statistical Society: Series A (Statistics in Society)*, 168(4), 701–722.
10. Chowell, G., Cazelles, B., Broutin, H., & Munayco, C. V. (2011). The influence of geographic and climate factors on the timing of dengue epidemics in Perú, 1994–2008. *BMC Infectious Diseases*, 11(1), 164.
11. Di Fatta, G., Blasa, F., Cafiero, S., & Fortino, G. (2011, December). Epidemic k-means clustering. In *2011 IEEE 11th International Conference on Data Mining Workshops* (pp. 151–158). IEEE, Vancouver.
12. Ben-Gal, I. (2005). Outlier detection. In Maimon, O., & Rokach, L. (Eds.). *Data Mining and Knowledge Discovery Handbook* (pp. 131–146). Springer, Boston, MA.
13. Yoon, K. A., Kwon, O. S., & Bae, D. H. (2007, September). An approach to outlier detection of software measurement data using the k-means clustering method. In *First International Symposium on Empirical Software Engineering and Measurement (ESEM 2007)* (pp. 443–445). IEEE, Madrid.
14. Pamula, R., Deka, J. K., & Nandi, S. (2011, February). An outlier detection method based on clustering. In *2011 Second International Conference on Emerging Applications of Information Technology* (pp. 253–256). IEEE, Kolkata.
15. Duan, L., Xu, L., Liu, Y., & Lee, J. (2009). Cluster-based outlier detection. *Annals of Operations Research*, 168(1), 151–168.
16. Loureiro, A., Torgo, L., & Soares, C. (2004, January). Outlier detection using clustering methods: a data cleaning application. In *Proceedings of KDNet Symposium on Knowledge-based systems for the Public Sector.* Springer, Bonn.
17. He, Z., Xu, X., & Deng, S. (2003). Discovering cluster-based local outliers. *Pattern Recognition Letters*, 24(9–10), 1641–1650.
18. Niu, K., Huang, C., Zhang, S., & Chen, J. (2007, May). ODDC: outlier detection using distance distribution clustering. In Washio, T. et al. (Eds.). *Pacific-Asia Conference on Knowledge Discovery and Data Mining* (pp. 332–343). Springer, Berlin, Heidelberg.
19. Hautamaki, V., Karkkainen, I., & Franti, P. (2004, August). Outlier detection using k-nearest neighbour graph. In *Proceedings of the 17th International Conference on Pattern Recognition, 2004. ICPR 2004* (Vol. 3, pp. 430–433). IEEE, Cambridge.
20. Pires, A., & Santos-Pereira, C. (2005). Using clustering and robust estimators to detect outliers in multivariate data. In *Proceedings of the International Conference on Robust Statistics ICORS 2005*, Finland.
21. Han, J., Pei, J., & Kamber, M. (2011). *Data Mining: Concepts and Techniques.* Elsevier, Amsterdam.

22. Lemnaru, C., & Potolea, R. (2011, June). Imbalanced classification problems: systematic study, issues and best practices. In *International Conference on Enterprise Information Systems* (pp. 35–50). Springer, Berlin, Heidelberg.
23. Singh, K., & Upadhyaya, S. (2012). Outlier detection: applications and techniques. *International Journal of Computer Science Issues (IJCSI)*, 9(1), 307.
24. Aleskerov, E., Freisleben, B., & Rao, B. (1997, March). Cardwatch: a neural network based database mining system for credit card fraud detection. In *Proceedings of the IEEE/IAFE 1997 Computational Intelligence for Financial Engineering (CIFEr)* (pp. 220–226). IEEE, New York, NY.
25. Brause, R., Langsdorf, T., & Hepp, M. (1999, November). Neural data mining for credit card fraud detection. In *Proceedings 11th International Conference on Tools with Artificial Intelligence* (pp. 103–106). IEEE, Chicago, IL.
26. Bolton, R. J., & Hand, D. J. (2001). Unsupervised profiling methods for fraud detection. *Credit Scoring and Credit Control*, VII, 235–255.
27. Nguyen, H. T., & Smeulders, A. (2004, July). Active learning using pre-clustering. In *Proceedings of the Twenty-First International Conference on Machine Learning* (p. 79). Springer, Cham.
28. Fawcett, T., & Provost, F. (1999, August). Activity monitoring: noticing interesting changes in behavior. In *Proceedings of the Fifth ACM SIGKDD International Conference on Knowledge Discovery and Data Mining* (pp. 53–62). ACM Press, New York, NY.
29. Cox, K. C., Eick, S. G., Wills, G. J., & Brachman, R. J. (1997). Brief application description; visual data mining: Recognizing telephone calling fraud. *Data Mining and Knowledge Discovery*, 1(2), 225–231.
30. Aggarwal, C. C. (2005, April). On abnormality detection in spuriously populated data streams. In *Proceedings of the 2005 SIAM International Conference on Data Mining* (pp. 80–91). Society for Industrial and Applied Mathematics, Kluwer Academic Publishers, Dordrech.
31. Barson, P., Field, S., Davey, N., McAskie, G., & Frank, R. (1996). The detection of fraud in mobile phone networks. *Neural Network World*, 6(4), 477–484.
32. Phua, C., Alahakoon, D., & Lee, V. (2004). Minority report in fraud detection: classification of skewed data. *ACM SIGKDD Explorations Newsletter*, 6(1), 50–59.
33. Brockett, P. L., Xia, X., & Derrig, R. A. (1998). Using Kohonen's self-organizing feature map to uncover automobile bodily injury claims fraud. *Journal of Risk and Insurance*, 65(2), 245–274
34. Donoho, S. (2004). Early detection of insider trading in option markets. In *Proceedings of the Tenth ACM SIGKDD International Conference on Knowledge Discovery and Datamining* (pp. 420–429). ACM Press, New York, NY.
35. Horn, P. S., Feng, L., Li, Y., & Pesce, A. J. (2001). Effect of outliers and non healthy individuals on reference interval estimation. *Clinical Chemistry*, 47(12), 2137–2145.
36. Campbell, C., & Bennett, K. (2001). A linear programming approach to novelty detection. In *Proceedings of Advances in Neural Information Processing* (Vol. 14). Cambridge Press, Cambridge.
37. Wong, W.-K., Moore, A., Cooper, G., & Wagner, M. (2003). Bayesian network outlier pattern detection for disease outbreaks. In *Proceedings of the 20th International Conference on Machine Learning* (pp. 808–815). AAAI Press, Menlo Park, CA.
38. Manson, G. (2002). Identifying damage sensitive, environment insensitive features for damage detection. In *Proceedings of the IES Conference*. Swansea.
39. Ruotolo, R., & Surace, C. (1997). A statistical approach to damage detection through vibration monitoring. In *Proceedings of the 5th Pan American Congress of Applied Mechanics*. San Juan, PR.

40. Hickinbotham, S. J., & Austin, J. (2000). Novelty detection in airframe strain data. In *Proceedings of 15th International Conference on Pattern Recognition* (Vol. 2, pp. 536–539). IEEE, Barcelona, Spain.
41. Brotherton, T., & Johnson, T. (2001). Outlier detection for advance military aircraft using neural networks. In *Proceedings of 2001 IEEE Aerospace Conference.* Big Sky, MT.
42. Augusteijn, M., & Folkert, B. (2002). Neural network classification and novelty detection. *International Journal on Remote Sensing*, 23(14), 2891–2902.
43. Cun, Y. L., Boser, B., Denker, J. S., Howard, R. E. Habbard, W., Jackel, L. D., & Henderson, D. (1990). Handwritten digit recognition with a back-propagation network. *Advances in Neural Information Processing Systems*, 2, 396–404.
44. Spence, C., Parra, L., & Sajda, P. (2001). Detection synthesis and compression in mammographic image analysis with a hierarchical image probability model. In *Proceedings of the IEEE Workshop on Mathematical Methods in Biomedical Image Analysis* (p. 3). IEEE Computer Society, Washington, DC.
45. Chen, D., Shao, X., Hu, B., & Su, Q. (2005). Simultaneous wavelength selection and outlier detection in multivariate regression of near-infrared spectra. *Analytical Sciences*, 21(2), 161–167.
46. Diehl, C., & Hampshire, J. (2002). Real-time object classification and novelty detection for collaborative video surveillance. In *Proceedings of IEEE International Join Conference on Neural Networks.* IEEE, Honolulu, HI.
47. Baker, D., Hofmann, T., McCallum, A., & Yang, Y. (1999). A hierarchical probabilistic model for novelty detection in text. In *Proceedings of International Conference on Machine Learning.*
48. Manevitz, L. M., & Yousef, M. (2000). Learning from positive data for document classification using neural networks. In *Proceedings of Second Bar-Ilan Workshop on Knowledge Discovery and Learning.* Jerusalem.
49. Manevitz, L. M., & Yousef, M. (2002). One-class SVMS for document classification. *Journal of Machine Learning Research*, 2, 139–154.
50. Allan, J., Carbonell, J., Doddington, G., Yamron, J., & Yang, Y. (1998). Topic detecion and tracking pilot study. In *Proceedings of DARPA Broadcast News Transcription and Understanding Workshop* (pp. 194–218).
51. Janakiram, D., Reddy, V., & Kumar, A. (2006). Outlier detection in wireless sensor networks using Bayesian belief networks. In *First International Conference on Communication System Software and Middleware* (pp. 1–6). IEEE, New Delhi.
52. Branch, J., Szymanski, B., Giannella, C., Wolff, R., & Kargupta, H. (2006). In-network outlier detection in wireless sensor networks. In *26th IEEE International Conference on Distributed Computing Systems.* Lisboa.
53. Subramaniam, S., Palpanas, T., Papadopoulos, D., Kalogeraki, V., & Gunopulos, D. (2006). Online outlier detection in sensor data using non-parametric models. In *VLDB '06: Proceedings of the 32nd International Conference on Very Large Data Bases. VLDB Endowment* (pp. 187–198). Seoul.
54. Chatzigiannakis, V., Papavassiliou, S., Grammatikou, M., & Maglaris, B. (2006). Hierarchical outlier detection in distributed large-scale sensor networks. In *ISCC '06: Proceedings of the 11th IEEE Symposium on Computers and Communications* (pp. 761–767). IEEE Computer Society, Washington, DC.
55. Albrecht, S., Busch, J., Kloppenburg, M., Metze, F., & Tavan, P. (2000). Generalized radial basis function networks for classification and novelty detection: selforganization of optional Bayesian decision. *Neural Networks*, 13(10), 1075–1093.

56. Crook, P., & Hayes, G. (2001). A robot implementation of a biologically inspired method for novelty detection. In *Proceedings of Towards Intelligent Mobile Robots Conference*. Manchester.
57. Shekhar, S., Lu, C.-T., & Zhang, P. (2001). Detecting graph-based spatial outliers: algorithms and applications (a summary of results). In *Proceedings of the 7th ACM SIGKDD International Conference on Knowledge Discovery and Data Mining* (pp. 371–376). ACM Press, New York, NY.
58. Ihler, A., Hutchins, J., & Smyth, P. (2006). Adaptive event detection with time-varying poisson processes. In *Proceedings of the 12th ACM SIGKDD International Conference on Knowledge Discovery and Data Mining* (pp. 207–216). ACM Press, New York, NY.
59. Ide, T., & Kashima, H. (2004). Eigenspace-based outlier detection in computer systems. In *Proceedings of the 10th ACM SIGKDD International Conference on Knowledge Discovery and Data Mining* (pp. 440–449). ACM Press, New York, NY.
60. Lu, C.-T., Chen, D., & Kou, Y. (2003). Algorithms for spatial outlier detection. In *Proceedings of 3rd International Conference on Data Mining* (pp. 597–600). Melbourne, FL.
61. Lin, S., & Brown, D. E. (2003). An outlier-based data association method for linking criminal incidents. In *Proceedings of 3rd SIAM Data Mining Conference.*
63. Sornalakshmi, M., Balamurali, S., Venkatesulu, M. et al. (2020). Hybrid method for mining rules based on enhanced Apriori algorithm with sequential minimal optimization in healthcare industry. *Neural Computing & Applications.* doi:10.1007/s00521-020-04862-2.
63. Ramasamy, L. K., Kadry, S., Jayanthi, M. A., Wei, W., & Rho, S. (2020). Handling failures in semantic web service composition through replacement policy in healthcare domain. *Journal of Internet Technology*, 21(3), 733–741.

3 Machine Learning Algorithms for Healthcare Data Analytics

G. Shyamala
Mahendra College of Engineering

Ilavendhan A.
Vel Tech Rangarajan Dr. Sagunthala R&D
Institute of Science & Technology

CONTENTS

DOI: 10.1201/9781003050827-3

3.1 INTRODUCTION

Cognitive functions of humans, and even animals and birds, are to approach different attributes of life and differentiate them in their own ways. Similarly, artificial intelligence (AI) is a new developing technology that allows a computer or machine to think for itself via algorithms by embracing the cognitive thinking method.

The subset or applied AI has a learning methodology that is used for autonomous acquisition of knowledge through computer programs known as machine learning, which is mainly developed for its key indicators that include general intelligence, learning, motion and manipulation, knowledge presentation, planning, reasoning, and processing with cognitive function. It also utilizes human interpretation whenever required.

ML integrates both the inductive and deductive learning strategies of AI. Inductive learning involves generalizing the input and opts for a result, whereas deductive learning deduces new information from known information and facts. It acquires the datasets by heuristic functions when the classic model fails to meet the output within the time frame. Its accuracy depends on the amount of data collected; the huge amount of data will help to get a better model with efficient output.

3.2 TYPES OF MACHINE LEARNING ALGORITHMS

Figure 3.1 depicts the broad categorization of the ML Algorithm [1].

3.2.1 Supervised Learning

The supervised ML algorithms can forecast future occurrences by applying previously learned data to fresh data using labeled examples. The first learning method analyses a known training dataset that includes test data and then generates an inferred function to predict output values. After sufficient training, the system may offer objectives for any new input. It may also use the learning process to compare its output to the future output and identify the mistakes to improve the model accordingly.

- Supervised learning may be applied in the real world for spam filtering, risk assessment, image classification, fraud detection, and so on.

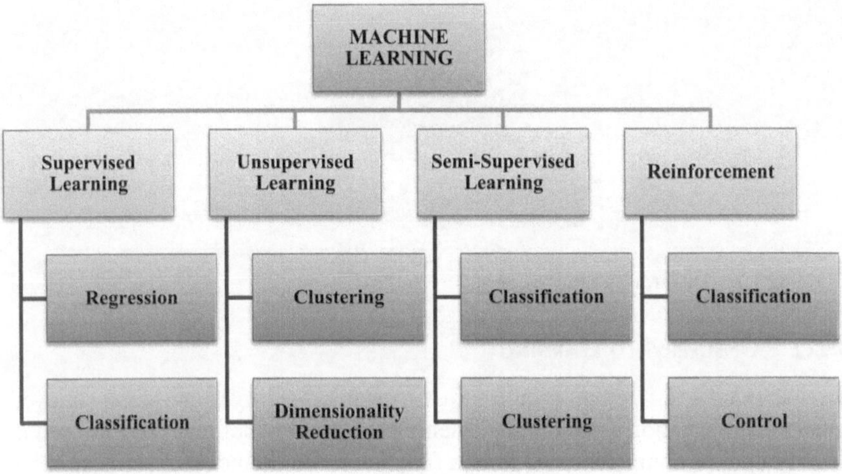

FIGURE 3.1 Flow diagram of types of ML.

- Steps involved in observed learning are as follows:
  - To begin, figure out what kind of training dataset is to be considered.
  - Gather training data that has been labeled.
  - Make a training dataset, a test dataset, and a validation dataset out of the training data.
- Determine the training dataset's input characteristics that will allow the model to properly predict the output.
- Choose an appropriate method for the model like support vector machine, random forest, linear regression, or a decision tree.
- Run the algorithm on the practice data. Validation sets, which are a subset of training datasets, are occasionally required as control parameters.
- Provide the test set to evaluate the model's correctness. The given model is accurate if it correctly predicts the result.

3.2.1.1 Algorithms for Supervised Machine Learning

3.2.1.1.1 Regression

It is a supervised ML technique in which the output is a continuous value concerning the number of given inputs, for example, weather forecasting and stock market analysis. If there is a link between the input and output variables, regression methods are utilized [2].

3.2.1.1.2 Classification

For classifying output variables, like Yes–No, True–False, and so on, classification methods are utilized.

- Spam Filtering and Random Forest are two of the categorization methods used.

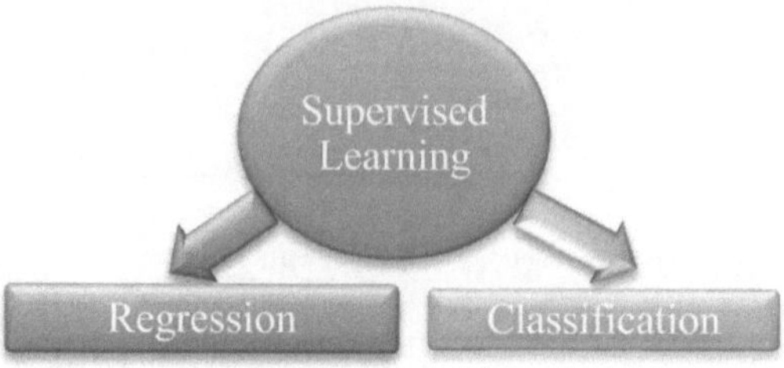

FIGURE 3.2 Illustration of types of supervised ML.

3.2.2 Unsupervised Learning

Unsupervised ML methods, in contrast to supervised learning, are employed when the information used to train is neither categorized nor labeled. It is used to investigate how computers may infer a function from unlabeled data to explain a hidden structure. To identify hidden structures from unlabeled data, the algorithm just looks at the data and derives conclusions from datasets. Unsupervised learning is a technique that allows a machine to learn without being supervised. The machine is taught given a collection of data that has not been labeled, classified, or categorized, and the algorithm is expected to operate on it without supervision. Unsupervised learning reorganizes the incoming data into new features or a collection of objects with the same patterns. There is no fixed outcome here. The computer tries to extract useful information from the massive volume of data. K-Means algorithm and Principal Component Analysis (PCA) are few examples of unsupervised learning.

3.2.3 Semi-Supervised Learning

Semi-supervised ML algorithm is the combination of unsupervised and supervised learning algorithms; it uses both the unlabeled data and labeled data for training. Usually, a minimum quantity of labeled data and the remaining maximum amount of unlabeled data is employed here.

3.2.4 Reinforcement Learning

Reinforcement learning is a feedback-based learning approach in which a learning agent interacts with the environment and receives a reward for each correct action and a penalty for each incorrect activity. Trial-and-error search-and-delayed rewards are the major aspects of reinforcement learning.

Reinforcement learning algorithm enables machines and software agents to automatically identify the best behavior for a given situation to enhance performance. For the agent to learn and pick the best action, it requires simple reward feedback, usually known as the reinforcement signal.

A learning agent must be able to perceive the state of its surroundings and perform actions that alter the state to some extent. In addition, the agent must have a purpose that is related to the condition of the environment. Any technique that is employed to solve such issues is regarded to be a reinforcement learning method. Markov decision processes are designed to contain only three aspects –: sensation, action, and objective – in their simplest feasible forms without trivializing any of them. It is frequently utilized in the present field of ML research [3].

A chess player, for example, makes a move. Both planning and predicting alternative responses and counter-responses, as well as quick, intuitive judgments about the desirability of certain movements, inform the decision.

3.2.5 Association Rules

The rule-based ML approach of association rules connects the differences between variables in a wide range of databases. Even with vast sets of inputs, bioinformatics uses these criteria to determine the sample's unique DNA and RNA pattern. There are several methods for creating association rules; some of them are as follows:

- A priori – It counts the support of item sets using a breadth-first search approach and a candidate generation function that makes use of support's downward closure characteristic.
- Eclat – It is a set intersection-based depth-first search method. It has locality-enhancing features and is suited for both sequential and parallel execution.
- FP-growth algorithm.
- OPUS.
- Lore.
- Multi-Relation Association Rules.

3.3 MODELS OF MACHINE LEARNING

3.3.1 Support Vector Machine

Support Vector Machine (SVM) is a familiar supervised learning model specialized for classification and regression analysis in ML. An SVM model method generates a decision boundary that divides n-dimensional space into classes, and a new data point is placed in the right category for further analysis using that line or boundary. A hyperplane is the optimum decision boundary utilized in SVM. Figure 3.3 gives the illustration of SVM.

The hyperplane is constructed using extreme points or vectors, which are known as Support Vectors, and hence the method is called SVM.

Figure 3.3 depicts how a decision boundary is utilized to classify two different categories.

Kernel trick is used to perform a nonlinear classification, and it implicitly maps the input into the high-dimensional feature spaces.

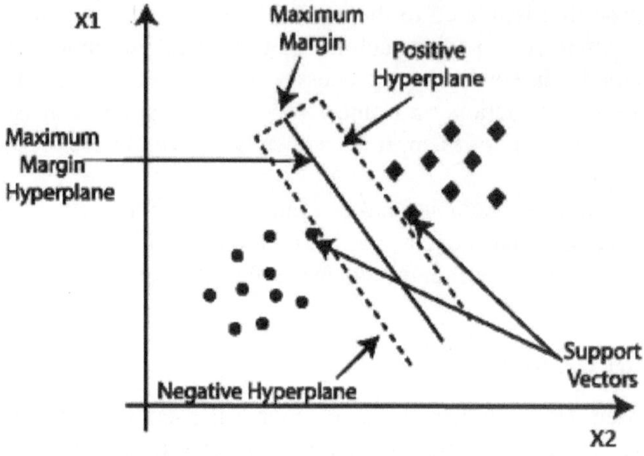

FIGURE 3.3 Illustrate SVM Model.

3.3.2 Bayesian Networks

A Bayesian network model uses a Directed Acyclic Graph (DAG) to describe a set of variables and their conditional relationships. It is otherwise called a Belief network, Bayes network, Decision network, Bayesian model, or probabilistic directed acyclic graphical model.

Bayesian networks are great for estimating the chance that any one of multiple potential known causes was a significant element after an event has occurred [4].

A Bayesian network, for example, might be used to depict the probability connections between illnesses and symptoms. The network is frequently used to compute the likelihood of the existence of various illnesses given symptoms.

In Bayesian networks, effective algorithms may infer and learn. Dynamic Bayesian networks are used to model sequences of variables such as speech signals or protein sequences. Influence diagrams are generalizations of these networks that can depict and solve decision issues under uncertainty.

3.3.3 Artificial Neural Networks

Artificial neural networks (ANN) are based on biological neural networks seen in animal brains. ANN "learn" to do tasks by looking at examples, rather than being coded with task-specific rules. Each connection may convey a signal to other neurons, similar to synapses in the human brain. An artificial neuron that receives a signal analyses it and then sends a signal to the neurons to which it is connected.

In the image recognition process, they input photos that can be recognized by examining sample images that have been manually tagged the name of the object and then applying the results to additional images. They do this without knowing anything about cats, such as the fact that they require hair, tails, whiskers, and cat-like features. Instead, they use the instances to automatically develop identifying features.

An ANN) is made up of a network of connected units or nodes called artificial neurons, which are modeled after the neurons in a biological brain. In ANN implementations, the "signal" at a connection may be real, and therefore the output of each neuron is computed by some nonlinear function of the sum of its inputs. Edges are the names given to the connectors. The weight of neurons and edges generally varies as learning continues. The intensity of the signal is modified by the weight at a connection. In neurons, there can be a threshold that can transmit a signal as long as the mixed signal passes through it. In layers, neurons are often organized. Various layers can alter their inputs in various ways. Signals can pass from the first layer (the input layer) to the last layer many times (the output layer) [5].

The ANN approach's original objective was to solve problems in the same way as a human brain would. However, as time passed, the focus shifted to specialized activities, resulting in biological abnormalities. The applications of ANNs vary from computer vision, voice recognition, social network filtering, machine translation, board, and video games to diagnostics and even skills formerly thought to be reserved for humans, such as painting.

3.3.3.1 Components of ANN Are

- **Input layer (neurons)**: It accepts input, uses an activation function to integrate it with their internal state of activation and an optional threshold, and then uses an output function to generate output.
- **Hidden layer (weights and connections)**: It allows one neuron's output to be fed as an input to another neuron. Each link is given a weight that indicates its relative importance.
- **Output layer (propagation function)**: This layer computes the input to a neuron as a weighted sum of the outputs of its preceding neurons and their connections.

Hyperparameter: Its value is determined before the start of the learning process, whereas the values of parameters are determined during the learning process. Learning rate, number of hidden layers, and batch size are examples of hyperparameters. The values of a few hyperparameters can sometimes be influenced by the values of other hyperparameters.

3.3.4 Regression Analysis

Regression Analysis describes the relationship between a dependent (target) and independent (predictor) variables with one or more independent variables in ML. It allows us to see how the dependent variable's value varies about one independent variable while the other independent variables remain constant.

3.3.4.1 Method

By building a network between the variables that suit the supplied data points, the ML model may make predictions about the data. Regression displays a line or curves on a target-predictor graph that passes through all of the given data points with the smallest vertical gap feasible between data points and the regression line [6]. A model's strong connection is determined by the distance between data points and the regression line. Regression analysis is used to predict continuous or real data.

- Regressions can be used to predict rain using weather forecasting or to identify temperature variations.
- Research industry trends and requirements.
- Road accidents are predicted as a result of reckless driving.

3.4 APPLICATION OF MACHINE LEARNING IN VARIOUS FIELDS

ML is being used in our daily lives without us realizing it, for example, in Google Maps and Google Assistant. Here are some examples of real-world ML applications.

3.4.1 Medical Diagnosis

In medical science, ML is used for disease diagnosis and can build 3D models, for instance, physicians can predict the exact position of lesions in the brain. Some other major healthcare applications that use ML are as follows [7]:

1. Computer-Aided Diagnosis, X-ray, MRI, Ultrasound, AI and computer vision are combined with radiology and pathology image processing, to detect tumors, breast cancer, and lung cancer.
2. The order of nucleotides in DNA in DNA sequencing has expedited biological and medical research and discoveries in medical diagnostics, biotechnology, forensic biology, virology, and biological systematics. When healthy and mutant DNA sequences are compared, they can be used to detect many illnesses, including malignancies, and to guide patient therapy.
3. The Brain–Computer Interface (BCI) is utilized to map, aid, enhance, and restore human cognitive and sensory-motor functions.
4. Computational anatomy is used in mathematical, statistical, and data analysis methods for modeling and simulation of biological structures like the Bayesian model of computational anatomy context of medical imaging.

5. Bioinformatics is the field of genomics where identification of candidate genes and single-nucleotide polymorphism (SNPs) is used to understand the genetic basis of disease.

3.4.2 Image Recognition Process

It is one of the most widely used ML programs. It's used to identify things like items, people, locations, geographical areas, and digital pictures. Automatic friend tagging recommendation on Facebook is a well-known example of image recognition, face detection, and recognition algorithm that was developed by the Facebook project "Deep Face," which is responsible for face recognition and person identification in photos [8].

3.4.3 Speech Recognition

It is also known as "Speech to text" or "Computer speech recognition," and it is the process of translating voice commands into text. ML techniques are now widely utilized in a variety of voice recognition applications, for example. When using Google, we come across a feature called "Search by Voice," which, like Google Assistant, Cortana, and Alexa, follows voice commands [9].

3.4.4 Predicting Traffic

Google Maps is the finest example of traffic predictions since it shows us the proper path if we visit a new location, allowing us to discover the shortest route, and forecasts traffic conditions.

- Google Maps app and sensors provide real-time car position.
- At the same time, the average time has been taken on past days.
- It also takes into account input from its database to enhance performance.

3.4.5 Spam and Malware Filtering in Email

The emails we receive are sorted into three categories: essential, normal, and spam. The ML technology allows us to get primary mail in our inbox with the important symbol and questionable emails in our spam box. Gmail uses the following spam filters:

- Content Filter
- General blacklists filter
- Filters based on rules
- Filters for permissions
- Filter for the headers

For email spam filtering and virus identification, ML methods such as Bayes classifier, Multi-Layer Perceptron, and decision tree are employed.

3.5 HEALTHCARE DATA ANALYTICS WITH MACHINE LEARNING

Artificial intelligence (AI) is defined as human intelligence as a goal and algorithm, in which data science is an approach. Many healthcare applications were associated with ML and AI. In this section, we are going to discuss the NLP.

3.5.1 Natural Language Processing

3.5.1.1 Introduction

NLP is the process of a machine automatically processing authentic human language. It is a field of AI that deals with the interpretation of human-generated data in the form of text or speech.

Substantive narrative tests such as fitness tests, physical laboratory reports, surgery, and discharge resumes are a substantial part of medical clinical information which, if specific text processing methods are not utilized, are unstructured and not understandable to a computer program. NLP addresses these issues by identifying a set of disease-relevant keywords in clinical notes using historical databases, which are then entered into a training dataset and used to improve structured data for clinical decision-making [10].

Initially, ML and natural language processing (NLP) in healthcare arose from the following two major use cases: understanding human speech and extracting its meaning as a future reference.

By mapping the fundamental ideas and values to unstructured data in databases and papers, clinicians may utilize this knowledge for making decisions and analytics.

3.5.1.2 Important Use Cases of NLP Are

- **Speech recognition**: Over the years, NLP allowed doctors to transcribe notes for Electronic Health Record (EHR) data input; NLP has expanded its use case in speech recognition. ML is utilized to manage clinical documentation, relieving physicians from the clerical and complicated structure of EHRs and allowing them to focus more on patient care.
- **Data mining research**: Companies are reducing subjectivity in decision-making and delivering relevant medical information by incorporating data mining into healthcare systems.
- NLP-driven computer-assisted coding (CAC) promises to increase the coder's accuracy in extracting information regarding procedures and treatments.
- **Automated registry reporting**: Regulatory reporting burdens many healthcare IT systems when measurements like ejection fraction are not maintained as discrete values, resulting in inaccuracies. It extracts

values for automatic reporting and stores them in a format that the organization's analytics platform can use.

- **Clinical trial matching**: Patients for clinical trials are identified as important use cases for NLP and ML in healthcare because NLP can automate and streamline the process.

3.5.1.3 Implementing Predictive Analytics in Healthcare

In healthcare, predictive analytics combined with NLP can be used to identify high-risk patients and improve the diagnostic process.

Emergency rooms require immediate access to comprehensive data. For example, if Kawasaki's illness is not diagnosed or treated properly, it might lead to serious consequences. When compared to a human examination of clinician's notes, an NLP-based algorithm identified at-risk patients with Kawasaki illness with a sensitivity of 93.6% and a specificity of 77.5% [11].

A group of French researchers created another NLP-based system to monitor, identify, and prevent hospital-acquired infections (HAI) among patients. NLP aided in the rendering of unstructured data, which was then utilized to identify early symptoms of sickness and alert physicians. NLP may be done using a variety of methods; we will go through a couple of them briefly here.

3.5.1.4 Workflow of NLP

STEP 1: Text pre-processing

TF-IDF is known as the Term Frequency-Inverse Document Frequency, and it serves as the basic algorithm for extracting keywords. It is a statistical measure of the importance of a word in a collection or corpus of documents. The importance of a word increases proportionally to how many times it appears in the document, but this is offset by the frequency of the word in the corpus.

In healthcare, TF-IDF is used in finding patients' similarities in observational studies and also in discovering disease correlations from medical reports. It is also used in finding sequential patterns in databases. Figure 3.4 gives an example of sample analytics using TF-IDF.

STEP 2: Mathematical and statistical analysis with EDA

The next step focuses on the EDA (Exploratory Data Analysis). It is the critical process of performing initial investigation on data and discovers patterns that identify variance and test hypothesis and check assumptions with the help of summary statistics and graphical representation.

a. **Input**: The input dataset is a corpus, a document matrix.
b. **EDA**: Summarizes the main characteristics of the data set, often using visual methods.
   i. Top words: find the most common words for each comedian
      - *Data*: Determine the format of the raw data that need to start.
        Example: Choose which data format do you think would be better.

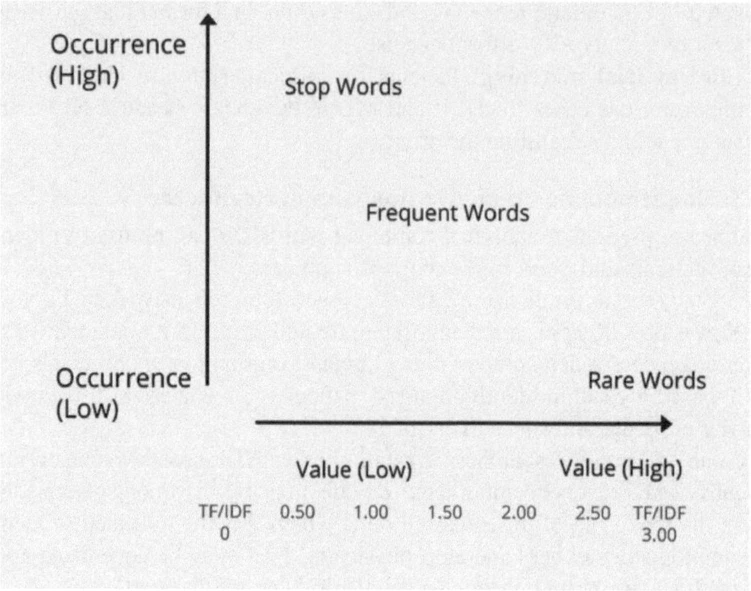

FIGURE 3.4 Sample analytics using TF-IDF.

Either the corpus or the document-term matrix.

- *Aggregate*: Figure out the aggregate data; i.e., for each trial (person who is a speaker), select the columns with the largest values.
- *Visualize*: Find the best way to visualize the data; i.e., let us say we found the top 30 words for each trial.
- *Insights*: Extract some key takeaways from the visualizations.

ii. Vocabulary: Take a look at the unique number of words used.

iii. Amount of profanity: Note the number of swear words said.

c. **Output**: Figure out the main trends in the data and if it makes sense.

STEP 3: Sentimental Analysis

Naïve Bayes

Naïve Bayes classifier acts as a base methodology for text categorization and predicting the category of the documents. It assumes that the presence of one feature in a class has no bearing on the presence of any other feature. Even if these characteristics are unrelated, there is a chance that they belong to a specific category [12]. For example, see Figure 3.5. Naïve Bayes is the most effective classification algorithm used for many medical problems such as the classification of journal articles and medical reports.

STEP 4: Topic Modeling

Topic modeling is the collection of the semantic structure of data, and it is collected with the popular technique called LDA (Latent Dirichlet Allocation).

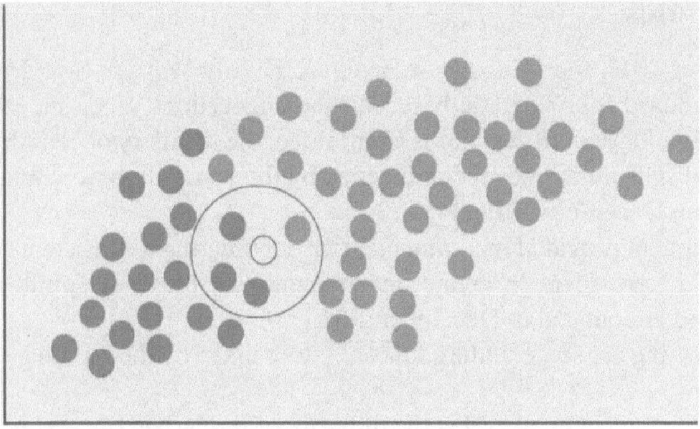

FIGURE 3.5 Example of Bayes Theorem.

STEP 5: Text Generation

The Text Generation is the regeneration of the words to text using Markov Chains. In the Markov chain, the data is predicted with the previous state.

The NLP is achieved with all the above steps and also can be modified further to produce the EHR or EMR with the vendor preference.

3.6 CASE STUDIES BASED ON EHR

3.6.1 Introduction to EHR

An EHR is an electronic version of a patient's medical history that is maintained by the treatment provider and may include all of the key administrative clinical data relevant to that person's care under a specific provider, such as medications, vital signs, past medical history, immunizations, laboratory data, and radiology reports.

The EHR streamlines the clinician's workflow by automating access to information regarding a specific patient. It also directly or indirectly supports other care-related tasks such as evidence-based decision support, quality monitoring, and outcomes reporting via multiple interfaces.

EHRs are the next stage in the evolution of healthcare, and they may enhance the interaction between patients and physicians. The data, as well as its timeliness and accessibility, will allow clinicians to make better judgments and deliver better treatment.

For example, the EHR may enhance patient care by:

- Minimizing the frequency of medical errors by increasing the quality and clarity of medical data;
- Making health information available, decreasing duplication of tests, reducing treatment delays, and informing patients to make better decisions.

3.6.2 Schemes

In September 2013, the Ministry of Health & Family Welfare (MoH&FW) proposed a standard for the EHR based on the suggestions recommended by the Electronic Health Record Standards Committee, the Ministry of Health & Family Welfare [13]. The guidelines recommended by the committee were amended and announced in December 2016 [14].

The document provided recommendations for creating a consistent framework for healthcare providers to create and maintain EHRs; these guidelines were amended and announced in December 2016 [13].

The following are some of the additional steps taken by the Indian government:

- In 2015, the National e-Health Authority of India (NeHA) was proposed to build an e-Health ecosystem in India, with the primary goal of establishing data management, privacy and security rules, guidelines, and patient health records following legislative regulations.
- The Healthcare Digital Information Act establishes national and state e-Health authorities and exchanges of health information; standardizes and regulates processes related to digital data collection, storage, transmission, and use; and ensures the confidentiality and security of digital health information; The National Institution for Transformation of India (NITI Aayog) proposes to generate digital health records for all inhabitants of India by 2022 for "The National Health Stack"; a futuristic digital framework.

3.7 ISSUES AND CHALLENGES

1. **Medical data transformation**: An ML algorithm makes use of the obtained quantity of data. In the real world, the medical data obtained are either in the form of sequences, trees, text descriptions, or combinations of any of these formats. The available qualitative data cannot be easily translated into the functional format, and processing is difficult, as numeric values are typically easily interpreted by the system. Hence, such algorithms that could transform the data easily are required under such a situation.
2. **Handling data stream**: Another problem with ML is fast data transfer and streaming. Though such data are readily available in abundance from several channels in the real world, including blood glucose monitoring, brain activity, electrograph, and MRI, there are still unresolved challenges regarding the effective analysis of fast-moving, large-scale memory consumption streaming data, feature selection, missing data, and computational complexity.
3. **Analyzing big data**: Another issue in learning is the availability of labeled data. Labeled data is expensive and time-consuming. Labeled data is required to contain all features required for the system for

training. Though the algorithm reuses data at every new data point and trains itself, it merely learns from experience. Some medical domains that want to function using ML become restricted due to the unavailability of labeling experts. Data Misuse, data sharing of sensitive details about patients, is another drawback under this domain.

4. **ML a stochastic approach**: ML is a stochastic approach, not a deterministic approach. It cannot differentiate between physical constraints though multiple researchers are looking out at eliminating such physical constraints today. The learning gets tougher with the presence of noisy data. Data augmentation works well only to a certain extend.
5. **Hardware standards for medical big data**: Learning using ML requires vast volumes of training data to work effectively. The medical data is very large, and the system needs to be built with enough processing power. High processor GPUs that are required for fast computing are impractical to be available in all medical institutions and hospitals where the actual data is captured and used. It becomes time-consuming and expensive to process the data by experts outside and transfer it to the system required. Though now-massive hardware designs have emerged to implement artificial neurons in a chip, such as Intel Curie, NuPIC, and software packages that have API for Python, Java, C++, and MATLAB, these are still under research.

3.8 CONCLUSIONS

Thus, we learned the basic types and modeling of ML that adopts the computational anatomy like mathematical, statistical, and data analytics and some of the algorithm that deploys the ML for healthcare analytics. That statistical computing will transfer the raw data into knowledge with the help of data mining and optimize the learning model as datasets. The information can be transferred from a remote area with the help of the Internet of Things (IoT). However, human interpretation is required for approving the analysis since healthcare is a dedicated and promising field in which one should not do any mistake that causes damage to life. We also discussed a few schemes and policies that the government had initiated to incorporate the EHR to maintain the health records, track them, help the patient to access their reports at any time, and help the clinicians make right decisions during treatment.

REFERENCES

1. Agrawal R, Imieliński T, Swami A. Mining association rules between sets of items in large databases. In *Proceedings of the 1993 ACM SIGMOD International Conference on Management of Data*, 1993 Jun 1 (pp. 207–216), San jose, CA.
2. Agrawal R, Srikant R. Fast algorithms for mining association rules. In *Proceedings of the 20th International Conference on Very Large Databases, VLDB*, 1994 Sep 12 (Vol. 1215, pp. 487–499) Santiago, Chile.

3. Wadhwa M. March 2020. CSD Working Paper Series: Towards a New Indian Model of Information and Communications Technology-Led Growth and Development.
4. Sutton RS, Barto AG. *Reinforcement Learning: An Introduction*. MIT Press; 2018 Nov 13. Trevor Hastie, Robert Tibshirani, Jerome Friedman. 2017. The Elements of Statistical Learning, 2nd Edition, UK.
5. Witten IH, Frank E. Data mining: Practical machine learning tools and techniques with Java implementations. *Acm Sigmod Record*. 2002 Mar 1;31(1):76–7.
6. Chatterjee S, Hadi AS. *Regression Analysis by Example*. John Wiley & Sons; 2015 Feb 25.
7. Kononenko I. Machine learning for medical diagnosis: History, state of the art and perspective. *Artificial Intelligence in Medicine*. 2001 Aug 1;23(1):89–109.
8. Svyrydov A, Kuchuk H, Tsiapa O. Improving efficienty of image recognition process: Approach and case study. In *2018 IEEE 9th International Conference on Dependable Systems, Services and Technologies (DESSERT)*, 2018 May 24 (pp. 593–597). IEEE, Ukraine.
9. Deng L, Li X. Machine learning paradigms for speech recognition: An overview. *IEEE Transactions on Audio, Speech, and Language Processing*. 2013 Jan 30;21(5):1060–89.
10. Khan W, Daud A, Nasir JA, Amjad T. A survey on the state-of-the-art machine learning models in the context of NLP. *Kuwait Journal of Science*. 2016 Nov 17;43(-4): 27–36.
11. Cohen IG, Amarasingham R, Shah A, Xie B, Lo B. The legal and ethical concerns that arise from using complex predictive analytics in health care. *Health Affairs*. 2014 Jul 1;33(7):1139–47.
12. Rish I. An empirical study of the naive Bayes classifier. *IJCAI 2001 Workshop on Empirical Methods in Artificial Intelligence*, 2001 Aug 4; 3(22): 41–46.
13. Mantri M. 2016. "Electronic Health Record (EHR) Standards for India (2016)," 57. https://www.itu.int/net4/wsis/archive/stocktaking/Project/Details?projectId=1515593078.
14. Ozair FF, Jamshed N, Sharma A, Aggarwal P. Ethical issues in electronic health records: A general overview. *Perspectives in Clinical Research*. 2015;6(2):73–76. doi: 10.4103/2229-3485.153997.

4 A Review of Challenges and Opportunities in Machine Learning for Healthcare

M. Arvindhan, D. Rajeshkumar, and Anupam Lakhan Pal
Galgotias University

CONTENTS

DOI: 10.1201/9781003050827-4

4.1 INTRODUCTION

Health issues affect human lives. Healthcare providers gather clinical data about each patient during medical treatment and use information from the common public to decide how to handle the patient. Consequently, data has a vital position in resolving health problems, and better knowledge is essential for improving patient care. Machine learning (ML) is an artificial intelligence (AI) application in which the optimization efforts are used for predictions or data, such as illustrations, experience, or guidance, figures patterns from data, and indicating future events based on indicators we provide. ML is increasingly being used throughout industries for a variety of reasons: vast volumes of data are now being detected and made accessible in a digital format; the scanning of large datasets seems to have become cost-effective due to the greater computing power now obtainable at competitive prices; and there are numerous free software applications, development tools, and community centers that could be used to develop and implement this [1–3].

ML has contributed to groundbreaking new advances, especially in healthcare, that could redefine cancer detection and management in the years ahead. ML will improve access to healthcare in developing economies that lack adequate medical specialists to treat those diseases, enhance sensitivity to diagnosis, add value to healthcare decisions, and help individualize treatment such that each patient receives the treatment that will be best for them. They would in many cases also contribute to the productivity of workflows in hospitals. The possibilities are endless [4–6].

4.2 IDENTIFYING DISEASE AND DIAGNOSIS

Health systems are increasingly being overburdened, inadequately resourced, and not prepared for the challenges they may face with increasing demographics and longer life expectancy. Scientists worked on ML models to predict susceptibility to infections or aid in premature diagnosis and treatment of diseases and disease. For what may be an incredible game-changer, a recent deep learning-based forecasting model has been urbanized by MIT's Computer Science & Artificial Intelligence Lab that really can estimate breast cancer growth for 5 years. Their system was built on mammograms and clinical follow-up data to recognize trends that individual clinicians might not consider apparent or even measurable [7,8].

4.2.1 Medical Imaging Diagnosis

ML algorithms can process large amounts of medical imaging at high speeds. And then they can be qualified to be exceptionally accurate in the detection of minuscule information in CT and MRI scans [9]. A deep learning coevolutionary human brain or CNN – designed and developed by Germany, France, and the US – can more reliably identify skin cancer than dermatologists can. The algorithm was able to correctly diagnose cancer in 95% of photographs of cancer moles and benign spots in a recently published test, while a team of 58 dermatologists was correct 87% of the time.

4.2.2 Robotic Surgery

By integrating real-time data, input from numerous previous operations, and data through electronic health records during the operation itself, AI- and ML-based techniques can increase the efficacy of surgical instruments [10]. This might help to alleviate human error and make a difference in general surgeons' performance of complex cosmetic procedures in resource-limited configurations where specialists are lacking.

4.2.3 Personalized Medicine

By applying AI and ML to multiple sources of data, viz. genetic data, patient records, sensor/wearable data, environmental, and genetic risk data, scholars have taken initial steps towards creating personalized therapies for diseases ranging from cancer to depression. IBM Watson Oncology is making great strides in treating cancer by using the medical history of patients to help create numerous treatment choices [11,12]. Similarly, a test called "Can Assist Breast" uses ML to classify a novel biomarker mixture that plays a key function in breast cancer recurrence.

4.2.4 Drug Development

ML may be practical at every segment of innovative drug development together with drug chemical/protein architecture design, target testing, drug safety review, and clinical study management. The idea is that the use of ML in drug discovery will not only help dramatically bring down the cost of bringing new medicines into the industry but will also make the experience of drug discovery quicker [13]. Most AI applications currently use the "deductive reasoning" model and therefore need to switch from "inductive reasoning" to "inductive reasoning."

4.2.5 Destructive Testing

As the name implies, destructive tests prepared to measure compression strength accomplish so with risking concrete structural's components possessions. The crushing test is the most commonly used tool for the hardness test.

4.2.6 Non-Destructive Testing

Concrete non-destructive techniques typically assess the identification of concrete compressive strength and damage. This research focuses on Rebound Hammer and Ultrasonic Velocity [14].

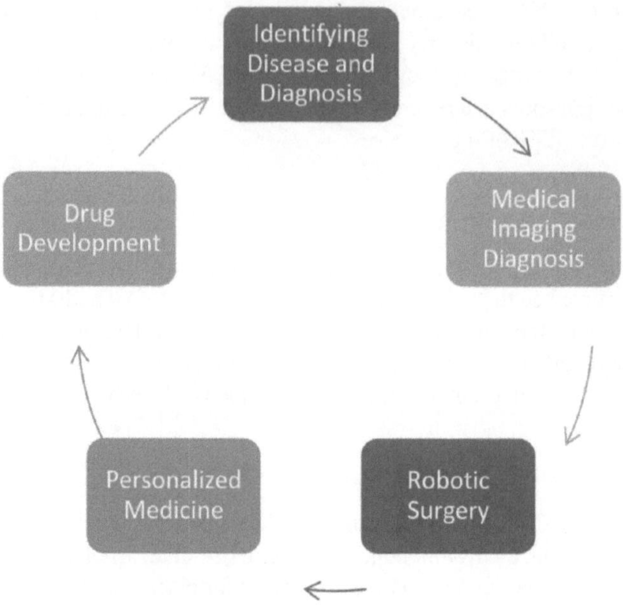

FIGURE 4.1 Five ways machine learning is redefining healthcare (based on lifecycle model of ML on healthcare system 2001).

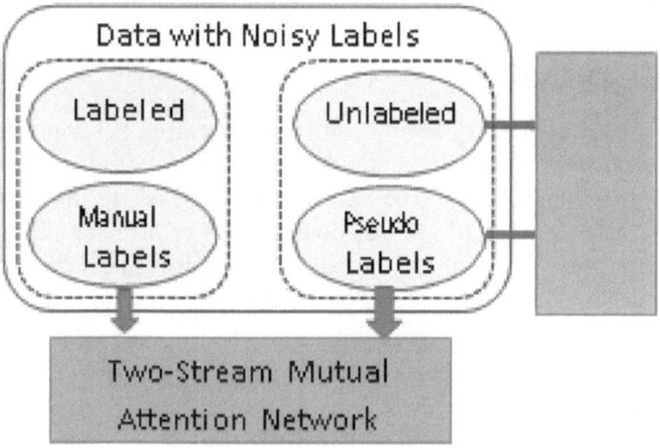

FIGURE 4.2 The pipeline of our self-training framework (structural model and designing format 2003).

4.3 SUPERVISED LEARNING TECHNIQUES

Those algorithms need external help. The dataset for inputs is split into training and testing datasets. The training database has changeable performance and a requirement to be estimated or graded. From the training dataset, the algorithm recognizes the pattern and adapts it all to the datasets for estimation. The well-known supervised machine learning (ML) approach selection is also handled as follows: Tree of Decision and Vector Machine Support [15].

4.3.1 Decision Tree Methods

A decision tree is a design that is capable of iterative process and is sophisticated such that each internal node connects in the direction of a check and each leaf (node) is associated with a classification. A decision tree is produced top-down using a simple recursive technique starting with targeted data collection. The fundamental idea of a decision tree is that it has a nearly split-up overarching rule that is interested in making clearer judgments, which might result in greater intelligible elucidation. It is indeed a high-quality results occurring in accordance with several forecast hypotheses. The usage of tree-based techniques was not widespread, especially in the non-destructive evaluation of tangible potency, though some experiments pay attention to the use of decision trees to evaluate concrete life cycle and some include try to determine dent to the structure through it. At the juncture of practical, non-destructive assessment and tree-based analysis, there seems to be a shortage of work to do [16]. We have been trying to resolve this in this research, particularly since the majority are non-destructive physical assessments.

4.3.2 Support Vector Machine

The SVM operates according to the theory of measuring margins. It draws, essentially, revenues and profits between training data. A dataset with a huge amount of investigational data is needed to predict exactly using SVM. The data illustrations are categorized as stated by the theory of maximizing menace design [5,17].

An array of data is provided by the ML-DB discussion tool. The important sections are as follows:

- patients
- observations
- medications
- conditions

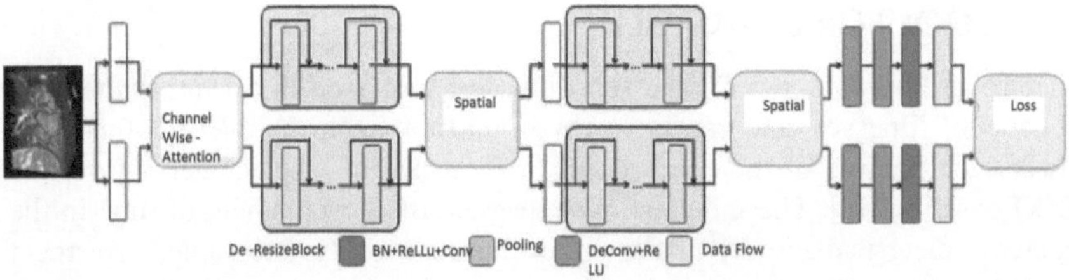

FIGURE 4.3 Channel-wise distribution of spatial data flow (distributed flow of data on channel-wise 2006).

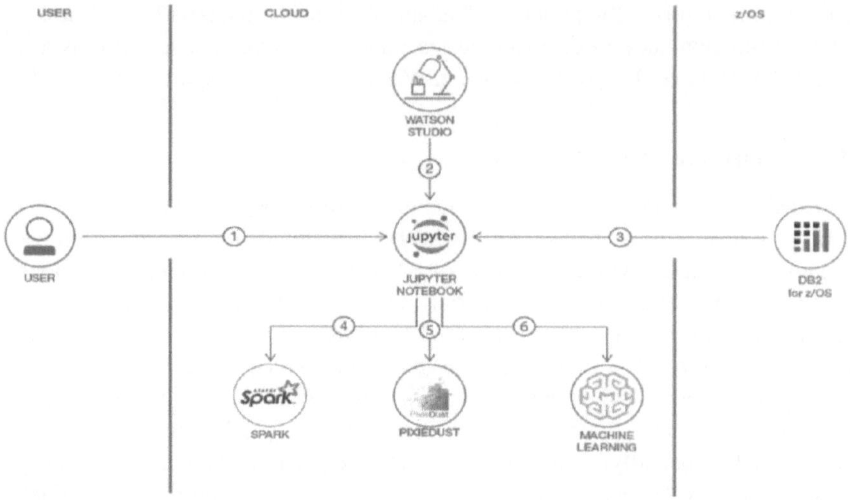

FIGURE 4.4 Method of storage on cloud platform (methods and platform variation on the structure storage cloud).

The ML-DB communication tool is called upon to produce a collection of CSV files that contain clean data on patient information care.

- Determining the maximum available number of patients in the DB2 for the z / OS data is called a JDBC program.
- The SQLite program was called to transform the ML-DB discussion tool-generated CSV folders to match the DB2 object model for z / OS.
- A JDBC software is called upon to load translated CSV files into the DB2.Health collection of code patterns for z / OS storage tables, which illustrates how cloud technology can access information stored on z / OS systems. To occupy the DB2 for the z / OS repository, we needed a way of generating a large amount of diagnosis healthcare data. We find an open-source tool named Synthea that produces synthetic data like this.

4.4 ELECTROCARDIOGRAPHY (ECG) TOOL FOR EVALUATION

Electrocardiography (ECG) is a semi-invasive tool used in cardiac anomalies, functional illnesses, and cardiac arrhythmia treatment and take-up. One may track each beating of the heart as just a sequence of displacements from the ECG baseline [18]. These deflections represent the development of time in the center of electrical activity that activates muscle activity. A complete heartbeat is also recognized as the cardiac cycle, the duration of instance starting through muscular contractions and ending using ventricular relaxation. A particular sinus (customary) cycle of ECG is usually labeled with the letters A, B, C, D, and E at each of its defining moments, equivalent to one heartbeat [19]. Because of atrial and ventricular depolarization and repolarization, we exploit the intention of an ECG signal as interpreted as a sequence of impressions: A wave pursued by a BCD collection and lastly an E wave. We classify certain particular waves in the unlabeled ECG signal of the person and refine the oppositional set of connections to produce waves identical to the arrhythmia shown in every patient although with a unique morphological cardiac signal of the identification issue throughout his customized arrhythmia. The signals garnered are most often utilized as ongoing training data for a profound beginner. Both types of heartbeats have an average duration of 700 ms, 300 ms earlier than the R max, and 500 ms after. The cardiac phases can be divided into unusual groups that can also reflect regular heartbeats or arrhythmias. The structures of the A, B, C, D, and E waves differ by rhythm class. We should focus on three types of heartbeats: Supraventricular ectopic beats, Ventricular ectopic beats, and combination of both [19].

4.5 FEATURE EXTRACTION

The Converter Loss function is characterized by an amalgamation of the traditional annoying entropy loss that attempts to produce cardiac pulses which trick the classifier with a Mean Square Error (MSE) function that attempts to create forged heartbeats that are much more morphologically close to actual heartbeats. The MSE loss mechanism restrains the produced heartbeat pulse values, A, B, C, D, and E to be as similar as possible to the phase values obtained in the training data from actual heartbeats [20].

$$\begin{aligned}\text{Loss } H(G) &= H((hbg,1)) = H((G(x),1)) \\ &= \left[-1 * \log D(G(y)) - (1-1)\log\left(1 - D(G(z))\right)\right] = -\log D(G(z))\end{aligned} \tag{4.1}$$

4.5.1 MSE Loss

To boost the generator's aptitude to produce artificial heartbeats that are further morphologically similar to actual heartbeats, we restrict the engender heartbeat wave ideals, A, B, C, D, and E, to be as identical to the wave principles extracted from actual heartbeats and in training evidence as possible. The values A, B, C, D, E, and T are calculated according to the mentioned algorithm. They add a restriction to the losses utility on behalf of how distant the wave is from both the transformer output was from the wave values taken as of the training data [21].

$$\begin{aligned}\text{Loss mse}(G) = &\ \text{MSE}(\text{Pdata}, \text{Pg}) + \text{MSE}(\text{Qdata}, \text{Qg}) + \text{MSE}(\text{Rdata}, \text{Rg}) \\ &+ \text{MSE}(\text{Sdata}, \text{Sg}) + \text{MSE}(\text{Tdata}, \text{Tg})\end{aligned} \tag{4.2}$$

Y is the matrix of practical component values being forecasted, then Pdata, Qdata, Rdata, Sdata, and Tdata are calculated from the training data.

$$\text{Loss}(G) = \text{Loss mse}(G) + \text{Loss } H(G) \tag{4.3}$$

With Adam optimizer, a processing rate of 0.0002 as recommended by training GAN models during optimization is achieved. The weights of the converter are educated two times every emphasis, whereas the discriminator is conditioned one time as proposed in.

4.6 SPATIAL (SA) AND CHANNEL-WISE ATTENTION (CA)

Because loss attention only extracts information from the two sub-networks' final predictions, we also exploit the same thematic information between the feature maps extracted by the two models. It presents two forms of attention models: spatial attention models and channel-wise attention models [22].

SA and CA during inference act as feature selectors, as well as gradient selectors during back-propagation.

4.7 ADDRESSING A HIERARCHY OF HEALTHCARE OPPORTUNITIES

Healthcare provides many elevated contact incentives. Goals must be clearly defined and confirmed as worthy of addressing before fitting the models. Here, we diagram future healthcare innovations into three high-level classifications: clinical process automation, clinical assistance, and clinical capability expansion. Definitions of how a technological solution is implemented may alter its purpose, so early involvement of clinical stakeholders is essential [23].

4.8 CLINICAL TASK AUTOMATION: AUTOMATING CLINICAL TASKS DURING DIAGNOSIS AND TREATMENT

There are several activities conducted by physicians offering low-hanging fruit to researchers in ML. Medical process automation covers a form of research currently being performed by clinicians [24]. These errands are quite well defined (i.e., established output and input spaces), and consequently involve the fewest number of action recognition and speculation required. Task substitution assessment is likewise straightforward performance and can be evaluated against existing criteria. Evaluation of the diagnostic picture is automatic. Medical imaging is an accepted resource for ML, as officials undertake rigorous training to record output (e.g., diagnosis) from a predetermined input space (e.g., images). Notable achievements in applying deep learning to image processing tasks are varied. Routine operations are automated. Similarly, automating clinical routine procedures means raising the stress imposed on clinical staff. For instance, prioritizing triage order is mostly left to personnel in the emergency room but could still be done algorithmically [25]. Summarizing the components of the paper medical records is also a time-consuming but useful task.

4.9 CLINICAL SUPPORT AND AUGMENTATION: OPTIMIZING CLINICAL DECISION AND PRACTICE SUPPORT

A similar problem with HACs is Adverse Drug Effects (ADE), which presents a significant challenge to patient safety, provider liability, and healthcare providers' financial status. Although all drugs on the marketplace are subject to rigorous clinical trials, a rigorous understanding of their ADEs is progressing for years after being accepted for therapy. The more data that can be processed easily, the easier it would be to avoid harm from medical procedures for individual practitioners as well as for the medical system as a whole. ML will help medical officials, insurance professionals, insurers, and other industry stakeholders deepen their understanding of this ocean of knowledge [26]. If the medications cause adverse reactions: Why? For what? ML is especially useful when there is an intersection of large quantities of data in various formats that need to be compiled and afterward extracted as information morphs and changes not just once, but continuously.

A group of Marshfield Clinic researchers developed a neural network in 2017 that examined 1.45 million medical journal articles and 420,000 social media messages in an attempt to uncover knowledge about ADEs. The findings in the journal papers yielded more than 12,000 ADE-related sentences and 181 ADE-related statements in social media posts. Researchers used that data to create a visualization, from weight gain to vertigo, that connected different drugs to different symptoms. Capper and colleagues agreed to concentrate on molecular knowledge that does not necessitate extensive visual considerations to be identified [27].We followed the introduction of functional groups to DNA using a ML approach to the classification of tumors caused by changes in DNA methylation and compared these diagnoses with those provided by pathologists using histological analysis. The computer used supervised ML to identify methylation patterns found in the pathologist-classified samples as well as unsupervised ML that involved the computer scanning the datasets for patterns it could use to assign samples to its computer-generated classification categories. After preparation, the computer was able to classify tumors into 82 distinct groups according to specific methylation profiles [28]. Among these, only 29 corresponded to a common tumor type as defined by the WHO, and 29 other tumor subclasses reported by the WHO. The authors used the device to identify 1104 cases of tumors that pathologists had identified using normal histological or molecular techniques (Figure 4.1). The computer-based classification was similar to the classification of the pathologist for 60.4% of these test cases and the computer and pathologist allocated the same category of tumor for 15.5% of the test cases but the computer could also classify the tumor to a subclass [29,30].

The machine diagnosis did not match the diagnosis of the pathologist in 12.6% of test cases. Remarkably, a more thorough study in this scenario, such as gene sequencing, resulted in the classification in 92.8% of these unexplained tumors being moved from the initial clinical diagnosis to the classification based on the computer [31,32].

In addition, 71% of reclassified tumors were given a separate category of the tumor, a categorization that may have consequences for prognosis or treatment. The machine could not identify the remaining test cases (11.5%). Additional computational research indicates that one-third of tumors in this category could be unusual tumors for which the machine has yet to find enough examples to produce a grouping of classifications [33].

It is helpful to obtain an extensive molecular summary of a tumor sampling, particularly when coupled with microscopic assessment, and it may be the way forward, while medical treatments turn increasingly tailored to the distinctiveness of a tumor. Nevertheless, histology remains indispensable for the diagnosis of diseases for now, as standard methods for specimen preservation and microscopic inspection provide the major open and consistent access point in the regular diagnostic process done in clinical laboratories throughout [33].

4.10 OPPORTUNITIES FOR NEW RESEARCH IN MACHINE LEARNING

Acknowledging the structure of healthcare resources provides multiple investment opportunities. Pertinently, healthcare workers and ML researchers also have balancing skills, and joint projects will only solve many of the high-impact issues. We mention many interesting research directions, specifically mentioning those that tackle non-constant data problems, model interpretability, and the discovery of correct representations [34]. New research works like the Internet of Things (IoT)-Based Humanoid Software for Identification and Diagnosis of COVID-19 Suspects [35] and Detection of diabetic retinopathy using a fusion of textural and ridgelet features of retinal images and sequential minimal optimization classifier [36] was carried out by few researchers.

4.11 ACCOMMODATING DATA AND PRACTICE NON-STATIONARY IN LEARNING AND DEPLOYMENT

Models are trained mostly on the biggest surface database in most current work and considered to be ready for deployment, i.e., systems do not continue to learn. This is troublesome in clinical practice, as patient communities and optional treatment processes will transform over time, leading to dishonored predictive efficiency as the target anthropogenic climate statistical properties change.

The preliminary model was an immense accomplishment, using Google exploration data to forecast influenza occurrence; nevertheless, in the following years with no updates, the model started to overrate influenza frequency as user explore patterns had changed. While the glide was accidental, the example encourages the necessity for constantly updating models [37]. Several factors affect predictive validity, including confined hospital procedures, dissimilar patient demographics, obtainable resources, and even the particular form of EHR one uses to transition from one EHR to the next that cause non-obvious mapping problems for apps. This problem should remain until the technology is more widespread for simple testing across numerous sites. Lack of such standardization creates incentives for data normalizing and model creation that is resilient to variations in data collection at different locations developing robust models to response loops. Models that learn from offered clinical practice can amplify the bias that is central to modern medicine. Although not yet observed in healthcare, these feedback loops were noted when predictive policing was implemented [38,39]. Such biases expressed in deployed projections can migrate into historic data on training, essentially creating a feedback loop that causes more bias. Operating in algorithmic justice should be taken into account as it empowers any need for systems that are sufficiently conscious to notify us of such inappropriate behavior. In fact, EHR-trained models may derive from current procedures rather than explicitly reflect best practice guidelines, which in certain clinical settings can be problematic due to low operational compliance. Executives in the world of ML have such a hard time placing a handle on whether ML makes its mark on primary care. We've written the story not to be a comprehensive list of potential applications but to illustrate a range of ongoing and prospective uses of medical ML. ML algorithms differ considerably in their structures, complexity, and also the conclusions their make about both the data to which they refer. To predict continuous (regression) or numerical (classification) variables, they may be linear or non-linear, but optimize various functions. Also, the performance of all ML algorithms depends on their hyperparameters being finely tuned. The result is that there are theoretically limitless variations of methods that could be taken out of the data to define associations. To reduce this complexity, it is necessary to develop tools that can automatically select the appropriate (a combination of) preprocessing and ML techniques to apply to a dataset to highlight relationships that are both generalizable and computer-efficient. Any real-world implementation of ML algorithms requires at least some customization level. The level of necessary automation for health data is very high for three reasons: the inherent complexity of the human body, the accessibility and usefulness of data sources, and incorporation into the current healthcare system.

With proper preprocessing of the data, simple models yield excellent results. Simple models performed well for us when we were developing ML algorithms to predict continuous vital signs as well as the physiology of the patient's state vector. Accurate physiological modeling makes a considerable difference and turns a feature that is poorly developed into one that is very relevant. Physiological modeling can also identify where a character may be appropriate and where it may not. That's one of the reasons we are using a more straightforward model ensemble and shying away from deep learning (Figure 4.8).

Personalized medicine will require mass customization of models that are trained at the hospital and cohort level and re-calibrated. Modern ML techniques focus not on generalization to different sites but on generalization beyond a training dataset. Transfer learning methods need further improvement to help resolve this problem and generalization needs to be done by localization in the meantime. Adapting a model to a local setting requires either the ability to collect and curate local datasets and retrain models at each location or the willingness to incorporate resources from outside.

4.12 PERSONALIZED MEDICINE

It's possible they would get Bactrim for a urinary tract infection (UTI). Few patients would ever get precisely the same dosage of any drug in the potentially not-too-distant future. If we know enough about the biology and history of the patient, perhaps the same medication can be prescribed to a few patients.

The promise of personalized medicine is a world where the health guidelines and therapies for diseases of all are customized to their medical background, genetic backgrounds, past illnesses, diet, stress levels, and more.

4.13 AUTOMATIC TREATMENT OR RECOMMENDATION

Although this can potentially extend to mild circumstances (i.e., offering someone a slightly lower dose of Bactrim for a UTI, or a completely new version of Bactrim designed to prevent side effects for a person with a different genetic profile), it is likely to have more of its initial impact in high-stakeholder situations (i.e., determining whether or not to undergo chemotherapy depending on an individual's age). In our article entitled "Where Healthcare's Big Data Comes From," we address issues related to personal medical data.

4.14 AUTOMATIC CARE OR GUIDANCE

In the video on diabetes produced by Medtronic and IBM, Medtronic's own Hooman Hakami notes that at some stage Medtronic wants its insulin control pumps to function autonomously, track blood glucose levels and administer insulin as required, without disrupting the everyday life of the patient.

This is a perfect reflection of a much more comprehensive view of autonomous diagnosis. Imagine a computer that could change the dosage of pain killers or antibiotics in a patient by monitoring blood, food, sleep, and stress data. Instead of relying on distractible humans to remember how many pills to take, a small kitchen table machine intelligence "agent" (think Amazon's Alexa) can dole out the pills, track how often you receive, then call a doctor if health seems to deteriorate or you've never followed its instructions.

The legal restrictions of putting so much control in an algorithm's "pocket" are not trivial. Like any other advancement in healthcare, free treatments of any type are likely to endure long journeys to prove their feasibility, protection, and superiority to other methods of treatment.

4.15 ERROR POTENTIAL AS APPLICABLE TO AI AND ML TO HEALTHCARE

Usage in genomics, where AI and machine learning are employed to classify patterns within a large volume of altered genetic datasets, could be confusing, with several possible choices and uncertain consequences. ML may theoretically assist clinicians in determining the most suitable and stable target sequences. In the future, avatars in AI-enhanced programs might be used as the first calling point in busy medical clinics. The application of remote medicines is essential for the future where access to medical and health facilities is often difficult in remote areas.

An online service could use an AI system to provide the public with first-hand information about minor ailments. Apps can allow patients to take control of their health data in the future and learn more about ways they can stay healthy.

Extracting medical data itself is a task, and the difficulty increases along with medical language precision and uncertainty. This discrepancy creates inherent differences between queries and records, which also make ineffective conventional IR approaches and ML algorithms. Such issues often impact dependencies in terms of grouping, hierarchies, and concept. For example, if a user enters how to treat a headache, a document containing the word acetaminophen may not be flagged as necessary to the system but is clearly so from a physician's or even an average human's point of view. The problem with this example is that to equate the words acetaminophen, can treat, and headaches, the model must be equipped with the appropriate domain knowledge. Additionally, these programs must be able to appeal to both individuals with and without the experience of medical terminology. The users may be physicians or patients, so technical terms need to be well meshed with ordinary, everyday words used.

This is an up and coming sector, as evident, with AI paving the way for medical advancements. The combination of powerful hardware and efficient algorithms makes it possible to integrate machines with medicine, with ML resurfacing in many domains. When reviewing papers for this study, only existing technologies were discussed that were either being used or being developed in the medical sector for real-world applications. Hypothetical applications were not considered for future work. This survey broadened both the area of computing science and the therapeutic area.

4.16 HEALTH DATA REQUIRES A HIGH LEVEL OF CUSTOMIZATION

Any real-world implementation of ML algorithms requires at least some customization level. The level of necessary automation for health data is very high for three reasons: the fundamental complexities of the human body, the accessibility and usefulness of data sources, and incorporation into the healthcare system. Organizations such as Spry Health are working tirelessly to address these problems: we already see the light at the end of the tunnel. When more data is

FIGURE 4.5 Machine learning database tool for array of data communication.

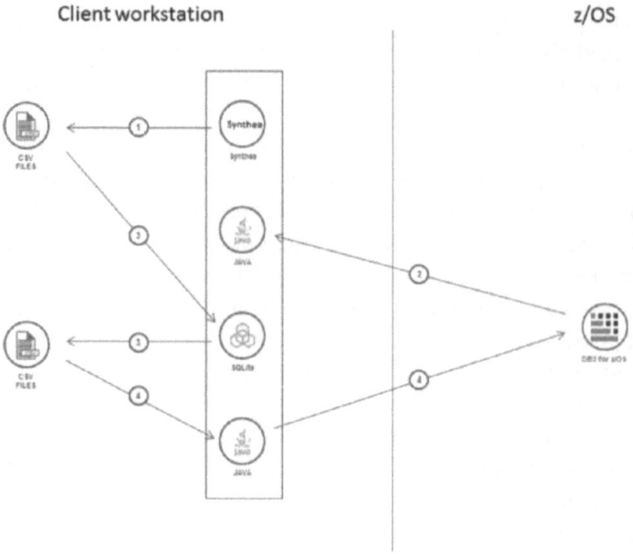

FIGURE 4.6 Transforming and loading big data CSV files into a DB2 for z/OS database.

available, barriers to the use of ML and AI should vanish. The major drawback of utilizing the AI and ML in healthcare is it requires huge data to perform analysis. Any learning pattern, person or computer, shows dependency on the quantity of data is used to train it. ML datasets that are publicly accessible, are scattered, and reasonably limited. We cannot compete with the comprehensive medical school training doctors undergo and the decades spent on study or treatment delivery. Some day, ML datasets may catch up. Still, we see a lot of value in integrating our models known medical and physiological rule sets and indicators as well as expert opinions. Although AI regulation of processes or devices directly linked to human life (insulin pumps, ventilators, etc.) is a long way off, researchers who are trying to find uses for such technologies have to tread carefully. ML algorithms are trained on relatively narrow datasets, and unlike humans, the broader context of a physician's needs or process improvement cannot be taken into consideration.

We will "play the game" and learn to produce outcomes that seem to be effective in the short term, which are against longer-term objectives. While they know, there are environmental and legal questions around how much "exploration" a ML technique can implement: with continually learning, the autonomous system will ultimately experiment with expanding the therapy limits in an attempt to explore new technologies that could potentially harm patients.

All of this encourages the very issue that should be solved by AI and ML: increased human intervention supervision. Designing AI in health by applying ML is a promising field of study. Still, the rapid pace of change, the variety of different technologies, and the variety of input variables make it challenging to get a clear picture of how effective such systems could be in clinical settings or how repeatable they might be in variable diagnostic contexts.

"This is aggravated by a lack of agreement about how ML studies can disclose possible bias that respondents found the Diagnostic Accuracy brings to the surface initiative may be a valuable starting point." "Scientists also need to explore how ML structures, such as empirical data sets, can indeed be approved and transmitted to promote the replication of study findings in various ways."

Data availability: strength and conditioning AI methods allow vast amounts of data from publications such as electronic health records, pharmacy records, extended warranty records, or consumer-generated details, including such wearable devices or background of purchases. Yet data on well-being is still troublesome. Usually, the data are distributed across several different structures. Also, except for the variety just listed, patients generally see various providers and switch insurance agencies, following multiple-system and multi-format data splitting. Such heterogeneity raises the risk of error, reduces the comprehensiveness of the datasets, and raises the expense of gathering data – which also restricts the types of organizations that can implement successful AI for healthcare.

Injuries and errors: The most apparent danger is that AI systems may be incorrect at times and can result in medical injuries or even other healthcare complications. If an AI program advises a patient with the wrong medication, fails to detect a tumor on a radioactivity scan, or assigns a hospital bed to one patient over another because it incorrectly determines what patient will gain more, the patient may be injured. For example, many accidents happen even due to medical negligence in the healthcare system, even without AI participation. For at least two reasons, AI failures are inherently distinct. First, care providers can respond differently to software-induced injuries than it was to operator error. Second, if AI were ordinary, an inherent failure in one AI might result in damage to hundreds of patients – rather than just the small number of patients wounded by the mistake of any given operator.

4.17 GENERATION OF DESIGN ECGs

1. **Custom LSTM-We designed a custom LSTM**

   Plan implementing a Learning Transfer System for all patients, we sequence the LSTM and apply expected ECG models from the actual easygoing. On average 800–1000 patient-specific beats are additional.

2. **Customized GAN**

   We establish the modified GAN approach to understanding the contribution of knowledge of ECG features heterogeneity during customized ECG synthesis. An LSTM model is being trained from a custom GAN model on all patients with additional produced ECGs. The GAN model is a regular GAN devoid of studying ECG morphology, which is performed by the ECG GAN. The customized GAN classifier is trained on every practice patient and additionally expected ECG samples are applied from the particular person (similar to the technique discussed in (i.e., with supplementary 800–1000 patient-specific beats) [40].

3. **A Customized DCGAN**

   We intend to research the contribution of the unique architecture of generators to the classification. DCGAN is deemed state of the art for producing pictures. Therefore, we build a model with a DCGAN architecture that would be close to customized GAN. That is, the LSTM is qualified using synthesized illustrations from a custom DCGAN [41].

4. **The method on PGAN**

   LSTM model qualified on every patient with additional manufactured ECG examples from our planned PGAN model. The quantity of unnaturally produced beats by the multiplicative linkages, which were added to the training set, is a constraint of the model. We experimentation with the following values: 0, 500, 1000, 5000, 8000, 10,000, 15,000, and 20,000.

4.18 CONCLUSION

As a consequence, there is a need for rigorous testing to ensure both healthcare providers that certain systems are accurate. It could be years before the implementation of deep-learning frameworks in hospitals and clinics. But many are better than actual-world use: at two hospitals in southern India, Google is now going to run clinical studies of its eye-scan system. Neural networks can be incredibly effective in the right setting, being able to recognize correlations in data that humans will never be able to find on their own. But even experts have trouble knowing why specific decisions are taken by these channels and how they educate themselves. Then, new knowledge was provided to the neural network, such as the symptoms of a patient as established during a physical exam. Eventually, it was able to create correlations between written reports and observable symptoms on its own. The program could match the output of experienced physicians when evaluated on unlabeled data. When diagnosing asthma, this was more than 90% precise; the accuracy of the research physicians increased from 80% to 94%.

REFERENCES

1. Yu, S., Ma, Y., Gronsbell, J., Cai, T., Ananthakrishnan, A. N., Gainer, V. S., et al. Enabling phenotypic big data with PheNorm. *Journal of the American Medical Informatics Association.* (2017);25(1):54–60.

2. Savova, G. K., Masanz, J. J., Ogren, P. V., Zheng, J., Sohn, S., Kipper-Schuler, K. C., et al. Mayo clinical Text Analysis and Knowledge Extraction System (cTAKES): architecture, component evaluation, and applications. *Journal of the American Medical Informatics Association.* (2010);17(5):507–513.
3. Ricci, Z., Cruz, D., Ronco, C. The RIFLE criteria and mortality in acute kidney injury: a systematic review. *Kidney International.* (2008);73(5):538–546.
4. Khwaja, A. KDIGO clinical practice guidelines for acute kidney injury. *Nephron Clinical Practice.* (2012);120(4):c179–c184.
5. Ghassemi, M., Wu, M., Feng, M., Celi, L. A., Szolovits, P., Doshi-Velez, F. Understanding vasopressor intervention and weaning: risk prediction in a public heterogeneous clinical time-series database. *Journal of the American Medical Informatics Association.* (2016);10:125–132.
6. Suresh, H., Hunt, N., Johnson, A., Celi, L. A., Szolovits, P., Ghassemi, M. "Clinical intervention prediction and understanding with deep neural networks," in: *Machine Learning for Healthcare Conference*; 2017. pp. 322–337.
7. Rajkomar, A., Oren, E., Chen, K., Dai, A. M., Hajaj, N., Liu, P. J., et al. Scalable and accurate deep learning for electronic health records. arXiv preprint arXiv:180107860. 2018.
8. Jha, S., Topol, E. J. Adapting to artificial intelligence: radiologists and pathologists as information specialists. *JAMA.* (2016);316(22):2353–2354.
9. Esteva, A., Kuprel, B., Novoa, R. A., Ko, J., Swetter, S. M., Blau, H. M., et al. Dermatologist-level classification of skin cancer with deep neural networks. *Nature.* (2017);542(7639):115.
10. Gale, W., Oakden-Rayner, L., Carneiro, G., Bradley, A. P., Palmer, L. J. Detecting hip fractures with radiologist-level performance using deep neural networks. arXiv preprint arXiv:171106504. 2017.
11. Capper, D. et al. Practical implementation of DNA methylation and copy-number-based CNS tumor diagnostics: the Heidelberg experience. *Nature.* (2018);555:469–474.
12. Bailey, P., Cushing, H. *A Classification of the Tumors of the Glioma Group on a Histo-Genetic Basis with a Correlated Study of Prognosis*, Lippincott Massachusetts, 1926.
13. Brat, DJ. Cancer genome atlas research network. *The New England Journal of Medicine.* (2015);372:2481–2498.
14. Eckel-Passow, JE. et al. New molecular classifications of gliomas established. *The New England Journal of Medicine.* (2015);372:2499–2508.
15. Sturm, D. et al. *Cell.* (2016);164:1060–1072.
16. Yan, H. et al. *The New England Journal of Medicine.* (2009);360:765–773.
17. Sturm, D. et al. Hotspot mutations in H3F3A and IDH1 define distinct epigenetic and biological subgroups of glioblastoma. *Cancer Cell.* (2012);22:425–437.
18. Louis, D. N., Ohgaki, H., Wiestler, O. D., Cavenee, W. K. (eds) *WHO Classification of Tumours of the Central Nervous System*, 4th edn, International Agency For Research on Cancer, 2016.
19. Aldape, K., Nejad, R., Louis, D. N, Zadeh, G. Integrating molecular markers into the World Health Organization classification of CNS tumors: a survey of the neuro-oncology community. *Neuro-Oncology.* (2017);19:336–344.
20. Kleppe, A. et al. Chromatin organisation and cancer prognosis: a pan-cancer study. *Lancet Oncology.* (2018);19:356–369.
21. Turcan, S. et al. IDH1 mutation is sufficient to establish the glioma hypermethylator phenotype *Nature.* (2012);483:479–483.
22. Wiestler, B. et al. Integrated DNA methylation and copy-number profiling identify three clinically and biologically relevant groups of anaplastic glioma *Acta Neuropathology.* (2014);128:561–571.

23. Sahm, F. et al. DNA methylation-based classification and grading system for meningioma: a multicentre, retrospective analysis. *Lancet Oncology.* (2017);18:682–694.
24. Kim, J., Kim, J., Ren, L., Fan, J., Kim, J., Drewniak, J. L. "Extraction of equivalent inductance in package-PCB hierarchical power distribution network," in *Proceedings of IEEE 18th Electrical Performance of Electronic Packaging and. System Conference*, Oct. 19–21, 2009, pp. 109–112.
25. Kim, J., Ren, L., Fan, J. Physics-based inductance extraction for via arrays in parallel planes for power distribution network design. *IEEE Transactions on Microwave Theory and Techniques.* (Sep. 2010);58(9):2434–2447.
26. Kim, J., Shringarpure, K., Fan, J., Kim, J., Drewniak, J. L. Equivalent circuit model for power bus design in multi-layer PCBs with vias arrays. *IEEE Microwave and Wireless Components Letters.* (Feb. 2010);21(2):62–64.
27. Archambeault, B., Kim, J., Connor, S., Fan, J. "Optimizing decoupling capacitor placement to reduce effective inductance," in *Proceedings of IEEE International Symposium on Electromagnetic Compatibility*, Aug. 14–19, 2011, pp. 179–183.
28. Ren, L., Li, T., Chandra, S., Chen, X., Bishnoi, H., Sun, S., Boyle, P., Zamek, I., Fan, J., Beetner, D., Drewniak, J. Prediction of power supply noise from switching activity in an FPGA. *IEEE Transactions on Electromagnetic Compatibility* (Jun. 2014);56(3):699–706.
29. Pan, S., Brice, A. "Optimization of power delivery network design for multiple supply voltages," in *Proceedings of IEEE International Symposium on Electromagnetic Compatibility*, Aug. 5–9, 2013, pp. 333–337.
30. Hansen, L. K., Salamon, P. Neural network ensembles. *IEEE Transactions on Pattern Analysis and Machine Intelligence.* (1990);12(10):993–1001.
31. Laine, S., Aila, T. "Temporal ensembling for semi-supervised learning," in *International Conference on Learning Representations (ICLR)*, 2016.
32. McDowell, L. K., Gupta, K. M., Aha, D. W. "Cautious inference in collective classification," in *AAAI Conference on Artificial Intelligence*, 2007, pp. 596–601.
33. Reed, S., Lee, H., Anguelov, D., Szegedy, C., Erhan, D., Rabinovich, A. "Training deep neural networks on noisy labels with bootstrapping," in *Proceedings of the 29th International Conference on Machine Learning (ICML)*, 2014.
34. Yu, L., Cheng, J. Z., Dou, Q., Yang, X., Chen, H., Qin, J., Heng, P. A. "Au- tomatic 3d cardiovascular mr segmentation with densely- connected volumetric convnets," in *International Conference on Medical Image Computing and Computer-Assisted Intervention*, 2017, pp. 287–295.
35. Karmore, S., Bodhe, R., Al-Turjman, F., Kumar, R. L., Pillai, S. IoT based humanoid software for identification and diagnosis of Covid-19 Suspects. *IEEE Sensors Journal.* (2020): 1–1, DOI:10.1109/JSEN.2020.3030905.
36. Ramasamy, L. K., Padinjappurathu, S. G., Kadry, S., Damaševičius, R. Detection of diabetic retinopathy using a fusion of textural and ridgelet features of retinal images and sequential minimal optimization classifier. *Peer Journal of Computer Science.* (2021);7: e456. DOI:10.7717/PEERJ-CS.456
37. Ilija Radosavovic, Piotr Dollr, Ross Girshick, Georgia Gkioxari, and Kaiming He. "Data distillation: towards Omni-supervised learning," in *The IEEE Conference on Computer Vision and Pattern Recognition (CVPR)*, 2018, pp. 4119–4128.
38. https://www.nature.com/articles/d41586-018-02881-7.
39. https://www.lung.org/lung-health-and-diseases/lung-disease-lookup/lungcancer/resource-library/lung-cancer-fact-sheet.html.
40. https://www.cancer.gov/types/lung/research/nlst.
41. https://www.cio.com/article/3194517/data-scientists-compete-to-createcancer-detection-algorithms.html.

5 Digitalizing the Health Records Using Machine Learning Algorithms

N. Pooranam, M. Diwakaran, and T. Vignesh
Sri Krishna College of Engineering and Technology

CONTENTS

5.1 INTRODUCTION

The machine learning techniques include three classes: generative, discriminative, and hybrid. In most of the applications such as neural networks, information processing including audio/speech, image/vision, language representation, natural language processing (NLP), and data recovery. Each application has a special feature that is illustrated with its architecture; classes are classified directly for the visible data. In some applications, the signal processing areas are defined by the matrix with the two axes of signal and processing analysis; the scheme uses machine learning algorithms. The main advantage in machine learning is to scale the CPU clustering using very high training data. In this proposal, the application is developed for digitalizing healthcare records using data analytic techniques and machine learning algorithms that helps in monitoring the patient's records and digitizing it efficiently. In this chapter, the discussion is fully focused on recognizing all the words; there are different types of handwriting processes: one is online and the other is offline. Online is inclined with the document in the form of softcopy, the process of converting this into recognition will be in image processing

DOI: 10.1201/9781003050827-5

steps using a machine learning algorithm. Another will be an offline process in which handwritten form is converted into text and stored smartly (Figure 5.1).

5.2 LITERATURE STUDY

In an experimental study, the performance measure for each data point is studied through various network layers in the entire data frame obtained in each dataset (Figures 5.2 and 5.3).

5.3 BACKGROUND

The modernization of digitalization in the machine learning environment is made by the integration of techniques. This section will exhibit digitally monitoring the

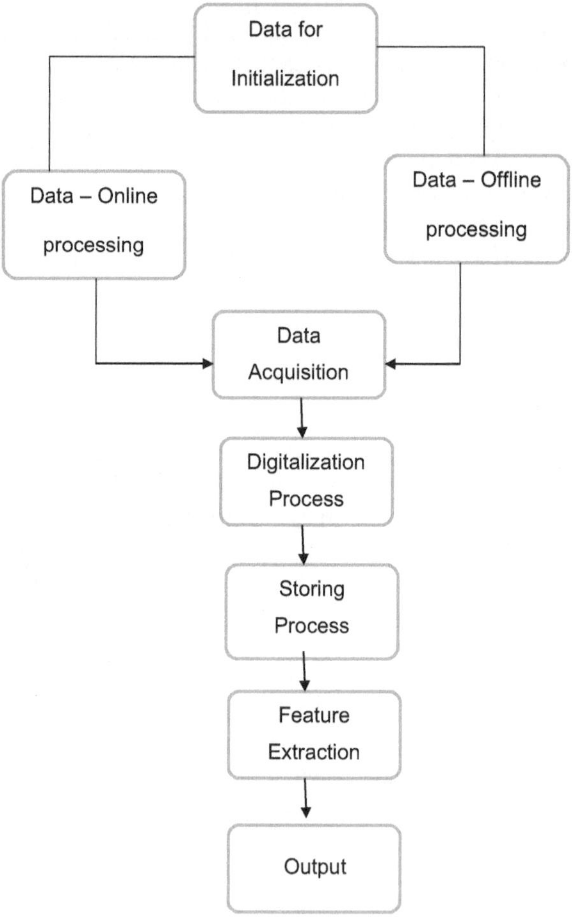

FIGURE 5.1 Process flow.

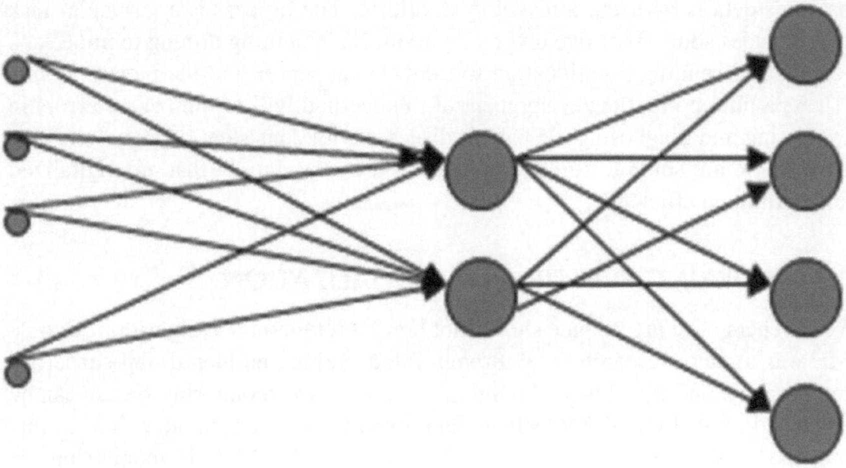

FIGURE 5.2 Layers of input and output networks.

FIGURE 5.3 Feedforward network layers.

patient's details by using a machine algorithm. The process is a particular task that includes some steps like text recognition, NLP, opining mining to mine each record, and sending a notification to the hospital server and the patient mobile [1]. As a final point, the development of this method will minimize the errors in evaluating and diagnosing the patient diseases with consistent data analysis process. There are some features to be extracted on the dataset that are digitalized and optimized efficiently.

5.4 WORKING PRINCIPLE OF DIGITALIZATION

Researchers have made some significant breakthrough on ML algorithm. Google ML was a major research breakthrough in analyzing and identifying cancerous tumors in an optimized way. Having this as a base, the monitoring system can be enhanced by making all the patient data digitalized. The technique used in this proposed system is to develop a smart application that helps in monitoring the data and analyzing those data in a fusion method.

The optimization technique will calculate the evaluation score of each text and give an opinion to the doctor that helps in diagnosing the data. The sentimental classification is based on the type of sentence; that is instead of classifying entire sentence, an approach called convolution neural network is used to divide the entire sentence into different types to provide a sentence-level sentimental classification. A model is developed to separate the sentence like subjective sentence, objective sentence, conditional sentence, target-based dependent sentence, etc. In machine learning, the sentence classification is through two different processes: first BiLSTM – CRF, which extracts the target sentence from the given opinion, and second, the CNN will extract predicted sentences into polarities and also connects different sentences like non-target sentences, one target sentence and even multi-target sentence. CNN is a connection model that acts as a learning agent. The agent will store information in the form of neurons [1]. The collected neurons classify each sentence into different types.

The major highlight of digitalizing is the behavioral approach of each text element through optimization techniques. The final development is to make the report digital data. The overall process is stored in open access, and it is authenticated by a particular person for future use, and the notification is generated through a time stamp [2] (Figure 5.4).

5.5 CASE STUDY ON DIGITALIZATION METHOD

In this discussion, a dataset is decided to find the pattern of each letter, each data frame in the dataset identifies the letter and should be digitalized in such a way that it can be an easy method if it is a classification process. In classification NN is the best algorithm to classify the data frame. Figure 5.5 describes the overall performance of classification, and each output step is clustered through the mapping function and the backpropagation approach is chosen for the effective output process.

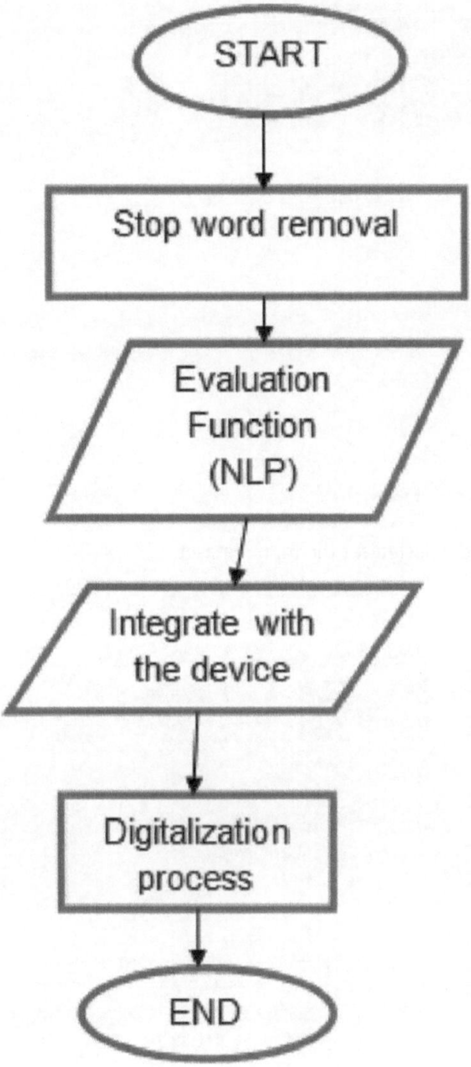

FIGURE 5.4 The workings of the system.

Each layer of output is linked with either the hidden layer or the input layer and the performance evaluation gets changed according to the evaluation value obtained in the output layer (Figure 5.6).

When the process of evaluation gets executed there a variation in the output layer, each single variation will make a huge difference in both the hidden and output layers, and the variation is shown as a separate screenshot. The different output layers will be obtained separately and the training state will be changed accordingly through the process that the algorithm matches the best output for

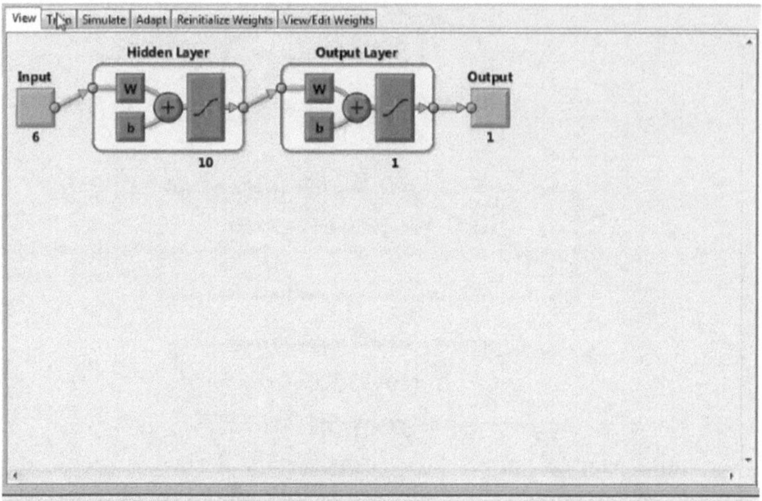

FIGURE 5.5 A model design for the input dataset.

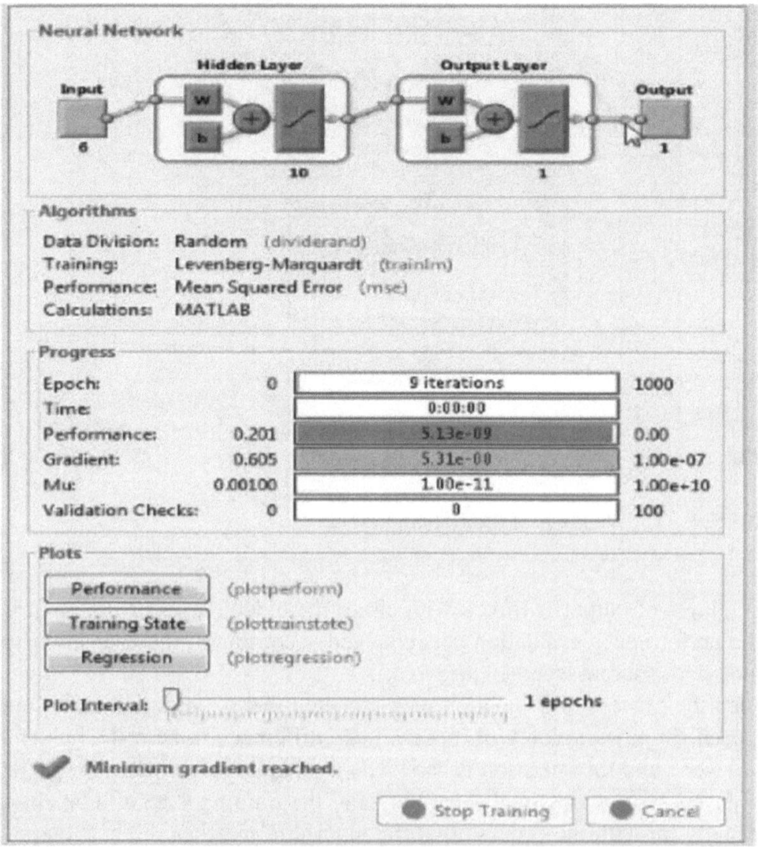

FIGURE 5.6 Performance measure for the model design.

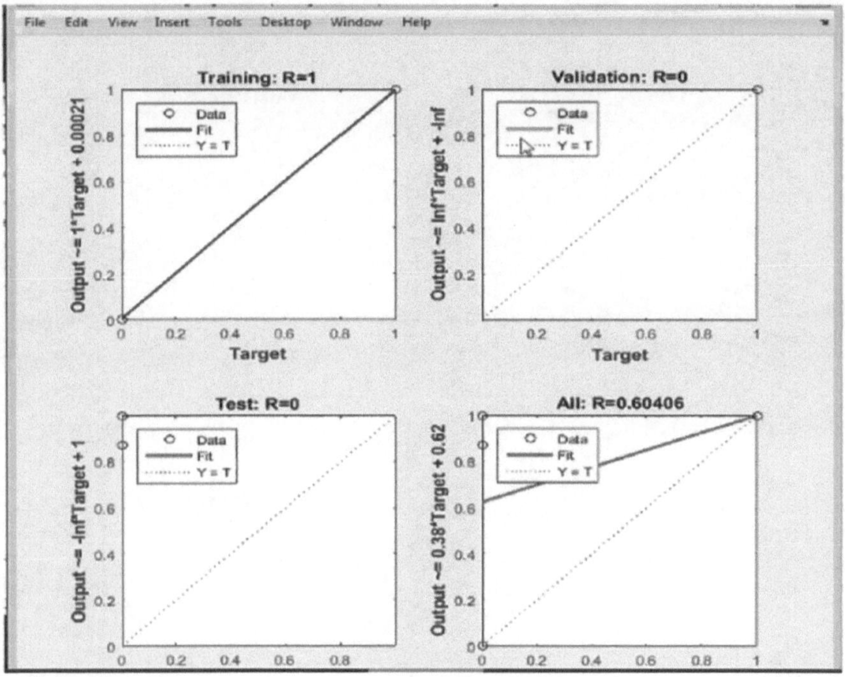

FIGURE 5.7 Performance measure for the first iteration.

the manipulation process. Performance is based on mean square error, and data division is a random method for evaluation (Figure 5.7).

Here training, validation, and testing charts are exposed to the output layer. Through target values, output values are obtained and classification process is done in an efficient way, each in parallel, and the data obtained is measured with the backpropagation method for the variation each is described in such a way that the process is scaled [3].

In Figure 5.8, there are some slight changes in performance evaluation, mainly in the training dataset; the chart has some variation while manipulating each data. In addition to the target value, there are several changes in the values; all the above changes are automatically done when there is some performance change [4]. A validation check is also done at the final stage.

5.6 CASE STUDY WITH IMAGE DATABASE

In this study, image dataset is taken as input and the feature extraction is made on each pixel value; different shapes are described using weight calculation through each person's handwriting being different from others. In this scenario, the weighted average is done on the input dataset; the value is obtained based on the weight calculation. The simulation is made on different classification algorithms; finally each output is defined by the target value, and graphical representation is

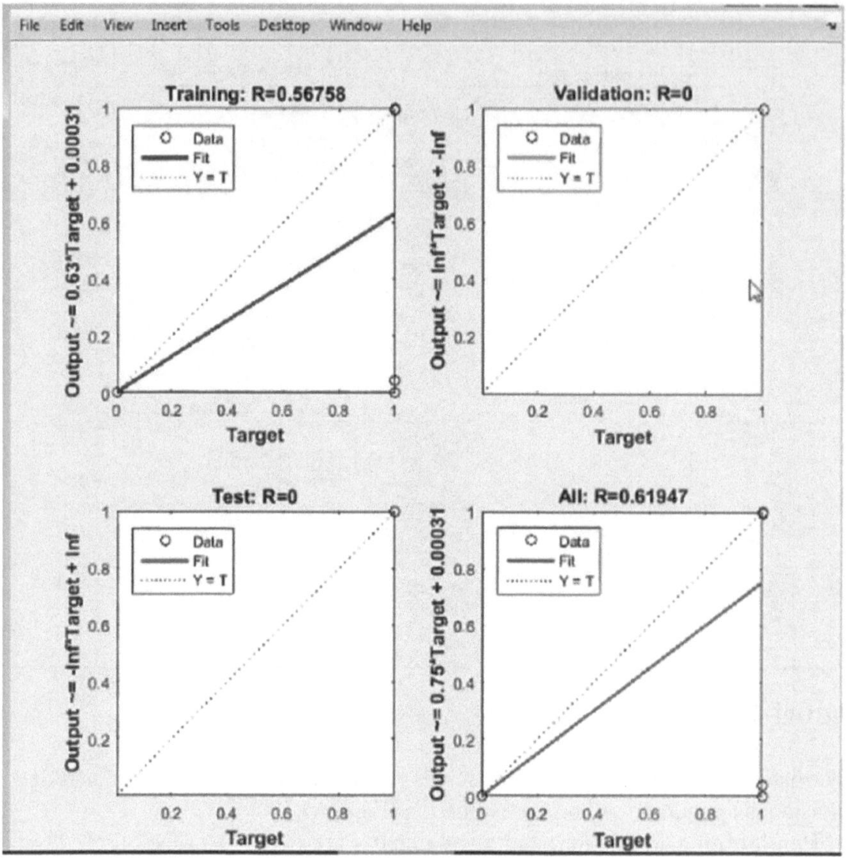

FIGURE 5.8 Performance measure for next iteration.

done on all the phases of writing the letter [5]. As we discussed in working principles, the flow of maintaining the database of each data layer will be described in such a way that it should produce a maximum output when compared with other data layers or hidden layer of the designed model [6]. However, the flow may vary with the size of data rather than input processing or target data maintained with some threshold value (Figure 5.9).

In Figure 5.10, network selection is made for training each data in the dataset. The input function is chosen with input data, target values are based on the target class, and finally, the training function is chosen carefully based on the trained data. Here the adaptive learning function is chosen by the user for obtaining the effective result. The network type is feedforward, which gives an effective result for the training method (Figure 5.11).

Here, the network layer should obtain new values based on weighted average property; the user can edit the weight based on training data [7], which has to be evaluated in a random method. Here, the first input layer is weighted with the bias

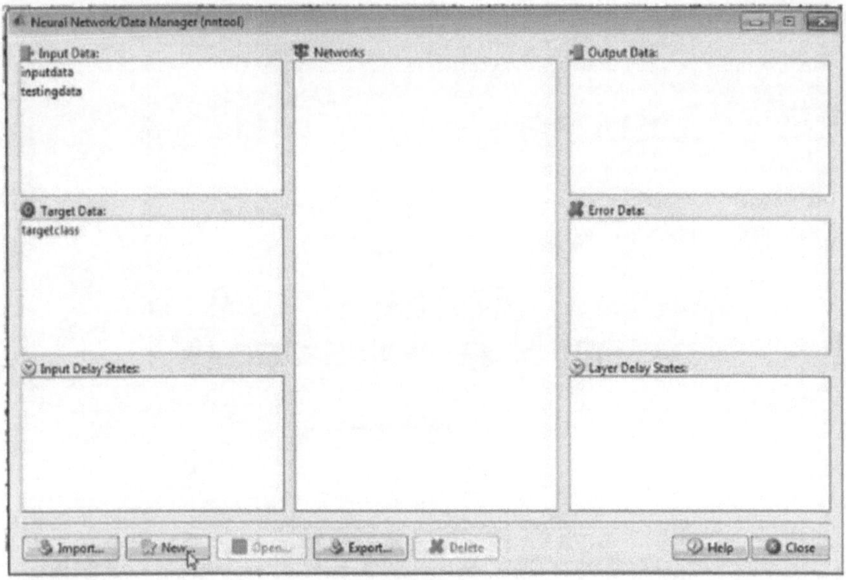

FIGURE 5.9 NN tool manager to maintain the database.

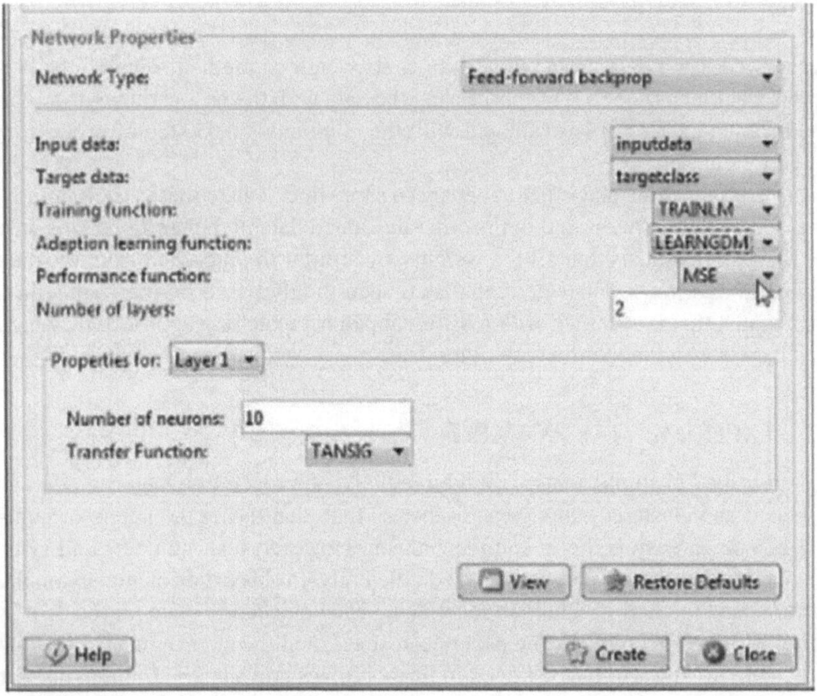

FIGURE 5.10 Network selection process.

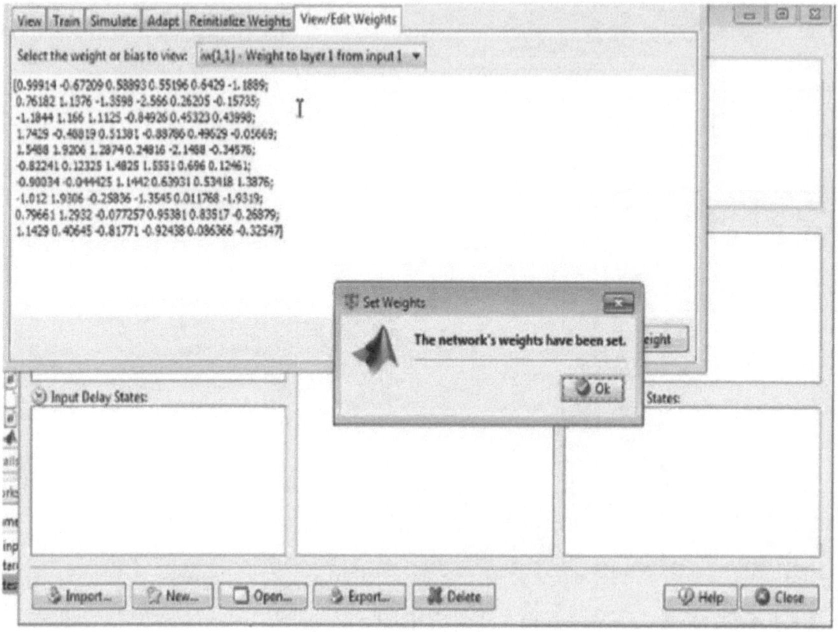

FIGURE 5.11 Network's weight estimation.

and viewed; each data processing unit has its weighted method. Finally, the network's weight has been set for processing the data with the best-estimated weight. The mean square error is evaluated with the continuous next step of processing the entire dataset.

If the target value and other value are in same line, it state that pattern is similar through NLP process and omits other unwanted data in the preprocessing step [8]. Figure 5.12 shows that all the weights are similar to other data from the data frame, though pixel values are calculated accordingly based on the change in a single value that is obtained with a different pattern matching approach by which each data is weighted with every other data.

5.7 EXPERIMENTAL EXAMPLE

The described example deals with the predictive approach in which each data is classified and clustered among each other so that identifying the letters or numbers can be an easy and efficient mechanism. However, text, numbers and symbols written by doctors can be digitized efficiently. In this experimental example, the proposed method will help in identifying each object and store all the data in digital form, which reduces the paper cost, space, and the environment. The flow of data processing will be explained in further topics now that the training dataset is preprocessed and stored for classification and clustering method. The prediction approach plays an important key role in identification. The nearest neigbor

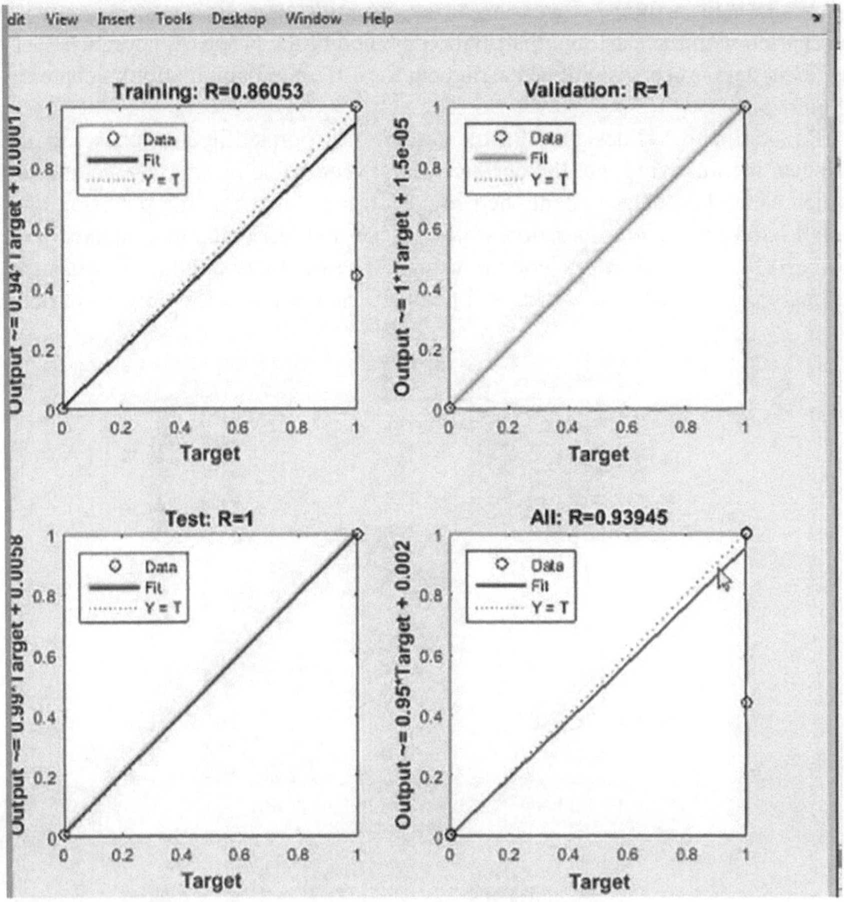

FIGURE 5.12 Final target values from the model.

algorithm evaluation function will describe each text using weights applied in identifying context. Training data will be exposed to the network layer where feedforward is chosen for processing, and symbols will have different weights when compared with other context data. Some preprocessing steps will be followed for reducing errors in the evaluation function. The data will be processed, and each time conversion will be carried for transferring it into a text document. In basic machine learning, the Chabot application will do the transformation from one form to another.

Predicting text content will be a difficult task for summarizing the data and analyzing data that can be semi-supervised data processing. Based on the shape of context data, the machine approach will be identified, classified, and clustered based on letter. All the trained data will be converted into some form of numeric representation so that the evaluation function will be working efficiently. In image processing, the context or symbols will be converted into pixels and data frames to help in future processing of data identification. The trained data will be charted with the class and distribution method of the proposed approach. The different data processes will be carried out for different manipulation of contexts (Figure 5.13).

Random model is designed for the dataset incorporated in each phase, while the data is transferred into the designed model confusion matrix is generated in which each class data is mentioned in performance metrics like precision and recall is taken into consideration while trained and test data are evaluated and a matrix is obtained in row-column design. The random method will estimate all feature extracted data, which will be mentioned as either loss or gain of data

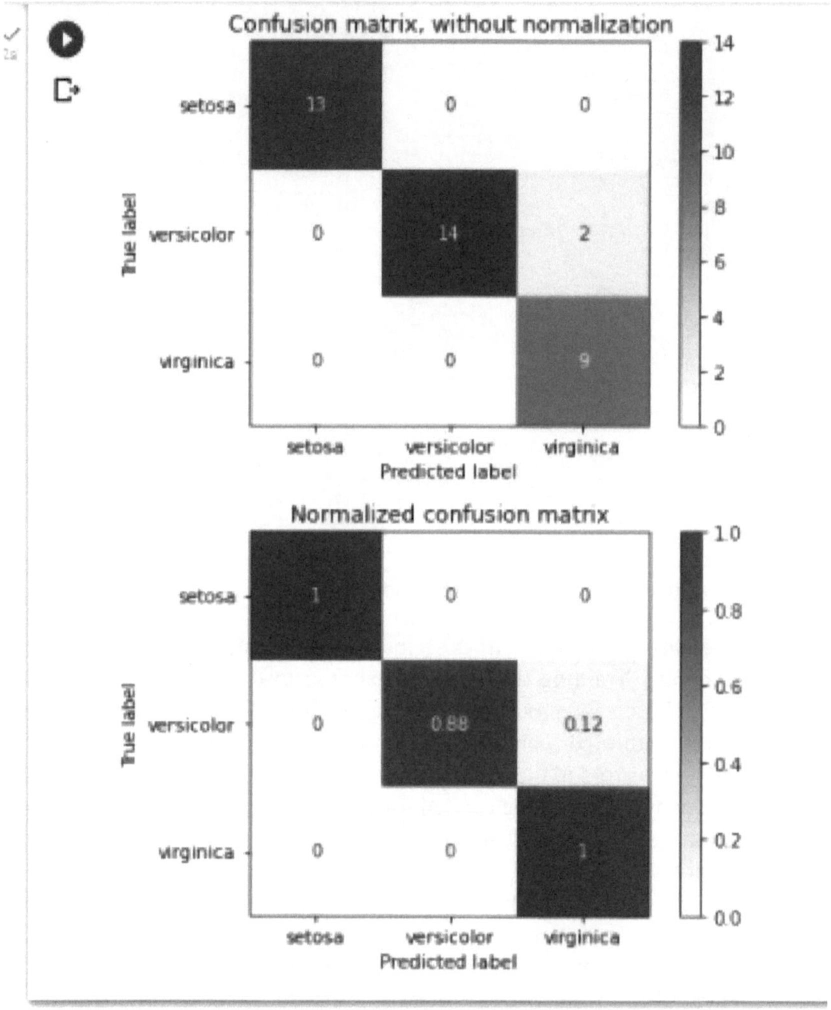

FIGURE 5.13 Matrix of the random model.

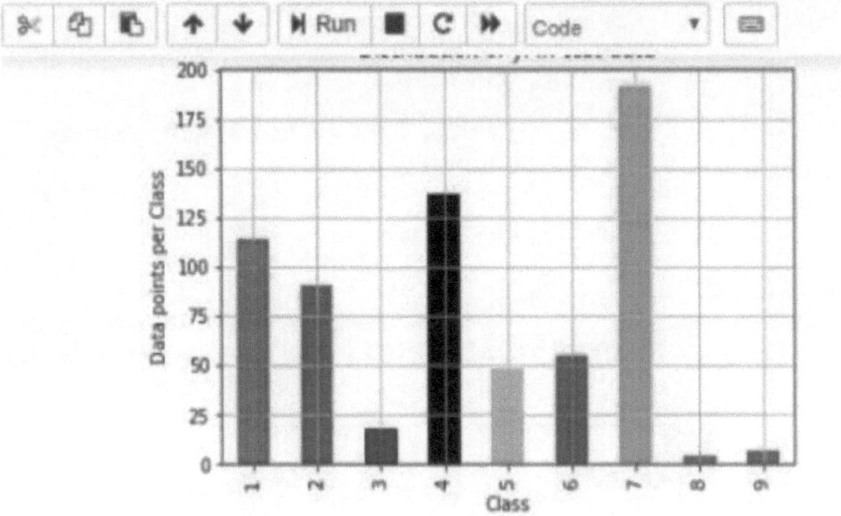

FIGURE 5.14 Precision of class and data points.

model designed [9]. Each color will describe each form of data processing in a random model: the dark color specifies the threshold value of the precision model, finally it describes the complex data which found its matrix representation and the output is processed by precision metrics which finds its gradual increase or decrease in data identification. The proposed method will find the best intermediate value for processing it with other future data fame.

In prediction finding of each text, mining process will be carried out till the final iteration process. Random forest method is built for selection of different data obtained through different data frames. Here, class is created using a model design by which each data point is classified accordingly (Figure 5.14).

For each class, data points may differ based on the input and target value generated using model design for the proposed approach. The performance measure is calculated through weights obtained for each training and test dataset. Here precision is obtained for each data point; different color shows the variation in the class creation process [10]. This example may be mapped for the case study described in the above section where each letter is classified in the same manner; the results are generated by the output value executed from each class.

Figure 5.15 describes the total data point generated in each data frame; here, classification is made on all three different data frames in which training data count is separated from other datasets like test datasets and also cross-validation data. Here the huge number of data is defined with all other data frames obtained for processing each category. Though there are some deviations in the number of data points, the count is stored for future calculation of dataset. There are some

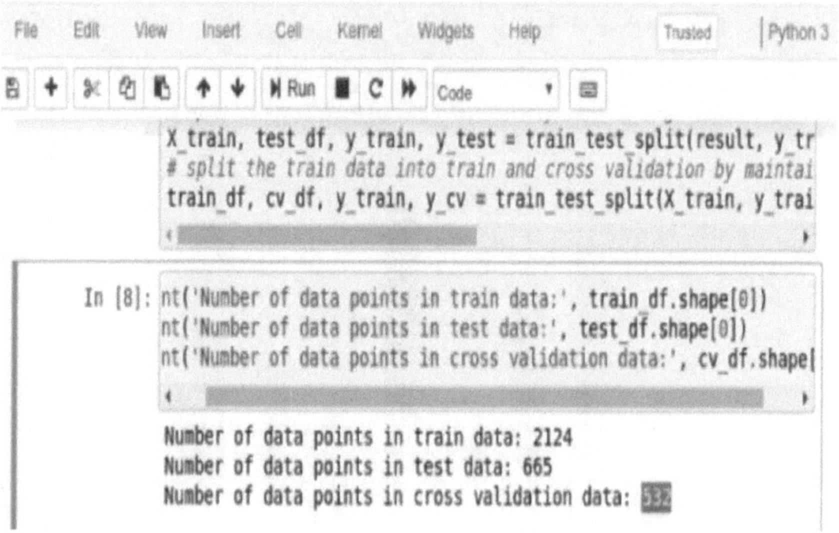

FIGURE 5.15 Number of counts in each dataset.

processing steps evaluated till the final iteration of the data point, which may still vary according to all other dataset used for the random model designed for future process. The separate test and cross-validation are maintained in all the classes designed for each set obtained by data frames.

5.8 DATA FLOW IN MACHINE LEARNING

Here the image processing on the handwriting dataset is classified according to some constraint. Each data in the dataset is iterated for more than a time for the best processing output value (Figure 5.16).

Diagnosis of diabetic retinopathy using multi-level set segmentation algorithm with feature extraction using SVM with selective features extraction process, which is described in the [11] (Figure 5.17).

5.9 FUTURE TRENDS AND CONCLUSION

The evaluation function which returns values based on the classification done by adding data analytic techniques will choose the locality value to initialize the position to cluster the data. After the evaluation process, the overall data will be synchronized with the mobile device of the end-user. The chapter concludes with the summary that the health record of each patient will be monitored and verified dynamically. There can be further research to develop deep learning techniques. In the future, it is planned to work on the Blockchain Security Framework for Electronic Health Records of patients [12].

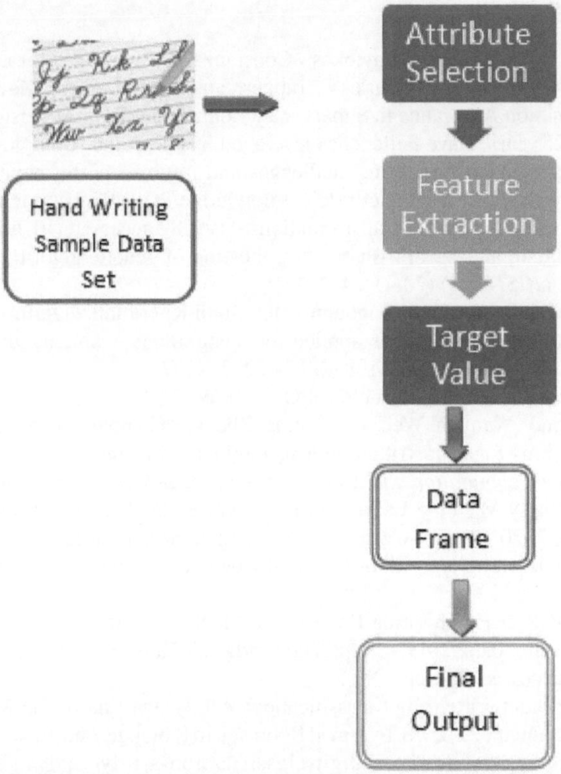

FIGURE 5.16 Data flow for machine learning process.

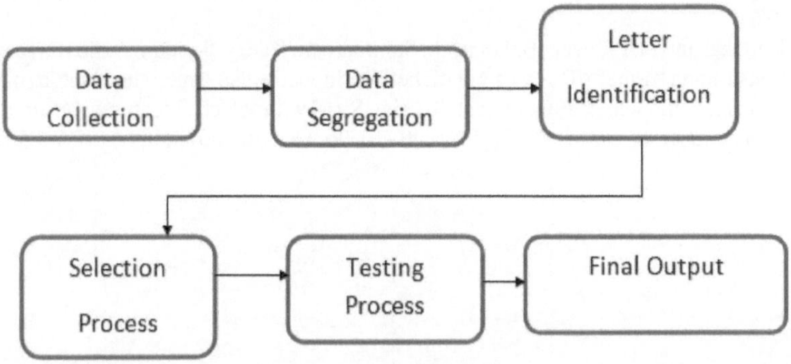

FIGURE 5.17 Data flow through individual letters based on text data.

REFERENCES

1. Alcalde, Richard, Peter Bayiokos, Constanza Cabrera-Mendoza, Sabrin Kaur Guron, Wildenslo Osias, Charles C. Tappert, and Avery Leider. "Machine Learning for Information Assurance in Supervisory Control and Data Acquisition Systems."
2. For over 25 years, Dave Fellers has leveraged business and technology solutions to help companies solve pressing challenges and improve performance. Since 1995, https://www.digitalistmag.com/cio-knowledge/2019/07/10/machine-learning-opens-pathway-for-digital-transformation-06199382 accessed (10-July-2019).
3. Bhasin, Harsh, and Surbhi Bhatia. "Application of genetic algorithms in machine learning." *IJCSIT* 2(5) (2011): 2412–2415.
4. Elleuch, Mohamed, Najiba Tagougui, and Monji Kherallah. "Optimization of DBN using regularization methods applied for recognizing Arabic handwritten script." *Procedia Computer Science* 108 (2017): 2292–2297.
5. https://www.youtube.com/watch?v=PO4hePKWIGQ.
6. Wu, Shixiao, Wanyun Wei, and Libing Zhang. "Comparison of machine learning algorithms for handwritten digit recognition." In *International Symposium on Intelligence Computation and Applications*, pp. 532–542. Springer, Singapore, 2017.
7. How To Apply Machine Learning to Recognise Handwriting by Great Learning, March 26, 2020 https://www.mygreatlearning.com/blog/building-a-framework-for-a-machine-learning-algorithm-that-can-recognize-handwriting/ accessed (March 2020).
8. 15 Best OCR & Handwriting Datasets for Machine Learning Article by Meiryum Ali, March 07, 2019, https://lionbridge.ai/datasets/15-best-ocr-handwriting-datasets/ accessed (March 2019).
9. Exploring handwritten digit classification: A tidy analysis of the MNIST dataset Posted on January 22, 2018, by David Robinson in R bloggers https://www.r-bloggers.com/exploring-handwritten-digit-classification-a-tidy-analysis-of-the-mnist-dataset/ accessed (January 2018).
10. Zeng, Yifan, Shouqiang Liu, Wei Zhang, and Yanliang Zhai. "Application of artificial neural network technology to predicting small faults and folds in coal seams, China." *Sustainable Water Resources Management* 2(2) (2016): 175–181.
11. Abunadi, Ibrahim, and Kumar, Ramasamy Lakshmana. "BSF-EHR: Blockchain security framework for electronic health records of patients." *Sensors* 21(8) (2021): 2865.
12. Kandhasamy, J. Pradeep, Balamurali, Saminathan, Kadry, Seifedine, and Ramasamy, Lakshmana Kumar. "Diagnosis of diabetic retinopathy using multi level set segmentation algorithm with feature extraction using SVM with selective features." *Multimedia Tools and Applications* 79 (2019):10581–10596. DOI:10.1007/s11042-019-7485-8.

6 Interactive Visualization for Understanding and Analyzing Medical Data

S. Suganthi
Cauvery College for Women

T. Poongodi
Galgotias University

CONTENTS

6.1 INTRODUCTION

Visualization is the graphical representation of raw data, and interactive visualization allows users to interact with the visualization itself to gain more insights from the data. Different aspects of data can be viewed through interactive visualization, which helps the user in gaining multiple levels of details. The user can converse with the data by giving inputs in real time and arrive at speedy solutions. Medical visualization systems require the users to have a deep understanding of the diagnosis methods, therapeutic decisions, and organizational and technical

DOI: 10.1201/9781003050827-6

constraints. Also, the interaction methods play a vital role in the interpretation and classification of data by exploring the data.

Healthcare data is voluminous, and such volumes of data can be effectively summarized by visualization [1]. The variety of data and the vast numbers of variables within each type pose a huge challenge in visualization. Data quality is another major issue that is affected by missing or invalid data. So, interactive visualizations must be designed in such a way they also handle issues related to volume, variety, and quality of data. The data handled by the visualization tools are mostly raw and unstructured. But analytic tool cleanses the data, evaluates the cleansed data, and displays the results, by using data mining algorithms, evaluation tools, and software tools. Visual analytics combines both these jobs of visualization and analytics in helping the users interact with the visual interface to arrive at the best feasible solutions. The data input for the interactive visualization requires the raw data to be transformed and cleaned before it is being imported into a tool. So a proper understanding of the data, such as its source, quantity, and type, is essential to make proper transformations.

The healthcare industry is complex and multidimensional, comprising of various components such as healthcare consumers (people who need services), healthcare providers (physicians, nurses, etc.), and healthcare organizations (hospitals, clinics, insurance companies, companies that provide technologies and medicines and equipment, public and private organizations that provide training for a career in the healthcare industry, research organizations, agencies that monitor the quality of healthcare services, etc.). So information coming from all these levels comprise a voluminous amount of data and needs to be organized and coordinated.

Healthcare data is growing at an exponential rate, and the revenue generated in the industry is also high. Due to the complex and huge amount of data in the healthcare industry, efficient technologies for visualizing data for analysis and interpreting the results are mandatory. Such visualization tools help in providing value-added services to the users and improve patient care.

6.2 SOURCES OF MEDICAL DATA

Medical (clinical) data generated from the healthcare industry refers to information related to the health of regular patient care or a clinical trial program. The medical data is more valuable, and it plays a prominent role in the diagnosis, treatment, and prevention of diseases and also in improving the healthcare services. The sources of medical data mainly include medical records, health surveys, claims data, disease registries, vital records, surveillance data, peer-reviewed literature, clinical trial data, and administrative records. These data are acquired, analyzed, stored, shared, and integrated by overcoming the difficulties existing in the incompatibilities concerning proprietary and privacy issues, hardware, medical equipment, and data formatting standards. Healthcare database systems are those in which clinical data are entered regularly and maintained. Electronic Health Records (EHRs) are the most commonly used form of healthcare databases, which store the person's medical information and history in digital format and are maintained by a healthcare provider. It includes data such as medications, demographics, diagnoses, procedures, lab values, radiology results, unstructured clinical notes, and genomic data [2]. The data are entered into the database by a transcriptionist, or automatically entered by a device that is used to measure or collect them, or patients themselves enter data with the help of interactive touch-sensitive screens, light pens, terminals with specialized keyboards, hand-held tablets with pen-based mechanisms, and mouse pointing devices.

6.3 TYPES OF MEDICAL DATA

Health data are usually acquired by health professionals such as doctors, nurses, office staff, admission personnel, physical or respiratory therapist, radiologist, laboratory personnel, and pharmacist. The data should be gathered and recorded properly as they form the basis for the historical and legal record, communication among providers, help determine health problems occurring in the future, record standard preventive measures, identify trends, and support clinical research [3]. Medical data can be broadly classified into structured or unstructured data.

Structured data: have a predefined data model and constitutes only 20% of the total data. They are easy to use and exist in a predefined format and include information such as discrete numeric values (e.g. height, weight, temperature, blood pressure), categorical values (e.g. blood type), nominal values (e.g. stage of the disease), patient demographics, coded health data, and medications. They can be stored in a relational table and require only SQL and supporting tools. They can be easily analyzed and interpreted.

Unstructured data: does not have a predefined data model and constitutes the remaining 80% of the data. They are generated in huge amounts and can be images, signals, or text. Images or medical video data are produced from medical imaging devices (e.g. endoscopes). The signal data are continuous analog signals (e.g. biosignals produced from intensive care units, electrocardiogram (ECG), or wearable health monitoring devices). The text or narrative data is the human-generated spoken or written data containing complex information including physician notes, nursing notes, dictation notes, or shorthand conventions used by the physicians that rely on Natural Language Processing (NLP).

Unstructured data contains a lot of medical information that is valuable but without a common structural framework [4]. They have to be transformed into structured clinical data before they are interpreted and evaluated. Structured data when combined with unstructured data provide a complete picture of the patient's medical history.

6.4 INTERACTIVE VISUALIZATION PROCESS OF MEDICAL DATA

Interactive visualization is an iterative process that implies complicated two-way communication between the visualization system and the user. The data is transformed into an image on the application of the visualization technique. The input can be given by interacting with the visualized objects and scenes, modifying the data to be visualized, adjusting various parameters, and extracting features from the original dataset [5]. Different visual representations are generated depending on the user's input. The data in EMR are processed by iteratively applying the processes of data collection, data preprocessing, data mining, evaluation, and knowledge application in an interactive manner, necessitating continuous feedback [4]. The structured data adheres to a certain predefined data model and is straightforward to analyze. Each field is discrete, and so they can be easily accessed and aggregated from different locations in the database. But unstructured data such as images, signals, and text need transformation into a structured format before they are processed.

6.4.1 Image Analysis

Medical imaging is essential in the diagnosis and treatment of diseases by exposing the features hidden inside the body. The amount of medical images obtained is voluminous, such that advanced techniques should be adopted to understand the relationships among the variables and identify their relative importance. The volume data are 3D entities that are characterized by a multidimensional array of scalar and vector data too voluminous to be represented geometrically. Volume visualization creates graphical representations of datasets that are defined on three-dimensional grids using interactive graphics and imaging.

The interactive volume visualization process of medical image data comprises various phases such as clinical image acquisition, image storage, image enhancement, image segmentation, registration, and visualization.

6.4.1.1 Data Acquisition

Medical images are obtained with the aid of a medical acquisition system. Advancements in technology have led to reduced acquisition time, which helps in providing quality images, as artifacts are reduced considerably. The problem with medical imaging is that the physical parameters in the image acquired cannot be verified for accuracy by comparing it with the actual true physical parameters. However, many image analysis techniques use prior knowledge of the physical parameters in overcoming the inaccessibility. Medical images are obtained from different modalities such as X-ray, computed tomography (CT), magnetic resonance imaging (MRI), ultrasound imaging, nuclear imaging, positron emission tomography (PET), or single photon emission computed tomography (SPECT).

There are four major techniques of image acquisition, as described in the following.

6.4.1.1.1 X-Ray Imaging

Wilhelm Roentgen discovered X-rays in 1895. He found that, except lead, this ray penetrated through many other materials and gave a blackening effect on the film. It is widely used and uses ionizing radiation to produce 2D images of the selected parts in different black and white shades due to varying absorption levels of radiation by the body. Ionizing radiation is the portion of the electromagnetic spectrum that possesses high potential energy, enough to traverse the matter and make physical changes in the electrons of an atom to convert to ions. In the medical field, when the radiations pass through the human body, they are either absorbed or scattered by the internal structures within the body, and the remaining radiation is transmitted to a computer screen or a film for recording and later analysis. It has a high spatial resolution in detecting hard tissues such as bones. However, there is difficulty in imaging soft tissues and the radiations can make damages to DNA that increases the risk of affecting the tissues and leading to the development of cancer later in life. However the risk is very minimal, and it depends upon factors such as age, body part where the radiation is given, amount of dosage, and sex.

The following are the imaging techniques that make use of X-rays:

Fluoroscopy is a kind of X-ray imaging technique in which a continuous X-ray image is showed on a monitor by injecting a contrast agent into the body. The contrast agent is the X-ray dye being applied to the location to be imaged which enhances the resultant images obtained. It is used in orthopedic surgery; for visualizing vascular systems such as blood vessels and organs, brain, and heart; to view the gastrointestinal tract and placement of devices within the body, such as stents.

Mammography is a type of X-ray imaging used specifically for breast imaging that detects cancer earlier by using low-dose X-rays. In this, monochrome X-rays are produced, which reduces scatter and increases the contrast, thereby producing images of high quality.

CT images possess higher spatial resolution when compared to X-ray images. The images are not directly obtained from the imaging devices, but apply projection measurements on the projection images obtained in different angles from X-ray imaging in getting the final image. In CT, narrow beams of rays are passed through the patient's body by the X-ray tube that rotates around the patient. The X-rays that leave the patient's body are captured by the X-ray detectors and transmitted to a computer. A 2D image slice of the patient is produced for one complete rotation and is sent and stored in the computer. Every time the bed is moved forward, the process is repeated to produce 2D image slices until the required amount of image slices is produced. The stored image slices can be viewed separately or combined by the computer to form a 3D image. A contrast agent can be used, which helps in enhancing the visibility of the internal structures. The risk of high exposure to radiation, greater acquisition times, and the cost of the equipment are the disadvantages of this method.

Radiographs are useful in identifying very small objects that have a high spatial resolution. Only distance ratios can be measured between the structures if the approximate distance to the radiation source is known [4]. Over- or underexposure may reduce contrast, and the images may be blurred due to motion or the size of the focal spot. Experience and extensive knowledge about the orientation of the organs and their relative positions concerning the projection direction are essential. CT scans overcome the above problem, but they have low spatial resolution when compared to radiograph [4].

6.4.1.1.2 Magnetic Resonance Imaging (MRI)

MRI is a non-invasive method used for producing detailed anatomical 3D images. In this, the protons within the body are made to align with the strong magnetic field produced by powerful magnets employed in the MRI machine. Later, when the radiofrequency field is passed through the body, the protons are pulled against the magnetic field. When the radiofrequency field is turned off, the protons try to realign with the magnetic field. Some energy is released during this process. The amount of energy released and the time taken for realigning can be measured using MRI sensors. This varies according to the type of tissues and the environment. Based on these magnetic properties, various tissues can be distinguished.

6.4.1.1.3 Ultrasound Imaging

Ultrasound imaging is a diagnostic imaging technique that uses high-frequency sound waves to look at internal organs and structures within the body and produces an image called a sonogram. They are non-invasive, free of radiation, economical, portable, and are used in real time to capture live and moving images from inside the body. Ultrasound imaging uses the acoustic impedance, the physical property of the tissues in determining the amount of resistance an ultrasound beam encounters when it passes through the tissues. Acoustic impedance is unique for each tissue and depends on the density of the tissues and the speed of the sound waves. Sound waves will be reflected at the boundaries between materials of different acoustic impedance. The transducer probe is the main part of ultrasound imaging in sending and receiving sound waves. It uses a principle called the piezoelectric effect that makes use of quartz crystals called piezoelectric crystals. These crystals have the property that when they receive electric current, they change shape rapidly, and this emits vibrations or sound waves and, similarly, when sound waves hit the crystals they produce electricity. The CPU sends electric signals to the probe that are then transformed to sound waves, transmitted through the patient's body, and the CPU receives back electric signals from the probe due to echoes. The processing of all the raw data is carried out in the CPU, and the resultant image is sent to the monitor for display. The user can interact by changing the scan mode duration and frequency of the sound waves depending upon the need.

6.4.1.1.4 Nuclear Imaging

Nuclear imaging is an imaging technique in which traces of radioactive material are injected into the patient's body before image acquisition. The distribution of the radioactive material is measured by the photons emitted by the radioactive material and captured as an image that is used to show the bodily functions, diagnosis, and treatment of diseases. The techniques that use nuclear medicine are explained in the following.

Scintigraphy imaging uses radioisotopes administered to specific parts or tissues inside the body and the gamma rays produced are captured by external detectors to produce 2D images. Here external rays are not given, and the isotopes are taken internally.

SPECT is an imaging technique in which gamma rays are used to measure the distribution of radioactive material. A 3D image is produced from the stacking of 2D projection image slices taken at different angles.

PET is a technique in which positron emitted during the decay of radioactive material is used in the imaging process. Positrons are small particles that have a mass equal to that of an electron but are oppositely charged. These positrons and electrons in the body annihilate to produce a small amount of energy that produces two photons which are used in the imaging process.

6.4.1.2 Data Transfer and Storage

The data from the healthcare industry constitute big data, and they have to be stored in a file format that is easily accessible for analysis and treatment. Medical data generated generally consists of image and non-image data, which are communicated and managed to comply with a universal standard format called DICOM (Digital Imaging and Communication in Medicine). The DICOM file format is usually used for image transfer, which has a ".dcm" extension. It integrates all medical imaging devices, network hardware, printers, scanners, archival systems such as PACS (Picture Archiving and Communication Systems), and all other types of equipment involved in the medical data transfer. Medical data contains extensive context information such as technical details regarding the image acquisition system, parameters used in the acquisition, demographic information about the person being examined, image sequences or images, details regarding the reports, etc. This leads to different image formats from conventional ones. Medical data can be exchanged between two entities if they follow the same DICOM standards and the file format. The digital medical images are usually stored in PACS of local workstations and are communicated with other workstations following the DICOM format and standards. Windows cannot recognize DICOM files as image files and so an additional "DICOM browser" is needed to view the files [19]. The non-image medical data such as reports can also be attached as PDF files during the sending process. Due to sensitive personal information in the medical data, the use of data is restricted and regulated.

The main components of PACS include

1. The image acquisition systems
2. A secure network for the transmission of medical data
3. Workstations for interpreting, analysis, and reviewing images
4. Archival system for the storage and retrieval of medical data

DICOM follows the client–server paradigm for communication between the entities and uses TCP/IP protocol for communication between devices. A DICOM file is composed of a header and image datasets all in a single file. The first few packets of information, which is the "header," consist of series of organized tags, where each tag, in turn, consists of groups of data elements and the corresponding value associated with each data element. Following the header is the image data itself which contains the pixel intensities of the image in series of 0s and 1s. The header and the image data are encoded in the DICOM file, so that all necessary information about the image data can be obtained from the header.

The communication of medical images is formulated by the DICOM Information Model in which the structure and organization of the information are defined. DICOM classes contain information entities in the form of data structures and services in the form of methods. Classes of information entities are templates and are called Information Object Description (IOD), which represent a group of real-world entities sharing common properties. An instance of IOD is called an information object which is uniquely defined by a Unique Identifier (UID), which is the name of a DICOM file. The UID is specified as a string or character vector with a root that uniquely identifies an organization. DICOM contains two different types of objects that can be exchanged. They are composite and normalized. Image is a composite entity consisting of text, numerical information, and the image itself. Normalized entities are single entities such as reports. DICOM defines several services associated with these entities.

DICOM is implemented by other imaging systems such as PACS, RIS (Radiology Information Systems), VNAs (Vendor Neutral Archives), 3D Visualization Systems, image printers, file scanners, CAD (Computer-Aided Detection/Diagnosis), image viewers, and display stations. Usually "Radiologist" is responsible for handling the image files which are stored on PACS. PACS are systems that archive digital image data on local workstations. The records in the RIS are associated with the corresponding images in the archives. Access to them is via the RIS or PACS interface. Communication is carried out through the DICOM standard and the RIS interface. Persons in the local workstations who have the access rights to RIS and PACS can access the data from anywhere within the hospital. When images need to be accessed from other workstations, DICOM client software is needed to PULL the images from other archives, and DICOM viewer places query to fetch the needed images and displays them in an organized way.

Medical images come from various modalities and are extremely in huge volumes. So these images have to be compressed for storage and communication over the network. The compression has to be done in such a way that the quality of the image should not be compromised. DICOM supports lossless compression and lossy compression. Lossy compression is not suitable with medical images as it has to be stored for a stipulated amount of time and later reviewed and analyzed. So, proper information regarding the image is needed. Lossy compression can be done for purposes where complete image details are not needed.

6.4.1.3 Image Enhancement

Image enhancement of medical data is the process of increasing the perceptibility of objects within the images by using software tools. Image quality is deteriorated due to noise, transfer, compression, processes in data acquisition, procedures used in imaging, etc. Image enhancement techniques are used to provide quality images for better analysis, diagnosis, and treatment, thereby enhancing manual or automated extraction of specific information. It should also properly convey the three-dimensional view of the object, reduce the illumination effects, find the partially hidden objects, track the objects, aid in detection, and help in delineation of structures within the objects [6]. The modality of the image and the specific parts or the organs to be examined determine the effectiveness of the specific enhancement technique applied. The advancements in the field of medical technology have led to automatic image enhancement in real time with low computational cost. There are several image enhancement techniques such as contrast enhancement, resolution enhancement, edge enhancement, and noise reduction.

- Contrast enhancement

  It deals with improving different intensities of the image by covering a wide range of image values to make features more perceptible for analysis. It is also used for highlighting specific regions of interes/t (ROI). Contrast is one of the most important characteristics that determines the accuracy of the results obtained and can be best depicted by a graphical representation of histogram which shows the frequency of pixel intensity values. Here the x-axis represents the gray-level intensity values, and the y-axis represents the frequencies of these intensities. A transformation function $T(f)$ is applied to the initial image (I_I) to get a transformed image (I_T). The gray-level values of each pixel are transformed from f to $T(f)$. The transformation function depends upon the original image and the desired final image. The transformations can be fixed functional forms such as linear and nonlinear stretching, or adaptive transforms such as histogram equalization, histogram specification, or adaptive histogram modification. The histogram should be made as flat as possible for contrast enhancement. For color images, the techniques can be applied separately for all the individual components in RGB or CMYK. The type of enhancement measure applied largely depends upon the physical attributes of the image (such as background, lighting, the content of the image, noise, and patterns), prior knowledge about the image, experience, etc. Contrast enhancement is related to contrast measures that assess the quality of the image. The contrast measures are chosen such that it best suits the attributes to yield a better informative image. Some of the contrast enhancement measures are global contrast, global variance, entropy, Gray-Level Co-occurrence Matrix (GLCM), contrast ratio, Mean Squared Error (MSE), Peak Signal to Noise Ratio (PSNR), Root Mean Square Error (RMSE), Normalization Co-efficient (NC), etc.
- Resolution enhancement

  Images are analyzed and processed to obtain an enhanced image. Interpolation is a technique used for resolution enhancement by increasing the number of pixels. It aims at finding intermediate pixels between the known pixel values in a low-resolution image row by row and column by column.

- Edge enhancement

  It is a prerequisite in image processing that enhances the edge contrast of an image or video. Edges represent discontinuities of the pixel intensity of an image. They characterize the boundaries of an object, and thereby edge enhancement plays a key role in identifying particular objects of interest. The structural properties in an image can be preserved by edge detection, which filters out useless information, thereby significantly reducing the amount of data to be processed. Edges represent the local change in intensity and so are related to intensity gradient. Different kernels such as Sobel, Gaussian, and Laplace are used for enhancement. The Laplace operator is most often used with a smoothing operator called Laplacian of Gaussian filter (LoG) as it is very sensitive to noise [6].
- Noise reduction

  It is the process of removing or reducing noise that deteriorates the quality of the image obtained through various modalities. It is due to the acquisition or transfer of images, or other environmental conditions. Noise leads to artifacts in the images leading to error in the perception and representation of any structures. There are several types of noise created in medical images most of which are Gaussian noise, Poisson noise, Speckle noise, Rician noise, Rayleigh noise, and Quantum noise. For an accurate diagnosis, the images should be free of noise and artifacts. Several denoising methods are used for noise removal or reduction depending upon the type of noise. In X-ray imaging, the photons are randomly received by the receptor and this depends upon the receptor sensitivity. As a result, it gives a grainy and snowy appearance on the film. This type of noise is called quantum noise or Poisson noise. Sometimes Gaussian noise is prevalent in X-ray imaging due to thermal activity. Fluoroscopic radiography is more prone to Gaussian noise. CT scan is more prone to Gaussian noise due to the high-speed computations needed. Gaussian noise is also more prevalent in PET and SPECT. In MRI, Rician noise, Gaussian noise, and Rayleigh noises are mostly seen. Random fluctuations caused due to electronic interferences, radiofrequency emissions, and measurement chain of MRI scanners reduce image contrast, resulting in reduced quantitative and qualitative evaluation. Speckle or Salt and pepper noise are dominant in ultrasound scans due to inherent granular interferences. They are also prone to Gaussian noise. Denoising methods such as low-pass filter, high-pass filter, median filter, histogram equalization, mean filter, and diffusion filter are used in the noise reduction process.

6.4.1.4 Image Segmentation

Image segmentation partitions the given image into multiple segments based on certain characteristics such as color, texture, brightness, and contrast. Each segment is comprised of a set of pixels with similar characteristics. The purpose of segmentation is to make the image more perceptible for the detection of specific objects of interest and enhance analysis. Interactive segmentation is a semi-automatic process of segmentation that allows users to interact with the computational system, by the way of giving inputs, checking the accuracy of the results, and manually controlling the segmentation process. The output of interactive segmentation largely depends on how the user interacts with the computational system. So, efficient interactive strategies have to be incorporated for accurate results.

The components of the interactive segmentation consist of the computational part, interactive part, and user interface. The computational part consists of the methods or algorithms used in the segmentation. The type of methods used to influence the accuracy of the results depends on the specific application. The user progressively interacts with the system by giving various inputs through the user interface until the desired result is achieved. The interactive part acts in between the computational part and the user interface in supplying the user input as parameters to the computational part and translating the results produced as visual feedback to the user. The inputs can be in the form of real parameter values, which can be continuous or discrete, spatial parameters or pictorial inputs, such as point, lines, or regions that focus specific regions of the image, and menu-driven.

Some of the techniques used in the interactive image segmentation process are as follows:

- Graph-cut method

  This method is semi-automatic and uses graph theory in which the pixel or voxel of an image represents the vertices of a graph and weighted edges connect these vertices. The relation between the corresponding edges determines the weight of the edges. The "cut" represents the partition of the set of vertices into two disjoint subsets. The cut-out is the set of all edges having one endpoint in each subset such that their removal results in the separation of the graph into one or more connected components. The seeds are given by the user by marking any of the pixels as "object" or the "background". The algorithm finds the disjoint sets and labels all pixels corresponding to one node as "object" and all pixels relating to the other disjoint node as "background". The region and boundary properties are included in the cost function and the method minimizes this energy function to provide a globally optimal solution.

  Some of the extensions of the graph-cut method are as follows:

  - *Grab cut*: It is an extension of the interactive graph-cut method done iteratively. The pixel colors are represented using the Gaussian Mixture Model (GMM) [7] and have less user interaction.
  - *Lazy snapping*: It has more speed and accuracy when compared to the interactive graph-cut method. To reduce computational speed, a graph is constructed with over-segmented pixels and accuracy is improved by adopting the Watershed algorithm.

- *Geodesic graph-cut method*: The edge information is combined with geodesic-distance region information in a graph-cut optimization framework for labeling the pixels [8].

- Random-Walk method

  It is an image segmentation method with multiple labels. The known labels are called "seeds" and the user interactively labels few pixels with "seeds". The random walk occurs on a weighted graph in which the pixels are represented by nodes and edges connecting the neighboring pixel. The weights represent the similarity between the pixels. In this algorithm, the probability that an unlabelled pixel first finds its way through the random walker to each of the seeds is calculated using linear equations. With probabilities calculated for each seed, the pixel that most likely reaches the seed is given the label of that corresponding seed. In a random walker, the segmentation is obtained by selecting the seed destination with the highest probability.

- Thresholding

  It is the simplest method of image segmentation based on a threshold value. Based on the pixel gray level, the classification of pixels is made [9]. It is the process of converting a grayscale image into a binary image by dividing them into two-pixel intensity levels by assigning either of the two values "0" or "1". The pixel intensities whose values are greater than or equal to the threshold specified are "foreground" and those below that are the "background". In interactive thresholding, the user selects the threshold value and varies until the desired structures of interest are met. There are various forms of thresholding, which are as follows:

  - *Local thresholding* is a technique in which an image is divided into several fixed-size sub-images with location and threshold value calculated for each image using a statistical method such as mean, standard deviation, and mean of maximum and minimum [10]. Finally, the results of thresholding from each image are combined. This technique is used when the background is not uniform and objects have diversified pixel values.
  - *Otsu's Thresholding* is a global automatic image thresholding technique that finds one optimal threshold value by reducing the intra-class variance and increasing the inter-class variance of black and white cluster pixels.
  - *Gaussian Mixture method* is a global automatic thresholding technique that finds the optimal threshold value by the average of the Gaussian mixture means found using the Expectation-Maximization (EM) algorithm.

- Region-Based methods

  In this method, regions are considered rather than pixels. Based on certain criteria, the image is segmented into similar regions. The preprocessing or the computation of initial regions for region-based segmentation is done first and then combined with the merging step, which helps in producing more efficient results. One of the most commonly used preprocessing algorithms for monochrome and color images are the watershed and mean shift algorithms, respectively, which are as follows:

- *Watershed algorithm*: It is the representation of a grayscale image as a topographic relief, flooded with water, in which watersheds are lines dividing areas of the water from different basins [11]. The value of each element of the image whether it is pixel (in 2D images) or voxel (in 3D images) represents height at this point. This algorithm aims to separate two regions of interest that are close to each other and find separate segments of interest and also the boundaries of the segments if needed. The flooding approach and rainfall approach are the two main approaches used. Oversegmentation is the main drawback of this method. It is widely used in medical applications.
- *Mean Shift algorithm:* It is a well-known technique for multimodal feature space analysis. It consists of two steps: the filtering of the original image data and the clustering of the filtered data. It is non-parametric, and the filtering consists of representing the highest data density points as modes and associating the corresponding data points which are close to these modes. Later, the data points are replaced with their corresponding modes and are grouped by clustering, which are less than one kernel radius apart resulting in segmentation of the filtered data.

Some of the techniques used in region-based segmentation are as follows:

- *Region Growing* is a region-based approach in which an initial seed point is selected and the unallocated neighboring pixels are compared iteratively and the region is grown.
- *Region splitting and merging* adopts a homogeneity criterion being defined and the image is split into quadrants. Based on the criterion, similar regions are merged, and this process is repeated until all regions satisfy the homogeneity criterion to get a segmented result.
- *Grow-Cut* is a region growing method in which the seed points are provided by the user. It is similar to graph-cut methods and extracts foreground images by removing the background. The user can enhance the results by refining the inputs. The interaction of the user and the image content determines the accuracy of the results. It can be applied for multiple object segmentation and can be used for 3D and 4D images.

- Edge-Based methods

These methods are used to detect the edges of images, which give significant information characteristics regarding the image. Edges are represented by significant changes in gray values of the image from one pixel to the next. Usually smoothing is performed as a preprocessing step before the edge detection techniques are applied. There are many methods for edge detection, which include the Gradient Edge Detection method, Laplacian method, Active Contour method, LiveWire, and Intelligent Scissors.

- Classification methods

  It involves training data that contains manually segmented images with target labels. It is a supervised learning method in which the classifier learns from the training data. The disadvantage is that the spatial information is not considered and the user has to manually segment images for training which needs experience. K-nearest neighbor, maximum likelihood, etc. are some of the most widely used classifiers.
- Clustering methods

  It is an unsupervised method of pattern recognition that does not involve training data. It works similarly to classification in grouping similar objects into one class using similarity measures. It consumes less time, and the disadvantage is that they do not consider spatial information. Some of the clustering algorithms include k-means, fuzzy C-means, mean shift, expectation-maximization, and hierarchical clustering.

6.4.1.5 Image Registration

Registration is a step involved in image analysis in which two or more input images are combined into a single image with a single coordinate system. In healthcare, the information acquired from separate images has to be integrated in a proper way for efficient decision-making [12]. This process of integration involves two steps which are *image registration* and *image fusion*. Image registration involves spatially aligning two or more image datasets taken at different times of the same scene. The image may be taken from different viewpoints or by different sensors. A coordinate transformation is done between the coordinates of multiple-image spaces which combine information from all forms of a particular patient into a single representation that gives the spatial and temporal relationship among all the images. Registration of medical images from the same modality taken at different time-frames is necessary to overcome the distortions due to movements of the object. Similarly, registration of medical images of the same structure obtained from different modalities is useful as they provide different representations of the same image. Integration of information from various images acquired from a single modality, taken at various time-frames and angles, is essential as they will not yield sufficient information when viewed separately [13]. Registration plays a significant role in radiotherapy, planning surgeries, and characterization of neurological disorders [14]. Combining images from different modalities of the same object, alignments of images from multiple subjects, aligning temporal sequences of images, and image guidance during interventions are some of the applications of image registration [15].

Registration can be classified based on the nine criteria formulated by Van Den Elsen et al., which are as follows:

1. **Dimensionality** of the registration method depends on the type and number of dimensions involved. Based on type, they are divided into spatial dimension (if all the dimensions are spatial only) or time series

with spatial dimension (if time can be added to the spatial dimension). In both cases based on the number of spatial dimensions, they are further subdivided into 2D–2D, 2D–3D, and 3D–3D.

2. **Nature of registration** categorizes the registration process into image-based and non-image-based. Image-based methods are further subdivided into *extrinsic* and *intrinsic* methods.

   Extrinsic methods are those in which external objects are attached to patients and which can be detected. Examples of common external objects attached to the patients include markers of screw-mounting and markers stuck on the patient's skin [16].

   Intrinsic methods use patient-generated information. They can be landmark-based registration, segmentation-based registration, and voxel property-based registration.

   - *Landmark-based registration methods:* Landmarks can be geometrical or anatomical. To define a transformation, elements such as curves, corners, points, and surface landmarks of one image are matched with the respective elements in the other image. By applying interpolation, correspondences of the remaining image volume along with the matched landmarks are inferred [17].
   - *Segmentation-based registration method:* Rigid or deformable models are used in the registration. Anatomically the same structures from both the source and the target images are extracted in rigid models, whereas in deformable models the structures are extracted from one image and are made to fit in the other image by elastically deforming them.
   - *Voxel property-based registration method:* In this method, the image intensity patterns are matched by directly operating on the image gray values, by using mathematical or statistical criteria [18]. The transformation is done by measuring the intensity similarity of the input image used for making adjustments in transformation for achieving maximum similarity.

3. **Nature of transformation** can be categorized into rigid, affine, projective, and curved. The transformation in which the entire 2D or 3D images are transformed only by translating and rotating is called rigid [12]. The transformation mapping parallel lines onto parallel lines is called affine, mapping of lines onto lines is called projective, and mapping lines onto curves is called curved.
4. **Domain of transformation** In a transformation, if the mapping parameters are applied to the whole image, it is called *global*, whereas if the mapping parameters are applied only to the sections within the image, it is called *local*.
5. **Interaction** can be automatic, interactive, or semi-automatic depending upon the relation between the user and the registration process.
6. **Optimization procedure** Registration is an iterative process that optimizes the transformation parameters by maximizing the similarity until

the target image is acquired. Transformation can be *parameters computed* in which parameters are computed from the available data directly or *parameters searched for* in which parameters are computed by an optimum function in the parameter space.

7. **Modalities** Depending upon the modalities used, the registration process can be *mono-modal* (two images are from the same modality), *multimodal* (two different modalities are used for obtaining images), *modality to model (*one image is used and the other modality is a model), and *patient to modality* (one image is used and the other modality is the patient).
8. **Subject** refers to the patients whose images are to be registered.
   - *Intrasubject* registration involves images from a single patient.
   - *Intersubject* registration involves two images from two different patients.
   - *Atlas* registration involves one image from a single patient, and the other image is made by imaging many subjects from an image information database.

   Such registrations of medical images are required for visualization, automatic detection, and quantitative analysis.
9. **Object-based** registration techniques involve parts of the human anatomy such as the head, abdomen, and limbs.

   Image fusion is the transfer of any imaging information from one dataset to another for the display of the data in an integrated way, which mainly concerns the proper visualization of useful information. For example, integration of CT scanning with MR imaging systems yields potential benefits as the CT scan is better suited for accurate computation of radiation results, while the latter is more suited for delineating tumor tissue. Computerized approaches and tools are used for the proper alignment of images and the display of the combined resultant image. How the images are compared and combined improves the efficiency and the potential benefits obtained.

6.4.1.6 Visualization

Visualization of medical data helps the physician in determining the structural (organs, tissues, and cells) and functional attributes (e.g. physiological, biomechanical) of the human body. The anatomical structures can be viewed by the physicians in real time with interactive and advanced display technologies that enable them to make decisions accurately and quickly. Visualization can be categorized into 2D and 3D visualizations. In 2D visualization, multi-planar reformatting is used in which visualization images are re-sliced into multi-planar reformatted images (MPR). Multi-planar reformatting consists of three important types; orthogonal, oblique, and curved planes. The 3D visualization can be divided into two types as surface rendering (SR) and volume rendering (VR). These rendering techniques can be direct or indirect. The direct rendering techniques can be classified as Direct Surface Rendering (DSR) and Direct Volume Rendering (DVR), while indirect rendering has Indirect Surface Rendering (ISR). VR is a collection of methods used to produce a 2D projection from a 3D dataset, enabling users to view three-dimensional scalar fields. VR techniques are developed to overcome the problems associated with SR. The intermediate geometric representations in SR are eliminated, thereby improving the surface integrity of each voxel. It also helps in assessing the underlying structures and provides a clear view of the anatomical structures.

The DVR technique assists in obtaining the 3D representation of the given large volume of data. There are two categories of direct volume rendering techniques, namely, object-space and image-space DVR. In object-space DVR, forward mapping strategy is used to obtain the 2D image, e.g. Ray Casting. In image-space DVR, a backward mapping strategy is followed to get an image, e.g. texture mapping, shell rendering, and splatting. The major components in DVR are as follows:

- Sampling
- Classification
- Composition

The techniques followed in DVR techniques are described below.

- Ray casting

  Ray casting is a classic DVR technique that is a variant of ray tracing. In these casting methodologies, rays are cast from the viewpoint via the image plane and dataset. The contributions in ray tracing are generally calculated at the intersection of rays. Based on the object material properties, the rays get either reflected or refracted. If both occur simultaneously, primary rays traverse in the new direction, while the secondary rays are cast from the point of intersection. Ray casting is an image ordering technique in which the ray is cast into sampling at certain time intervals. The sample is reconstructed/interpolated from the voxel at every sampling location.

  The voxels in the volumetric dataset are considered for computing the intersections. The samples in the dataset may (if opacity values are greater than zeros) or may not (if opacity values are zero) contribute depending on certain properties. The location of samples in the data volume depends on the sampling rate and direction, which manages the distances among neighboring samples. In most cases, sample values are computed using the interpolation method instead of locating positions straight away at voxel positions. The significant difference between ray tracing and ray casting is the absence of secondary rays; the rays are not reflected or refracted at the sampling points. Moreover, in ray casting, at least a single ray is to be cast per pixel in the image plane. According to sampling theory, more rays per pixel are required for casting, which automatically maximizes the computational load. In general, a precise intersection cannot be computed in discrete ray casting. If isosurfaces are rendered into ray casting that appears smaller and few characteristics may also get lost if the sample count is not adequate. In most medical applications, the orthographic or parallel project is used, where the complete rays will traverse in a parallel direction with the volume dataset.

- Shear Warp

  This algorithm combines the advantages of both object-space and image-space DVR methodologies. The shear matrix converts the complete viewing rays; hence, the sheared object surface permits the image and volume to pass through simultaneously. The main objective of this algorithm is to reduce the volume sampling by minimizing the cost incurred for memory access. In a shear warp, the rays are cast from a base plane that is generally parallel to the data volume. Now, the resultant ray is perpendicular to the relevant volume face, which facilitates the data access. Initially, a shear warp was proposed in [20–22] to simplify the voxel access patterns, and then it was extended by [23]. The latest shear warp is based on Lacroute's approach and provides better caching of data access in accessing voxels. Subsequently, it can be simplified by sampling the rays using a base plane. Run-Length Encoded (RLE) representation permits rapid processing and loading of volume datasets. Shear warp is considered the fastest rendering technique in accessing regular voxel patterns. However, the quality is not assured due to the fast rendering process. In the proposed shear warp algorithm [23], a single ray is cast per pixel on a base plane. It becomes inadequate for the largest view plane images; the base-plane image is enlarged during the warping step, resulting in heavy blurring. Bi-linear interpolation is exploited in sharp-wear for computing the ray samples in one slice; in turn, it resulted in lower quality when compared to trilinear interpolation, particularly for anisotropic datasets.

- Splatting

  Voxels are projected on an image plane in a splatting algorithm and are based on object-space. Voxels are referred as tiny points of scalar value, and it is to be convolved using volume filter, a typical Gaussian function. However, it functions well for isotropic grids in which the entire voxel distances are the same and the footprint screen projection is a circle. It significantly improves the computation speed of DVR; however, the accuracy rate is low. The data points are accumulated by directing kernels to each voxel grid in the drawing plane. The composition of footprints forms a visual representation of an image in the drawing plane. The kernel size and shape determine the quality of the resultant image.

  Ref. [24] presented a hybrid splatting approach that solves the undersampling problem that occurred in [25] splatting approach along with cone-beam ray casting. A pre-classification rendering scheme is followed in splatting; the voxels are classified first and resampled by the sheet buffer and filter kernel. The main benefit of splatting is in the utilization of pre-computed footprints via a lookup table. The footprints are commonly rasterized for generating the visual contribution. The rasterization is an expensive step in the splatting approach and, subsequently, researchers attempted to minimize the cost of it. In splatting, all nontransparent voxels that exist on the image plane should be projected even if there is no contribution of it.

- Texture mapping

  In 2D texture mapping, the volume is categorized into three perpendicular object-aligned polygons. In 3D texture mapping, the volume is uploaded in the graphics memory area as a single texture and the image is textured using trilinear interpolation. In this context, a volume dataset is primarily loaded in the texture memory and resampled using texture mapping functions and mapped into a rasterized polygon that appears on the screen. Texture mapping gathers all relevant samples to be placed in the object-space; hence, it is based on the object-space approach. 3D texture mapping can be employed in DVR for resampling the volume dataset. Interpolation functions can be exploited for computing the individual texels. For 2D textures, volume resampling has to be arranged differently. 2D textures could not be resampled arbitrarily; instead, Cartesian grid is exploited for arranging images in the volume dataset. Moreover, storing data volume in the Cartesian grid comprises three orthogonal orientations that eventually increase the memory requirement. Another drawback in texture mapping VR is the lack of shading that limits the perception significantly.

- Shell Rendering

  VR occurs based on the data structure generally referred to as nontransparent voxels along with the set of attributes interconnected with each voxel that represents the visualization. The major aim of shell rendering relies on fuzzy boundaries; these commonly refer to voxels that represent a cuboid. The voxels of the shell are maintained in a linked list format that permits sorting and fast addressing of voxels to facilitate both forward and backward traversal. Later, the voxels are projected as in splatting, and the voxel rendering has been significantly improved by employing better splatting techniques. The main advantage of shell rendering lies in choosing the voxels on a boundary; the count is significantly less than the total count in the volume dataset. This approach allows rapid rendering once the boundary is chosen.

6.4.2 Unstructured Text Analysis

Traditional preprocessing technology applied to structured data cannot be directly applied to unstructured text. They have to be converted into a structured format readable by the computer. Medical text mining is employed for this process of transformation, which uses NLP technologies. The burden of manually organizing text notes can be reduced with the help of machine learning in the form of text mining, which uses sophisticated algorithms. The text mining process consists of four stages, namely, information retrieval, information extraction, knowledge discovery, and knowledge application [4]. Information retrieval is the process of collecting the desired text. Information extraction is the preprocessing of the desired text. Knowledge discovery is the process of extracting new knowledge from the text, and knowledge application is the application of facts inferred from the texts into practice. The technologies involved in the process include named-entity recognition (NER) and relationship extraction (RE).

NER is the process of identifying required medical entities and classifying them into predefined categories. The differences in doctor's writing styles, different forms of medical terms such that the same disease can be represented by two names, ambiguity in short notations used by the physicians, and others reduces the effectiveness of finding the required entity. Several approaches such as rule-based NER, dictionary-based NER, and machine-learning NER approaches are being used for this process.

RE is the process of establishing semantic relationships among the selected entities. Different approaches are used for extracting entity-relationship such as pattern-based, concurrence-based, and machine learning-based approaches.

6.4.3 Signal Data Analysis

Bio-medical signals are the signals collected from the body to diagnose and detect pathological or physiological conditions at the organ level, molecular level, or cellular level. This information, which may include glucose levels, heart rate, and blood pressure or body oxygen levels, is captured through physiological instruments and processed using various algorithms. The signals are analog in nature and are available at all instants of time (continuous-time signals). Hence, to be processed and analyzed by the computers, the analog data have to be converted to digital signals by the sampling of the analog data at equal time intervals. These processes are carried out by the digital signal processing system (DSPS). The bio-medical signals are most advantageous in a critical care setting where the biosignals from a patient's body are continuously recorded, monitored, and analyzed in real time.

6.5 VISUALIZATION METHODS

Visualization techniques are playing a significant role in visualizing medical data by analyzing the massive number of clinical events. Innovative visualization techniques are needed for handling medical data that contains complex data with multiple variable types and data points. The utility of patient medical data stored in EHRs can be improved by using Interactive visualization tools (IVT) that helps in exploring and efficiently visualizing the data. The visualization techniques facilitate interaction by focusing on parts of the displayed information via zooming, filtering, and scaling, which gives more detailed information. Patient data consists of multidimensional information space consisting of both structured and unstructured data and this has to be given as a graphical summary, which is an integration of both graphical and textual information. Visualization design is to be such that all medical data such as numerical, text, image data, and signal data have to be reproduced in a visual view with enhanced interactivity.

Some of the techniques for visualizing quantitative data, incidents, and planning of different scenarios are given below:

- *Simple Charts* are simple representations that give a clear idea about the data being represented. Charts such as line charts, bar charts, pie charts, and histograms are used very commonly in many applications for finding patterns. However, they are not efficient for depicting periods that are longer and not suited for periodic data.
- *Graphical Patient Record Summary* gives an overview of the patient record containing all known parameters included in one sheet of paper excluding more detailed information. They are used for representing quantitative data and recording incidents. They are used for seeing developments and combining values. In these, different levels of significance are given for the data values. For example, the data values for the current week are given more magnification than the data of the year before and so on.
- *Spirals* are used for displaying data that are close in terms of temporal dimension and for identifying intervals.
- *Shinkansen Timetables* are used when many incidents have to be tracked as in this many numbers of objects can be represented on single screen space. It is mainly used for observing trends and patterns.
- *Timelines and LifeLines* Timelines are an intuitive and powerful way of representing temporal activities. This horizontal line is drawn at a particular period to depict an incident. This basic conceptual idea is further extended as LifeLines. It is a two-dimensional Timeline mainly used to visualize specific patient records or history. In LifeLines, horizontal lines drawn at different sets of intervals depict the duration and the actual temporal location of the data for different kinds of scenarios. The "Facets" are introduced in an extended version, the vertical segments that support grouping similar events. It is more flexible in providing information to the user in different aspects without re-organizing the display.

  LifeLines evolved and have been extended to LifeLines2, which enables multiple patient records to be visualized. The users can view both numerical and categorical data which can be analyzed by comparing and aggregating the data to provide higher levels of detail. But scalability is the main issue related to LifeLines2.

  LifeFlow is developed, which surpasses the scalability problem by allowing users to view millions of patient records on a single page and which summarizes the patterns.

EventFlow is an interactive graphical tool used for viewing, reviewing, and summarizing data from records based on the event sequences. The temporal patterns can be viewed and analyzed with EventFlow. The data from EHRs can be used to analyze the treatment patterns along with their outcomes.

- *Time cubes* are used to represent the spatiotemporal data, and it is an extension of timelines [26]. The change in incidents can be tracked easily and is used for finding patterns.
- *Gantt charts* are used in planning a medical action and they use Timelines as a basis for representation. The hierarchies can be easily displayed using Gantt charts, but the same task with different facets cannot be displayed.

6.6 CONCLUSION

Interactive visualization in healthcare aids the user in exploring the data for better understanding and analysis, which would yield fruitful results for arriving at the best feasible solutions in the diagnosis, planning, and treatment of diseases. Interaction in addition to viewing capabilities incorporated in the design helps the user in interrogating the dataset, and this can be enhanced by considering perceptual and cognitive theories. In this chapter, a brief introduction is given about interactive visualization and its requirement in the healthcare industry. Later, various sources of medical data are discussed along with its types. Then analysis techniques related to unstructured medical data which consists of image, text, and signal data are discussed. Finally, the methods mainly used in the medical field for interactive visualization are discussed briefly. This chapter gives an overall idea about how medical data are acquired, stored, analyzed, and visualized using several algorithms and methods.

REFERENCES

1. D. Gotz and D. Borland. 2016. Data-driven healthcare: Challenges and opportunities for interactive visualization, *IEEE Computer Graphics and Applications*, 36: 90–96.
2. C. Silva, E. Groeller, and H. Rushmeier (eds.), 2005, VAN WIJK J.: The value of visualization. *Proceedings of IEEE, Visualization*, 79–86.
3. E.H. Shortliffe, G.O. Barnett. Medical data: Their acquisition, storage, and use. In: Shortliffe E.H., Perreault L.E. (eds) *Medical Informatics. Health Informatics*, Springer, New York, 2001.
4. W. Sun, Z. Cai, Y. Li, F. Liu, S. Fang, and G. Wang. 2018. Data processing and text mining technologies on electronic medical records: A review, *Journal of Healthcare Engineering*, 2018: 9.
5. E. Zudilova-Seinstra, T. Adriaansen, and R. van Liere, *Overview of Interactive Visualisation*, Springer-Verlag London Limited, 2009, 3- 15.
6. K.D. Toennies, *Guide to Medical Image Analysis, Advances in Computer Vision and Pattern Recognition*, Springer-Verlag London Limited, 2012, 147–169.

7. J. He, et al., *Interactive Segmentation Techniques*, SpringerBriefs in Signal Processing, Singapore, 2014.
8. B.L. Price, B. Morse, and S. Cohen. 2010, June. Geodesic graph cut for interactive image segmentation. In 2010 IEEE computer society conference on computer vision and pattern recognition (pp. 3161–3168). IEEE.
9. J.S. Weszka. April 1978. A survey of threshold selection techniques, *Computer Graphics and Image Processing*, 7(2): 259–265.
10. A. Norouzia, M.S. Mohd Rahima, A. Altameemb, T. Sabac, A. Ehsani Rada, A. Rehmand, and M. Uddin. May_Jun 2014. Medical image segmentation methods, algorithms, and applications, *IETE Technical Review*: 199–213.
11. A.S. Kornilov and I.V. Safonov. 2018. An overview of watershed algorithm implementations in open source libraries, *Journal of Imaging*, 4: 123.
12. J.B.A. Maintz and M.A. Viergever, *An Overview of Medical Image Registration Methods*, Utrecht University Repository (Book), Nederland, 1998.
13. F. Alam, S.U. Rahman, A. Ud Din, and F. Qayum. 2018. Medical image registration: Classification, applications and issues, *JPMI*, 32(4): 300.
14. J.S. Duncan. January 2000. Medical image analysis: Progress over two decades and the challenges ahead, *IEEE Transactions on Pattern Analysis and Machine Intelligence*, 22(1): 85–106.
15. D.L. Hill, P.G. Batchelor, M. Holden, and D.J. Hawkes. 2001. Medical image registration. *Physics in Medicine & Biology*, 46(3): R1.
16. H. Mitchell, *Data Fusion*, Springer, Berlin, 2012.
17. T. Lindeberg. 1998. Edge detection and ridge detection with automatic scale selection, *International Journal of Comput Vision*, 30(2): 117–156.
18. F.E.-Z.A. El-Gamal, et al., 2016. Current trends in medical image registration and fusion, *Egyptian Informatics Journal*, 17: 99–124.
19. D.R. Varma. February 2012. Managing DICOM images: Tips and tricks for the radiologist, *Indian Journal of Radiology and Imaging*, 22(1): 4.
20. C.G. Cameron and P. Undrill. Rendering volumetric medical image data on a SIMD architecture computer. In *Proceedings of Eurographics Workshop on Rendering*, pp. 135–145, 1992.
21. R. Yagel and A. Kaufman. "Template-based volume viewing." Computer Graphics Forum. Vol. 11. No. 3. Edinburgh, UK: Blackwell Science Ltd, 1992.
22. P. Schroder and G. Stoll. Data parallel volume rendering as a line drawing. In *Proceedings of IEEE/ACM Symposium on Volume Visualization*, pp. 25–32, 1992.
23. P. Lacroute and M. Levoy. (1996, October). Fast volume rendering using a shear-warp factorization of the viewing transformation. In Proceedings of the 21st annual conference on Computer graphics and interactive techniques (pp. 451–458).
24. K. Mueller and R. Yagel. 1996, October. Fast perspective volume rendering with splatting by utilizing a ray-driven approach . In Proceedings of Seventh Annual IEEE Visualization'96 (pp. 65-72). IEEE.
25. L. Westover. 1990 September. Footprint evaluation for volume rendering. In Proceedings of the 17th annual conference on Computer graphics and interactive techniques (pp. 367–376).
26. R. Kosara and S. Miksch. 2002. Visualization methods for data analysis and planning in medical applications, *International Journal of Medical Informatics*, 68: 141–153.

7 Heart Disease Prediction Using Tableau

R. Indrakumari, Priyanka Shukla, and Akanksha Sehgal
Galgotias University

CONTENTS

DOI: 10.1201/9781003050827-7

7.1 INTRODUCTION

The heart is the smallest but strongest muscle in the human body. The heart starts to beat before birth in the uterus after 21–28 days of conception. The heart beats at a rate of 70 beats per minute, and with each beat pumping the blood to each part of the body. The rate at which the heart beats can be different depending upon the kind of exercise and type of emotional state a person is in. Blood is pumped through the left chamber of it, is transported through arteries and then through capillaries, and then reaches to tissues and body organs. After supplying oxygen to all organs and after collecting waste, the blood is supplied back to the right chamber of the heart. This complete heart system is vulnerable to break down because of many factors. Some factors that can affect heart and blood vessel systems are inflammation, drugs, alcohol, toxins, unhealthy diet, and lack of exercise; depending on these factors, heart diseases can be divided into coronary heart disease, rheumatic heart disease, congenital heart disease [1]. Dietary, population, and environmental-specific risk factors are believed to be linked with coronary heart disease [2]. Another factor that is associated with chronic heart diseases is dental diseases. Dental disease is a risk for coronary heart disease and mortality [3]. Coronary heart diseases can be controlled by using a low-fat diet, exercise, no smoking, and decreasing emotional stress levels [4].

Unsaturated fatty acids can be replaced by saturated and Trans-fats, and consumption of omega-3 should be increased to lower the risk of heart diseases. Along with these regular exercises, weight control can be considered for decreasing the risk of heart diseases [5]. The personality traits that can be considered as facilitators of heart diseases are anger, hostility, aggression, depression, extroversion, and anxiety. It is also considered as a person with a more negative thought process is more prone to heart diseases. If the fetus was undernourished, then also he is at risk of heart disease in later stages of life. Fetal health is also found to be related to the risk of heart diseases. Sixteen

Fetal Nutrition and Cardiovascular Disease in Adult Life. Healthy emotional life, regular workouts, no to alcohol and smoking, a high fiber diet, and reduction in body weight can be considered as some factors by which heart diseases can be controlled.

7.2 TYPES OF HEART DISEASE

Heart disease is a very common disease nowadays. There are numerous types of heart disease. All have their causes and symptoms. The types of heart disease, causes, and symptoms are as follows.

7.2.1 Congenital Heart Disease

This type of disease is present at birth, like a hole in the heart, and this defect in the heart may cause inhaling issues in the body, and the blood flow through various chambers of the heart is partially or blocked. Recognition of congenital heart diseases can be identified using ultrasound carried out on pregnant women when there is a problem in the heartbeat of fetal [6].

7.2.1.1 Causes and Symptoms

When a woman takes alcohol or a proscribed drug that is not good, during pregnancy, one of the causes may be excessive blood sugar level, which might also affect childhood development [7]. Symptoms may include difficulty in feeding; breathing problems; lips, fingers, and toes being bluish; chest pain; and delay in growth of a child.

7.2.2 Coronary Heart Disease

Coronary artery disease is another name for coronary heart disease. The coronary arteries pump blood to the heart, supplying oxygen and nutrients. The vessels transport blood to the heart. If these veins contract, the inflow of oxygen and nutrients to the heart may be compromised. As a result, the body may experience chest discomfort. As a result, the heart may not function effectively [8]. When the level of cholesterol in the human body rises, the risk of coronary artery damage rises.

7.2.2.1 Causes and Symptoms

The main cause is smoking; also, eating too much unhealthy food will increase the cholesterol, which may cause heart failure. Hypertension, high body mass index, and a family history of coronary artery disease also cause this disease. When the blood lipid level is either too high or too low, then it also causes coronary artery disease [9]. Symptoms of CAD are a pain in the legs while walking or standing, pain and tightness in the chest, breathlessness, upset stomach, nausea and vomiting, and swelling in legs, ankles, feet, stomach, and veins of the neck.

7.2.3 Arrhythmia

In this, the abnormal pattern of heartbeat occurs, this may not threaten the life of a person if it is cured properly. There are lots of ways that heartbeat gets abnormal: after doing exercise, the heartbeat increases. This is called tachycardia. The other is when the heartbeat is very slow called bradycardia; this can be diagnosed by different procedures and tests like blood test, ECG, ultrasound, X-ray, etc. [10]. The most common test for arrhythmia is ECG [11]. ECG processes electrical signals. There are some steps to the process: preprocessing, feature extraction, feature selection, and then the classification of signals. There are many algorithms used for the classification of electrical signals such as K-nearest neighbor, SVM, and Neural Network [12].

7.2.3.1 Causes and Symptoms

When the heart tissue and activity changes, this may cause arrhythmia, and this may happen because of a harm or may be present in family history [13]. There are not as many symptoms of this particular disease. One of the symptoms is either dizziness or facing problems in breathing [14].

7.2.4 Dilated Cardiomyopathy

As a result of cardiac muscle weakening, the heart chambers widen and are unable to pump blood adequately. The most prevalent cause is a lack of oxygen reaching the heart muscle as a result of coronary artery disease. Typically, this affects the left ventricle.

7.2.4.1 Causes and Symptoms

Some causes of dilated cardiomyopathy are diabetes, obesity, hypertension, alcohol consumption, cancer medication, the difficulty of late-stage pregnancy, exposure to toxins (such as lead, mercury, and cobalt), an infection caused by viruses or bacteria, etc. Dilated cardiomyopathy has symptoms such as feeling breathless when a person is active, not being able to do much exercise, swelling in leg ankle and feet, pain in the chest, and heart murmurs.

7.2.5 Myocardial Infarction

This is sometimes referred to as a heart attack, a cardiac infarction, or coronary thrombosis. An interruption in blood flow harms or kills a portion of the heart muscle. This is most usually caused by a blood clot developing in one of the coronary arteries, although it can also occur when an artery suddenly narrows or spasms. Hypertrophic cardiomyopathy is a genetic condition [15].

7.2.5.1 Causes and Symptoms

A person who does not have a proper diet may suffer from MI; fat, cholesterol, excessive salt intake also contribute, as do smoking, nicotine, alcohol, drugs, blood pressure, and not doing exercise every day [16]. The symptoms of MI include chest pain, which travels from left arm to neck, shortness of breath, sweating, nausea, vomiting, abnormal heart beating, anxiety, fatigue, weakness, stress, depression, and other factors [17].

7.2.6 Heart Failure

Congestive heart failure is another name for heart failure. One of the reasons for heart failure is when the body does not receive adequate blood flow from the heart. The heart might be impacted on either the left or right side, and rarely both sides. Coronary artery disease or excessive blood pressure can cause the heart to stiffen or weaken over time, making it unable to fill and pump adequately [18]. An increase in blood flow within the organs or to another organ can also induce heart failure.

7.2.6.1 Causes and Symptoms

One of the major origins of heart failure disease can be the flow of blood is restricted to tissues. Dyspnea is a condition in which a person has a problem in breathing. Angina, or chest pain, is the most prevalent symptom of coronary artery disease. Angina is characterised by a feeling of discomfort, heaviness, pressure, hurting, burning, fullness, squeezing, or pain in the chest [19]. It's easy to confuse it with indigestion or heartburn. Angina can affect the shoulders, arms, neck, throat, jaw, and back.

7.2.7 Hypertrophic Cardiomyopathy

Hypertrophic cardiomyopathy is a hereditary condition characterized by thickening of the left ventricular wall. As a result, pumping is difficult. The majority of unexpected deaths may occur among athletes. A child has a 50% chance to get this disorder from their parent who is already having hypertrophic cardiomyopathy [20].

7.2.7.1 Causes and Symptoms

Some symptoms of hypertrophic cardiomyopathy are syncope, palpitations, shortness of breath, fatigue, and chest pain after having a meal, doing exercise, or doing any physical activity [21].

7.2.8 Mitral Valve Regurgitation

It is also known as mitral regurgitation, mitral insufficiency, or mitral incompetence. This happens when the valve between the left atrium and the left ventricle of the heart is not closed properly because blood starts flowing back into the heart when it should leave. So, blood may not pass through the heart or body efficiently. One having this disease feels tired and also feels breathlessness.

7.2.8.1 Causes and Symptoms

Disease progression is very slow; so, for many years, one may not get any symptoms. Symptoms are abnormal heart sound (heart murmur) heard through a stethoscope and sensations of a rapid, fluttering heartbeat.

7.2.9 Mitral Valve Prolapse

The valve that connects the left atrium to the left ventricle does not close completely and bulges upwards or back into the atrium. Most people do not have a life-threatening ailment that requires treatment. Some patients may require treatment, especially if their disease is characterized by mitral regurgitation.

7.2.9.1 Causes and Symptoms

Mitral regurgitation is the most common cause of mitral valve prolapse. Fluttering or fast heartbeats, known as palpitations, is a symptom, as is shortness of breath, particularly during activity. Mitral valve prolapse symptoms include dizziness, passing out, fear and anxiety, numbness or tingling in the hands and feet, and so on.

7.2.10 Pulmonary Stenosis

Because the pulmonary valve is overly tight, it becomes difficult for the heart to pump blood from the right ventricle into the pulmonary artery. To overcome the impediment, the right ventricle must work harder. A baby suffering from severe stenosis may become blue. In most cases, older children will not exhibit any symptoms. If the pressure in the right ventricle is excessively high, treatment is required, and a balloon valvuloplasty or open-heart surgery may be performed to relieve a blockage [22] (Figure 7.1).

7.2.10.1 Causes and Symptoms

The causes of this disease are not known. One of the reasons may be that during eighth months of pregnancy, pulmonary valve does not form correctly [23]. Symptoms are dizziness, unconsciousness, chest pain, swelling in the legs or abdomen, cyanosis, or high red blood cell count [24].

7.3 MACHINE LEARNING ALGORITHMS USED FOR PREDICTING HEART DISEASE

Machine learning is an interdisciplinary field. Nowadays, machine learning algorithms are in demand for the detection, prevention, and prediction of heart disease. The factor which affects the heart disease are BMI, age, smoking per day, etc. [25]. There are many machine learning algorithms for classification [26] (Figure 7.2).

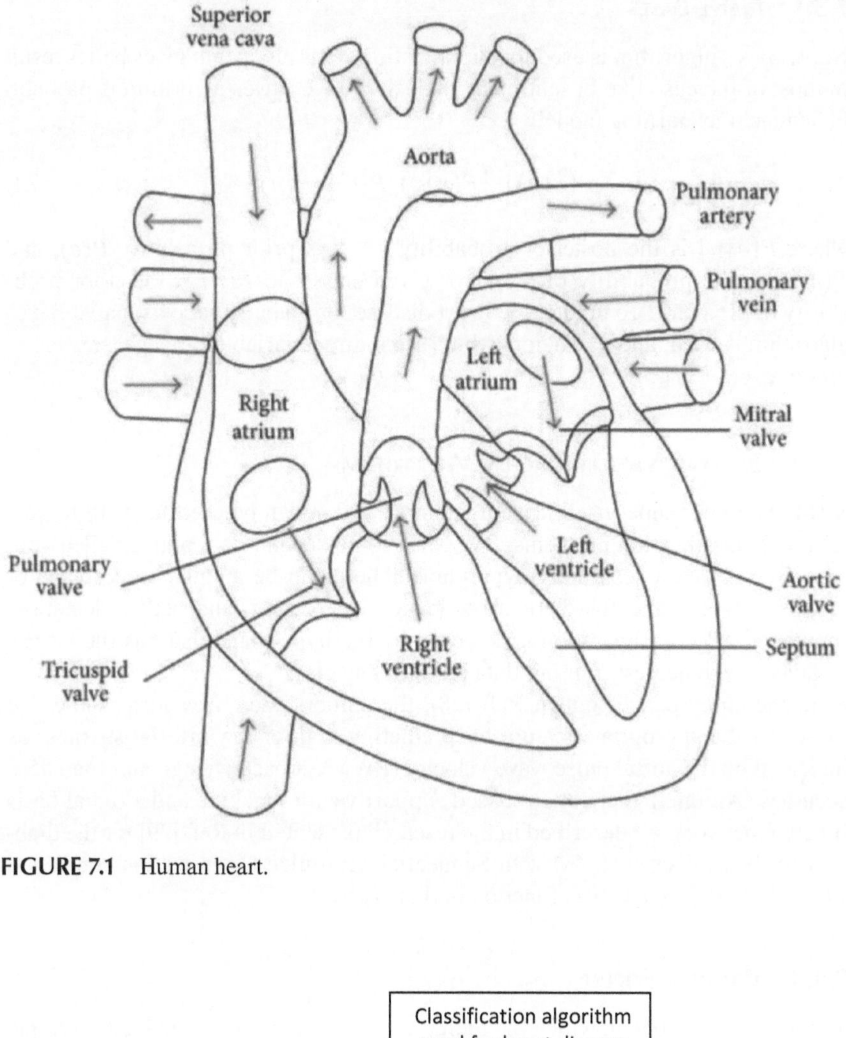

FIGURE 7.1 Human heart.

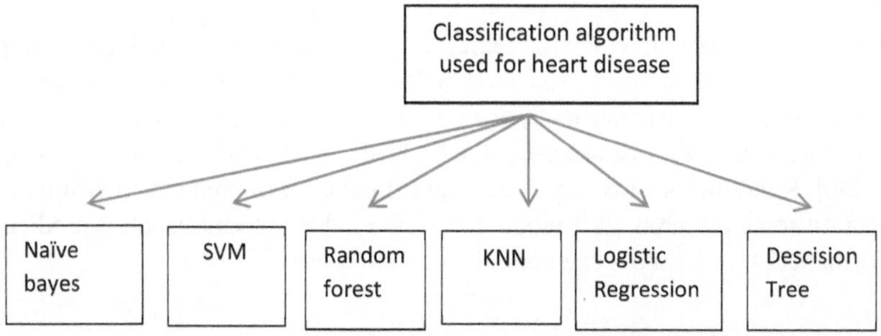

FIGURE 7.2 Classification algorithms used for heart disease prediction.

7.3.1 Naïve Bayes

Naïve Bayes algorithm is used for classification. This algorithm gives better result in most of the case like in sentiment analysis. Naïve Bayes algorithm is probabilistic machine learning model.

$$P(c \mid n) = \{P(n \mid c) \cdot P(c)\} / P(n) \tag{7.1}$$

where $P(c \mid n)$ is the posterior probability of class prior probability $P(c)$, and $P(n \mid c)$ is the probability of predictor given class and $P(n)$ is the prior probability of predictor. To predict the heart disease for diabetic patients, naïve Bayes algorithm is used, and the factors which are indirect variables are age, sex, blood pressure, etc. [27].

7.3.2 Support Vector Machine Algorithm

SVM is another supervised learning algorithm that can be used for both regression and classification problems. SVM has some special data points called support vectors and a separating hyperplane. There can be a single hyperplane or group of hyperplanes used for classification, regression, and outlier detection. Intuitively, a good separation is achieved by the hyperplane that has the largest distance to the nearest training-data point of any class.

In the study carries out in Ref. [28], the authors used Waveform and SVM to extract the appropriate feature for predicting high or low arterial stiffness as indicated by the aortal pulse wave velocity (PWV). It has given greater than 85% accuracy. An intelligent system-based support vector machine and a radial basis function network are described in the research presented in Ref. [29] for the diagnosis of heart disease. SVM with Sequential Minimize Optimization conducted similarly to the radial basis function in this case.

7.3.3 Random Forest

Random forest is one in which the multiple decision trees are combined. Multiple algorithms are used for feature selection. SVM, random forest, and JRip algorithms are used for the performance analysis of a selection of features in cardiac arrhythmia. In that, random forest has given the highest accuracy with 85.58% using gain ratio [30]. Random forest is one of the algorithms used to predict arrhythmia in a patient. In some research, the random forest is suitable for multiple class predictive performance. It can handle categorical and continuous variables.

7.3.4 Decision Tree

The decision tree is one of the classification algorithms also used for diagnosing heart diseases. There are different types of decision tree for information gain such as Gini index and gain ratio [31], as explained below.

7.3.4.1 Information Gain (ID3)

To maximize the information gain, the entropy of a selected attribute is calculated, and the attribute with the highest information gain is chosen for splitting. The information gain is calculated using the following formula [32]:

$$E(s) = \sum_{n=1}^{k} p_n \cdot \log_2 p_n \tag{7.2}$$

where s is the current state and P_n is the probability of an event n of state s.

7.3.4.2 Gain Ratio (C4.5)

Gain ratio is the extended form of information gain. As in information gain, there is a constraint that the large amount of data should be there to get a better split which is overcome by this method. The gain ratio is calculated by giving a formula, where IG is information gain and SI is the split information:

$$\text{Gain ratio} = \text{IG}/\text{SI} \tag{7.3}$$

7.3.4.3 Gini Index (CART)

Gini index is another type of decision tree. The formula used for calculation is given below [16]:

$$G = 1 - \sum_{n=1}^{k} (P_n)^2 \tag{7.4}$$

7.4 K-NEAREST NEIGHBOR

K-Nearest neighbor algorithm is one of the classification algorithms. It is also known as a lazy learner because it does not have the training dataset. It takes all dataset as training data. Then it chooses any integer value of k and finds the nearest neighbor by calculating the distance using Euclidean which is the most commonly used formula. With a smaller distance, the emphasis will be placed on that class. The number of classes is represented by k. The distance is determined using a particular mathematical model if the first set of values is (x_1, x_2, x_3) and the second set of values is (y_1, y_2, y_3), then the distance is calculated by using a given mathematical model:

$$= \sqrt{(x_1 - y_1)^2 + (x_2 - y_2)^2 + \ldots} \tag{7.5}$$

7.5 LOGISTIC REGRESSION

This algorithm is a very popular algorithm for classification. This algorithm is suitable for discrete values like the classification into yes or no. And this can also be used for multiple classes as well. In this, the value nearest to 0 means it belongs to that class and the calculated value nearest to 1 means it belongs to this particular class in this algorithm. In the healthcare sector, numerous machine learning techniques including as SVM, KNN, LR (Logistic Regression), Random Forest, Decision Tree, and Gaussian nave Bayes are used to forecast illness risk.

7.6 CASE STUDY: EXPLORING HEART DISEASE FROM THE GIVEN DATASET

Exploratory data analysis is an effective method to explore the data to make predictions and decisions from the huge volume of data generated by the healthcare industry. In this study, we use Tableau, a data visualization tool to explore data to predict whether a person is having heart disease or not. Here, the Cleveland Heart Disease dataset from the UCI Repository is used to perform the analysis.

7.6.1 Dataset

The Cleveland Heart Disease dataset contains 303 specific data with 15 columns. The explanation of the attributes is defined below [33].

1. **Age**: It contains the age of the person.
2. **Sex**: This column contains the gender of the individual with a numeric code 1 for male and 0 for female.
3. **Chest pain type**: Indicates the chest pain type experienced by the persons using the numeric code: 1=typical angina, 2=atypical angina, 3=non-anginal pain, and 4=asymptotic.
4. **Resting blood pressure**: It contains the resting blood pressure value of an individual in mmHg (unit).
5. **Serum cholesterol**: It contains the serum cholesterol in mg/dl (unit).
6. **Fasting blood sugar**: compares the fasting blood sugar value of an individual with 120 mg/dl. If fasting blood sugar > 120 mg/dl, then: 1 (true) else: 0 (false).
7. **Resting ECG**: 0=normal, 1=having ST-T wave abnormality, and 2=left ventricular hypertrophy.
8. **Max heart rate achieved**: displays the max heart rate achieved by an individual.
9. **Exercise-induced angina**: 1=yes and 0=no.
10. **ST depression induced by exercise relative to rest**: displays the value which is an integer or float.
11. **Peak exercise ST segment**: 1=upsloping 2=flat 3=downsloping.
12. **Number of major vessels (0–3) colored by fluoroscopy**: displays the value as integer or float.
13. **Thal**: displays the thalassemia: 3=normal, 6=fixed defect, and 7=reversible defect.

14. **Diagnosis of heart disease**: Displays whether the individual is suffering from heart disease or not: 0 = absence and 1, 2, 3, 4 = present.

7.6.2 Percentage of People Having Various Chest Pain Type

The populations are classified according to the chest pain type. The percentage of people suffering from microvascular angina is 47.52%, stable angina is 7.59%, unstable angina is 16.50%, and variant angina is 28.38%. The data are explored using Tableau software without writing a program (Figure 7.3).

Exercise can bring on angina; hence, an unsupervised program of exercise is not safe. To strengthen the heart, one should start the exercise slowly and gradually. Table 7.4 shows the number of males and females affected by exercise-induced angina. From the table, it is inferred that 77 females and 22 males are affected by exercise-induced angina (Figure 7.4).

7.6.3 Resting Blood Pressure Histogram

The following figure shows the histogram of the resting blood pressure details. From the Figure 7.5, it is inferred that the resting BP from 120 to 140 mm Hg is widely distributed.

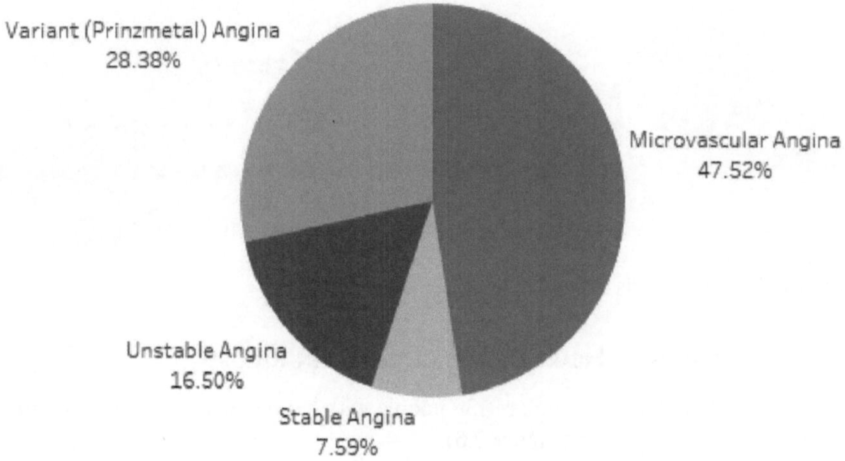

FIGURE 7.3 Percentage of people having various chest pain type exercise-induced angina vs. sex.

	Sex	
Exercise ind..	Female	Male
no	129	75
yes	77	22

FIGURE 7.4 Exercise-induced angina vs. sex.

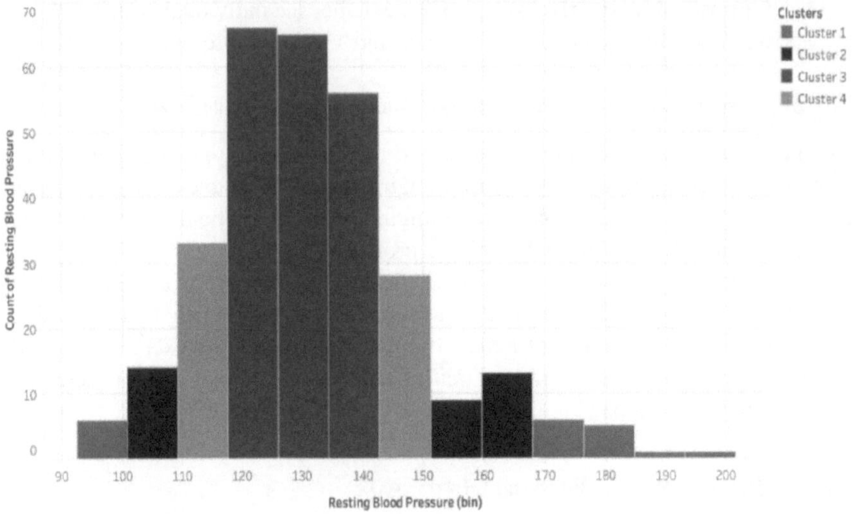

FIGURE 7.5 Resting blood pressure histogram.

FIGURE 7.6 Diagnosis of heart disease in the given record.

7.6.4 Diagnosis of Heart Disease in the Given Record

From the given database, there are 139 people with heart disease, and the remaining 164 are healthy people (Figure 7.6).

7.6.5 Chest Pain-Based Heart Disease Count

Figure 7.7 depicts that a total of 105 people are affected by microvascular angina-type chest pain, 7 persons are affected by stable angina-type chest pain, 9 persons are affected by unstable angina, and 18 persons are affected by variant angina.

Figure 7.8 shows the impact of blood pressure and cholesterol on heart disease. Persons with high blood pressure and high serum cholesterol are affected by heart diseases.

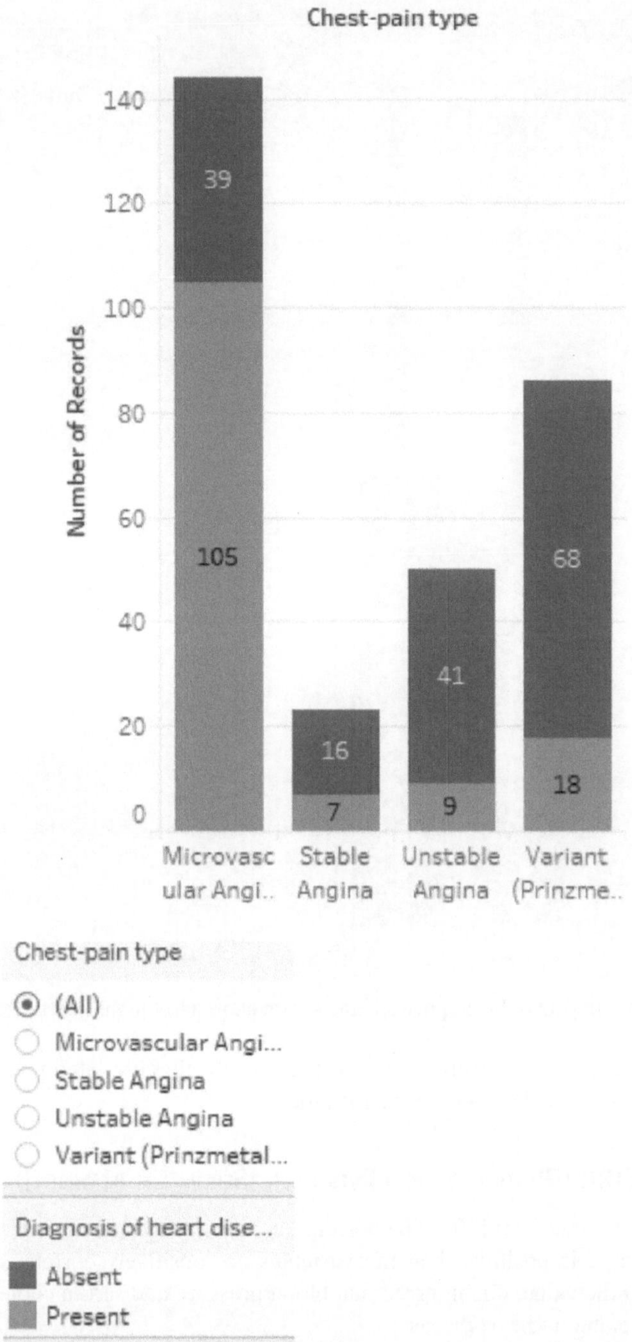

FIGURE 7.7 Chest pain-based heart disease count.

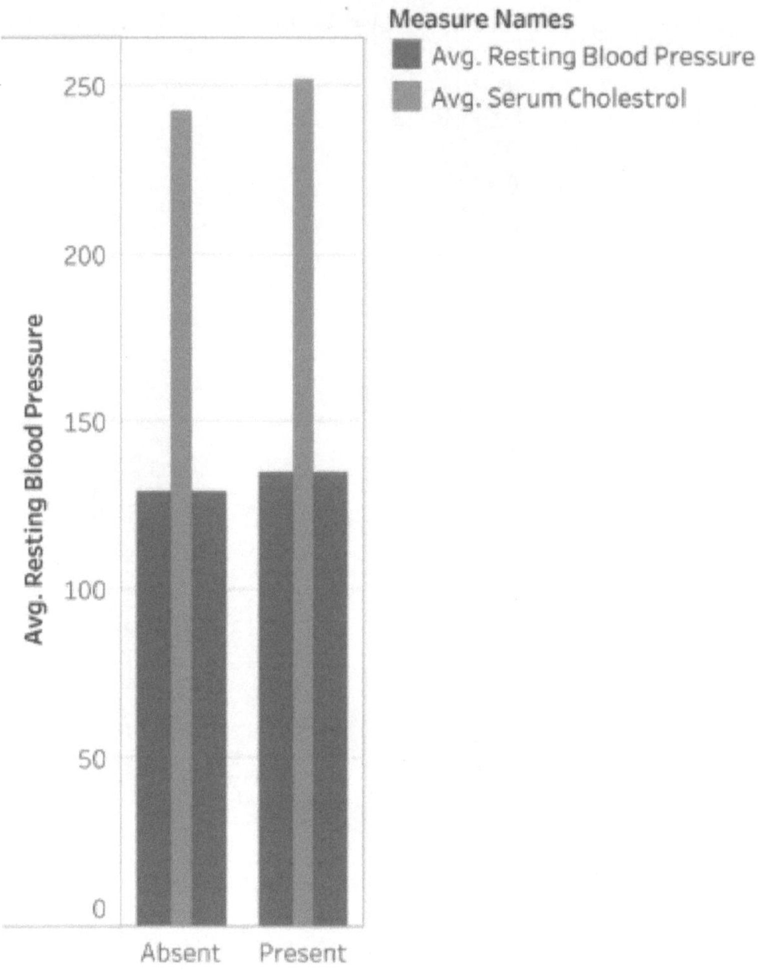

FIGURE 7.8 Impact of blood pressure and serum cholesterol in the heart disease.

Blood pressure and serum cholesterol are positively correlated to each other. The correlation is calculated by the formula.

7.6.6 CORR ([Resting Blood Pressure], [Serum Cholesterol])

The value obtained is 0.1301. The correlation value ranges from –1 to +1. If the obtained value is positive, then the variables are positively correlated to each other. From the value, it is inferred that blood pressure and serum cholesterol are highly correlated to heart disease.

Correlation value can be obtained manually by taking the square root of the R^2 value. From the line intercept, the R^2 value is 0.0169312. The sqrt (0.169312) is 0.1301 (Figure 7.9).

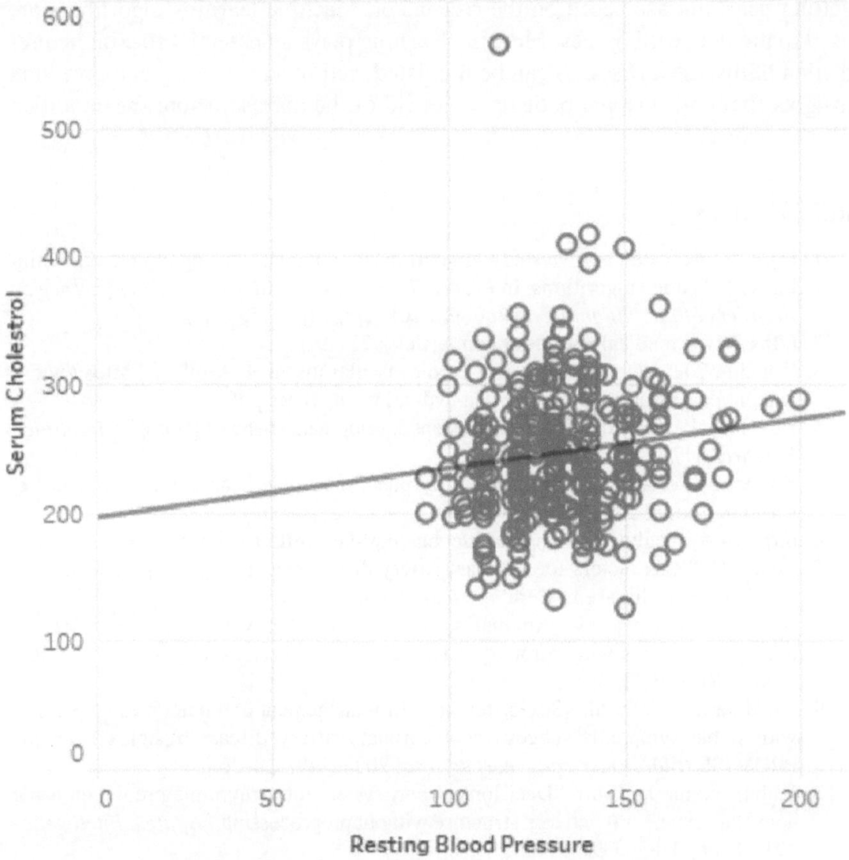

FIGURE 7.9 Correlation of blood pressure and serum cholesterol.

P-value:	0.0234965
Equation:	Serum Cholesterol = 0.382802 * Resting Blood Pressure + 196.282

Coefficients

Term	**Value**	**StdErr**	***t*-Value**	***p*-Value**
Resting blood pressure	0.382802	0.168127	2.27686	0.0234965
Intercept	196.282	22.3369	8.78736	<0.0001

7.7 CONCLUSIONS

Exploratory Data Analysis (EDA) is extracting knowledge from the data by summarizing the vital characteristics by presenting them visually. Heart disease is a serious disease in society. It is very hard to determine the odds of the chances of getting heart disease based on the risk factor. Machine learning algorithms are used to predict health issues. Machine learning plays an essential role in the prediction of diseases. Diseases can be predicted well in advance and can give vital insights to physicians and patients to get rid of the disease before the condition become worse.

REFERENCES

1. Enriko, I. K. Comparative study of heart disease diagnosis using top ten data mining classification algorithms. In *Proceedings of the 5th International Conference on Frontiers of Educational Technologies* 2019 Jun 1 (pp. 159–164).
2. https://www.medicalnewstoday.com/articles/237191#types.
3. Who.int, 2021. [Online]. Types of cardiovascular diseases. Available: https://www.who.int/cardiovascular_diseases/en/cvd_atlas_01_types.pdf.
4. Dassanayaka, S., and S. P. Jones. "Recent developments in heart failure." *Circulation Research* 117.7 (2015): e58–e63.
5. Kupper, N., et al. "Determinants of dyspnea in chronic heart failure." *Journal of Cardiac Failure* 22.3 (2016): 201–209.
6. https://www.healthline.com/health/congenital-heart-disease#prevention.
7. James, C. "Risk factors for coronary artery diseases: a study among patients with ischemic heart disease in Kerala." *Heart India* 1 (2013): 7–11.
8. Mozaffarian, D., et al. "Anginal symptoms consistently predict total mortality among outpatients with coronary artery disease." *American Heart Journal* 146.6 (2003): 1015–1022.
9. Pagidipati, N. J., et al. "Sex differences in management and outcomes of patients with stable symptoms suggestive of coronary artery disease: insights from the PROMISE trial." *American Heart Journal* 208 (2019): 28–36.
10. Fujita, H., and D. Cimr. "Decision support system for arrhythmia prediction using convolutional neural network structure without preprocessing." *Applied Intelligence* 49.9 (2019): 3383–3391.
11. Singh, N., and P. Singh. "Cardiac arrhythmia classification using machine learning techniques." *Engineering Vibration, Communication, and Information Processing, Processing ICoEVCI 2018*. Springer, Singapore, 2019, 469–480.
12. Bhirud, B., and V. K. Pachghare. "Arrhythmia detection using ECG signal: a survey." *Proceeding of International Conference on Computational Science and Applications*. Springer, Singapore, 2020.
13. https://www.nhlbi.nih.gov/health-topics/arrhythmia.
14. Ciuca, C., et al. "Dilated cardiomyopathy in a pediatric population: etiology and outcome predictors–a single-center experience." *Future Cardiology* 15.2 (2019): 95–107.
15. Rodrigues, L. L. R., et al. "Machine learning in coronary heart disease prediction: structural equation modeling approach." *Cogent Engineering* 7.1 (2020): 1723198.
16. Han, J. and M. Kamber. *Data Mining Concepts and Techniques*, Morgan Kaufmann Publishers, San Francisco, CA (2006).
17. Shouman, M., T. Turner, and R. Stocker. "Using decision tree for diagnosing heart disease patients." *Proceedings of the Ninth Australasian Data Mining Conference – Volume 121*, Ballarat, 2011.
18. Parthiban, G., A. Rajesh, and S. K. Srivatsa. "Diagnosis of heart disease for diabetic patients using naive Bayes method." *International Journal of Computer Applications* 24.3 (2011): 7–11.

19. Swain, D., et al. "An efficient heart disease prediction system using machine learning." *Machine Learning and Information Processing, Proceedings of ICMLIP 2019*. Springer, Singapore, 2020, 39–50.
20. Morrow, D. A. "The fourth universal definition of myocardial infarction and the emerging importance of myocardial injury." *Circulation* 141.3 (2020): 172–175.
21. Lu, L., et al. "Myocardial infarction: symptoms and treatments" 72.3 (2015): 865–867. doi: 10.1007/s12013-015-0553-4.
22. Khan, F. W., and M. S. Çiçek. "Pulmonary valve stenosis." In: Raja, S. (ed.), *Cardiac Surgery* (885–889). Springer, Cham (2020).
23. Mackay, J., and G. A. Mensah. *The Atlas of Heart Disease and Stroke*. World Health Organization, Geneva (2004).
24. Barker, D. J. P. "Fetal origins of coronary heart disease." *BMJ* 311.6998 (1995): 171–174.
25. DeStefano, F., et al. "Dental disease and risk of coronary heart disease and mortality." *BMJ* 306.6879 (1993): 688–691.
26. Ornish, D., et al. "Can lifestyle changes reverse coronary heart disease? The lifestyle heart trial." *The Lancet* 336.8708 (1990): 129–133.
27. Hu, F. B., and W. C. Willett. "Optimal diets for prevention of coronary heart disease." *Jama* 288.20 (2002): 2569–2578.
28. Elliott, P., and W. J. McKenna. "Hypertrophic cardiomyopathy." *The Lancet* 363.9424 (2004): 1881–1891.
29. Saini, I., Singh, D., and Khosla, A. (2013). "QRS detection using k-nearest neighbor algorithm (KNN) and evaluation on standard ECG databases." *Journal of Advanced Research*, 4.4, 331–344.
30. Shouman, M., T. Turner, and R. Stocker. "Applying k-nearest neighbour in diagnosing heart disease patients." *International Journal of Information and Education Technology* 2.3 (2012): 220–223.
31. Ghumbre, S., C. Patil, and A. Ghatol. "Heart disease diagnosis using support vector machine." *International Conference on Computer Science and Information Technology (ICCSIT 2014)*, Pattaya, 2011.
32. Alty, S. R., S. C. Millasseau, P. J. Chowienczyc, and A. Jakobsson, "Cardiovascular disease prediction using support vector machines." *2003 46th Midwest Symposium on Circuits and Systems*, Cairo, 2003, 376–379, Vol. 1.
33. Kaggle.com. *Cleveland clinic heart disease dataset*, 2021. [online] Available at: https://www.kaggle.com/aavigan/cleveland-clinic-heart-disease-dataset.

8 A Deep Learning Framework Using AlexNet for Early Detection of Pancreatic Cancer

Geraldine Bessie Amali, Gaurav Ramtri, Anukriti Kacker, and Siddharth Menon
Vellore Institute of Technology

CONTENTS

DOI: 10.1201/9781003050827-8

8.1 THE PANCREAS AND ITS FUNCTION

The pancreas is an organ located behind the lower side of the stomach across the back of the abdomen. It is 6 in. long, flat pear-shaped, and surrounded by the liver, gallbladder, stomach, spleen, and small intestine. It is both an exocrine as well as an endocrine gland, which plays two crucial roles: digestion of food and regulation of sugar in the blood. Almost 95% of pancreatic cells are exocrine in nature. Exocrine cells help in the digestion of food by producing digestive enzymes into the pancreatic duct. A few of the enzymes used for digestion include the breakdown of carbohydrates by amylase, the breakdown of proteins by trypsin and chymotrypsin, and the digestion of fats by lipase. Endocrine cells produce hormones to control or regulate specific functions of the body. The islet of Langerhans cells are endocrine cells that produce and secrete two main hormones of the pancreas, that is insulin and glucagon, into the blood. Insulin decreases whereas glucagon increases the sugar level in blood. Both of them work simultaneously to maintain a proper balance of sugar levels in the blood [1] (Figure 8.1).

8.2 CANCEROUS CELL AND ITS EFFECT ON THE HUMAN BODY

Human bodies are formed from trillions of cells that converge to form tissues and organs. The genetic material inside a cell regulates the activity of the cells. These genes are also responsible for the growth and division of cells. Cancer cells are formed when there is a mutation in the gene of the healthy cells, giving rise to the abnormal growth of cells without any control. These gene mutations could be inherited, or developed over time, or by engaging in activities like alcohol, UV radiation from the sun, and smoking could also damage the DNA. Cancer cell divides rapidly, forming a tumor that then undermines the immune system and provokes other changes which hinder the normal functioning of the body. The property of cancer cells to proliferate locally or regionally to nearby organs, lymph nodes, tissues, or far-flung organs is called metastasis [2].

8.3 PANCREATIC CANCER

Pancreatic cancer majorly falls under two categories: First is "Pancreatic Neuroendocrine Tumor," which affects the endocrine pancreas, and the second one is "Pancreatic Adenocarcinomas," which affects the exocrine pancreas, this develops from the cells that line the ducts which deliver digestive enzymes to the intestine and is more commonly found (Figure 8.2).

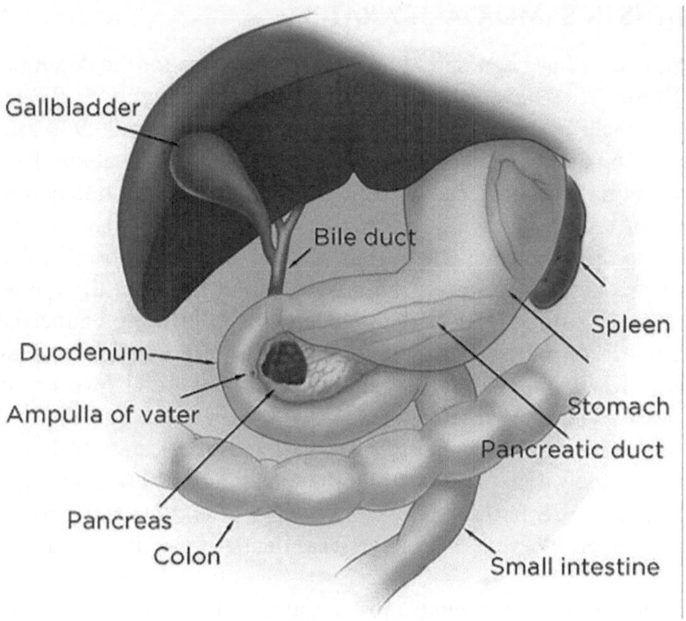

FIGURE 8.1 Organs surrounding pancreas in the body.

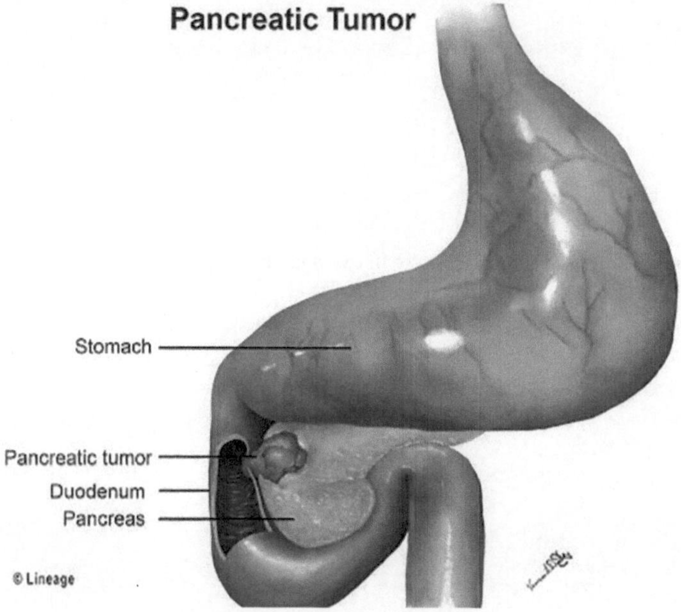

FIGURE 8.2 Tumor inside the pancreas.

8.4 STATISTICS (MORTALITY RATE)

Pancreatic cancer has been one of the primary reasons for the death of people in the US. Statistics have estimated that there were roughly 56,770 cases of pancreatic cancer reported in the US in 2019, out of which 29,940 are males, and the rest 26,830 are females. The data predicts that only about 9% of pancreatic cancer patients can recover from this deadly disease. It is believed that up to 93% of the people who are prone to this disease will have the tumor developing in the exocrine cells in the pancreas that is responsible for generating enzymes that are required for the digestion of food in the human body whereas the remaining (7%) of the people develop it in the endocrine gland which is also referred to as the neuroendocrine tumors (NETs). Pancreatic cancer is a low-incident but extremely lethal disease. However, in 2020 approximately 57,600 adults, of which 30,400 were men, and the remaining 27,200 women in the US, were confirmed with pancreatic cancer. It accounted for 47,050 deaths, which included 24,640 men and 22,140 women. Among all the patients in the US in the year 2020, only 3% of all the cancer patients had pancreatic cancer, whereas 7% of the total deaths were due to cancer in the pancreas.

Pancreatic cancer is the tenth most frequent form of cancer in males and ninth in women. It is the fourth major cause of death in both men and women [3] (Figures 8.3 and 8.4).

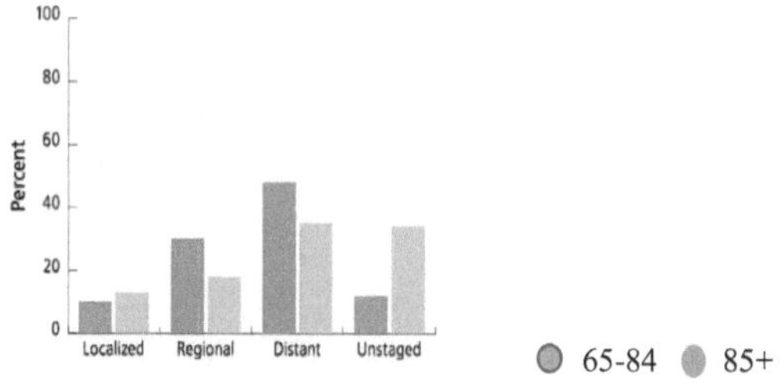

FIGURE 8.3 The four stages of cancer in two age groups [4].

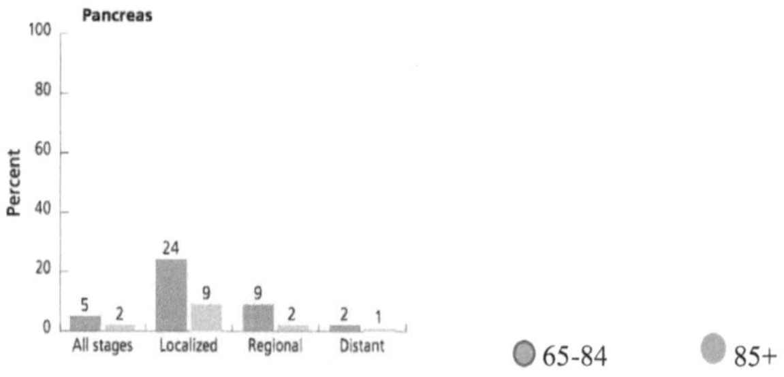

FIGURE 8.4 Five-year survival rate [5].

8.5 SYMPTOMS

Symptoms of pancreatic cancer vary based upon the extent of the tumor and the cell types induced by the tumor. The most common symptom of early warning signs of pancreatic cancer is jaundice. Jaundice is a medical condition in which the skin and the whites of the eyes turn yellowish hue due to the accumulation of excess bilirubin, which occurs by the blockage of the bile duct by the tumor. Some other symptoms include dark-colored urine and pruritus (itchy skin). Pancreatic cancer gives rise to gastrointestinal problems, which include abnormal stools, nausea or vomiting, bloating in the abdomen, loss of appetite, and indigestion. Other common symptoms include pain in the upper or middle abdomen, which may start radiating to the back. Apart from this, unintentional or significant weight loss and extreme fatigue are also common symptoms. Diabetes caused by the destruction of insulin-making cells by the tumor gives rise to an increase in blood sugar. Most pancreatic patients are diagnosed with advanced disease, which is cancer that spreads from the pancreas to different sections of the body. The advanced pancreatic cancer state can be identified through symptoms like swelling of the upper abdomen or pain in the back, especially after eating or lying down, bedsores, and depression [6,7].

8.6 CAUSES AND RISK FACTORS ASSOCIATED WITH PANCREATIC CANCER

Though the exact origin of pancreatic cause is still not known, there exist some risk factors associated with it, which may increase the chances, namely, conditions like obesity, tobacco use, smoking, a history of diabetes, or a sedentary lifecycle. One of the most significant and unavoidable risk factors is aging; people who are above 60 years old are at a higher risk.

It can be observed from the graph below that 10% of pancreatic cancer is due to smoking [8,9] (Figure 8.5).

8.7 DETECTION AND TREATMENT

8.7.1 Previous Work

There have been various attempts to detect pancreatic cancer early on; these include methods like multispectral endoscopy which end up being very complex and a pain to be implemented. As proposed in [10], LSS (light scattering spectroscopy) is used along with light gating to detect whether the cancer is malignant in the pancreatic ducts or not. The vital information to help diagnose if a person has cancer can be extracted through spatial or polarization gating along with a fiber optics probe.

Harmonic motion imaging is another revolutionary step in the detection of tumors. They adopted the method of elastic imaging and the system consisted of an ultrasound transducer.

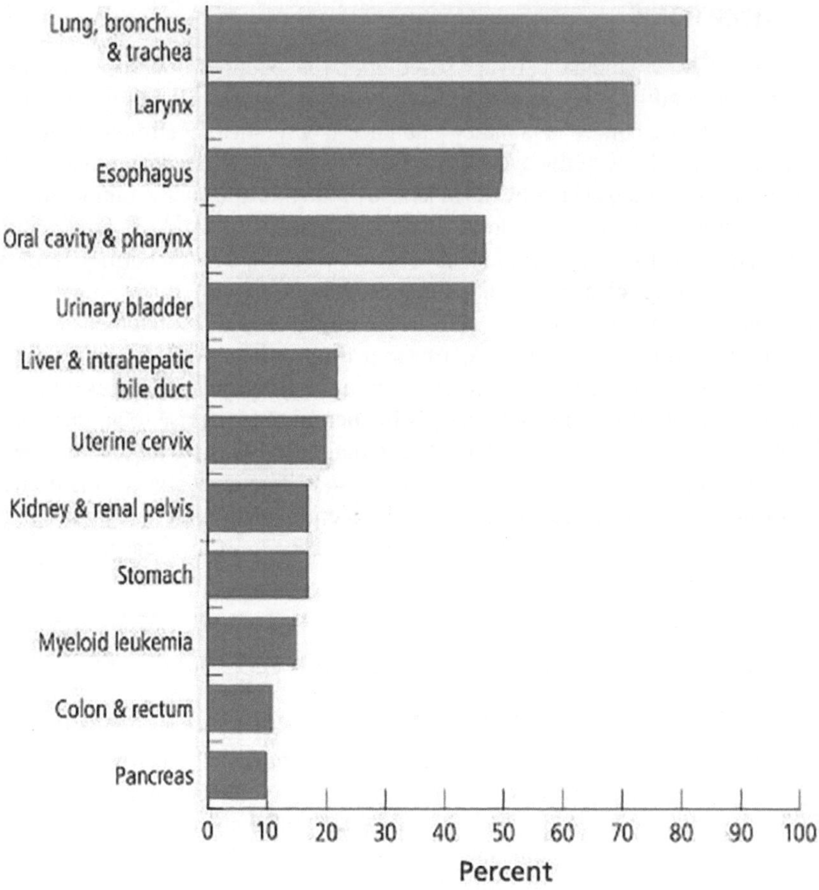

FIGURE 8.5 Types of cancer smoking can lead to [9].

Another usage for pancreatic tumor ablation is called monitoring wherein the concentrated ultrasound transducer is supposed to be operated at greater acoustic power and longer pulse length than the traditional length used for detection of the tumor. This was done to induce HIFU thermal ablation and oscillatory tissue displacement side by side which allowed HMI monitoring without disrupting the tumor ablation [11].

High-frequency focused ultrasound is another commonly found unique method to detect pancreatic cancer as explained by [12]. It does not only help with pancreatic tumors but with various other solid tumors be it breast, prostate, hepatic, etc. HIFU uses powerful ultrasound waves which induce thermal effects in the respective tissue. The only major side effect that can be caused by High-Frequency Focused Ultrasound is that it can induce cell apoptosis through hyperthermia which is a sub-lethal thermal injury. In this case, the nucleus starts self-destruction, leading to the rapid degradation of DNA by endonucleases. The tissue that is next to the tissue, which is being examined, can be damaged [12,13].

Since pancreatic cancer has led to a significant number of deaths in recent years and has the poorest prognosis among all the other tumors, it is essential to find the root cause of pancreatic cancer because a major reason is it being hereditary. So data regarding the family background will play a very major role in understanding if the cancer is hereditary and it will help in understanding the genes and starting the treatment way ahead. As proposed in [14], the registry of each family should be maintained in the dataset and this will allow a follow-up of the prospective population who are at a high risk of pancreatic cancer. This registry will give recognition to studies such as genome sequencing etc.

8.7.2 Significance of Early Detection

If cancer is diagnosed in the early stages, then surgical eradication of the tumor is feasible. A 5-year survival rate term is used, which tells us the percentage of people who will be alive for at least 5 years after the detection of cancer. These survival rates are established on many factors in conjugation with the distinct stage of disease when it is detected. 9% is the general 5-year survival rate of pancreatic cancer. If the cancer is diagnosed in an initial stage, and it is possible for surgical eradication of tumors, then the 5-year rate would increase to 37%. If cancer has advanced to neighboring organs or tissues, then the 5-year survival rate will be reduced to 12%. The 5-year survival rate is 3% for those patients who are diagnosed when cancer has already advanced to different parts of the body [7,15].

8.7.3 Challenges of Early Detection

Cancer in the pancreas is usually challenging for diagnosing because there are no particular, cost-efficient tests that can conveniently and accurately find early-stage cancer with patients having negligible symptoms. The pancreas is located deeply in the abdomen; therefore, the presence of a tumor cannot be felt during a physical examination. Tests that analyze pancreatic cancer sometimes fail to detect pre-cancers, small lesions, and early-stage cancer. There are no proven biomarkers that can detect the tumor early. This shows that pancreatic cancer is mostly not found until the last stages when cancer cannot be removed by surgery or has traveled to distant parts of the body [7].

8.7.4 Existing Approaches in Detecting Cancer

The initial tests for pancreatic cancer include knowing the medical and family history of the patient, risk factors associated with it, and some common symptoms of pancreatic cancer like jaundice or swelling up of liver or gallbladder. If these symptoms are positive, then various imaging tests will be done to examine for symptoms of cancer recurrence, areas that might be suspicious for cancer, how far it has spread, and to check if treatment is successful or not.

8.7.5 Imaging Tests

This test makes use of magnetic fields, X-rays, radioactive substances, or sound waves to generate detailed illustrations of the organs inside the body. Some of the most commonly used imaging tests are CT scans, which make detailed cross-sectional images of the body using multiplex-rays images. If a possibility of pancreatic cancer is suggested by a doctor, then two specialized types of CT scans also known as a pancreatic protocol CT scan or multiphase CT scan can be used. CT scans can also be adapted for guiding the biopsy needle into the speculated tumor. Another is Magnetic Resource imaging (MRI), which makes use of radio waves and robust magnets to make accurate images of the body. There are two particular sorts of MRI scans for people with a high risk of cancer in the pancreas: MR Cholangiopancreatography (MRCP), which examines pancreatic and bile ducts, and MR angiography (MRA) which examines the blood vessels. An ultrasound test uses harmless sound waves which are reflected by the organs to create images of the organs. The two most frequently used ultrasound tests for diagnosing cancer in the pancreas are abdominal ultrasound and endoscopic ultrasound, which are more accurate and very significant for detecting this particular cancer. This is performed with a tiny probe on the tip of an endoscope, which is used to view the digestive tract and to obtain specimens of a tumor for biopsy. Cholangiopancreatography is a test that views the pancreatic and bile ducts to find out if they are narrowed, dilated, or blocked. This test is suitable for telling if the tumor is blocking a duct or not. There are three ways in which this test can be performed: endoscopic retrograde cholangiopancreatography (ERCP), magnetic resonance cholangiopancreatography (MRCP) (a non-invasive test), and percutaneous transhepatic cholangiography (PTC). Another imaging test is a positron emission tomography (PET) scan in which slightly radioactive sugar is infused into the veins and is consumed by the cancer cells. This inspection helps in determining the degree of spread of pancreatic cancer. The last one is the angiography test, which takes X-ray images of the blood vessels. An angiogram is useful in providing information regarding the blockage of blood vessels caused by the tumor [7,15].

8.7.6 Other Tests Available

Different types of blood tests can be performed for the diagnosis of pancreatic cancer. As jaundice is a primary indication of cancer, a blood test to assess liver function is done to find out the exact cause of the disease. Another test could be based on the presence of specific tumor markers which can only be seen in the blood of a patient having cancer. Tumor markers that may help identify pancreatic cancer are CA 19 and carcinoembryonic antigen (CEA). Tumor markers test is not self-sufficient to tell if a person has pancreatic cancer or not but can tell us how efficient the treatment is and the signs of cancer recurrence. An individual's imaging test, medical history, and physical exam results may actively propose cancer in the pancreas, but the only method to determine is to extract a miniature fragment of the tumor and observe it under the microscope. This method is known as a biopsy. Biopsies can be performed in various methods like percutaneous (through the skin) biopsy, endoscopic biopsy, and surgical biopsy. Genetic counseling and testing can be another method. People with genetic mutations (such as BRCA mutations) in all the cells of their body are at a heightened possibility for cancer in the pancreas, and so testing for these mutations will be helpful to know the treatment method [7,15].

8.8 TREATMENT

The treatment method is based on various factors, including the type and stage of cancer, the patient's predilection, overall health, and probable side effects. Different treatment methods available are surgery, radiation therapy, and systemic therapy. Surgery includes eradication of all parts of the pancreas in which the tumor is present. Only 20% of people who have pancreatic cancer can have surgery as they were diagnosed at early stages. Several types of surgery are there, which depend on the section of the pancreas where the tumor is present. Pancreaticoduodenectomy, also termed as the Whipple procedure, is executed when the tumor is found in the upper-side of the pancreas, distal pancreatectomy to eradicate the left side of the pancreas, and total pancreatectomy for extraction of the entire pancreas. Radiation therapy utilizes high-energy X-rays for the destruction of cancer cells. In some instances, proton therapy is used, which shows fewer side effects when compared to standard radiation techniques. Systemic therapy incorporates different types of medication for the destruction of cancer cells; these include chemotherapy, immunotherapy, and targeted therapy. Chemotherapeutic drugs mostly inhibit the growth and division property of cancer cells. Chemotherapy can be used at any stage: before surgery, after surgery, and for advanced stages. Chemoradiation is chemotherapy combined with radiation therapy used for treating cancer that has not spread to other organs. Targeted therapy makes use of different targeted drugs, for example, Erlotinib, an EGFR inhibitor, Olaparib, a PARP inhibitor, Larotrectinib, and entrectinib, an NTRK inhibitor to target cancer's specific genes and proteins which contributes to the growth and survival of cancer. Immunotherapy improves the body's immunity to fight cancer. Pembrolizumab drug, a PD-1 immune checkpoint inhibitor, can be an option for treating pancreatic cancer [15].

8.9 PREREQUISITE KNOWLEDGE REQUIRED

8.9.1 Pancreas Image Classification

Classifying whether a person has pancreatic cancer or not is a supervised learning algorithm solution as we need to know some data of people who already have it. We will run this algorithm on 200 images in the dataset where 100 people have pancreatic cancer while 100 people do not have it. After training the model, we would be able to predict whether the person can be a cancer patient or not (Figure 8.6).

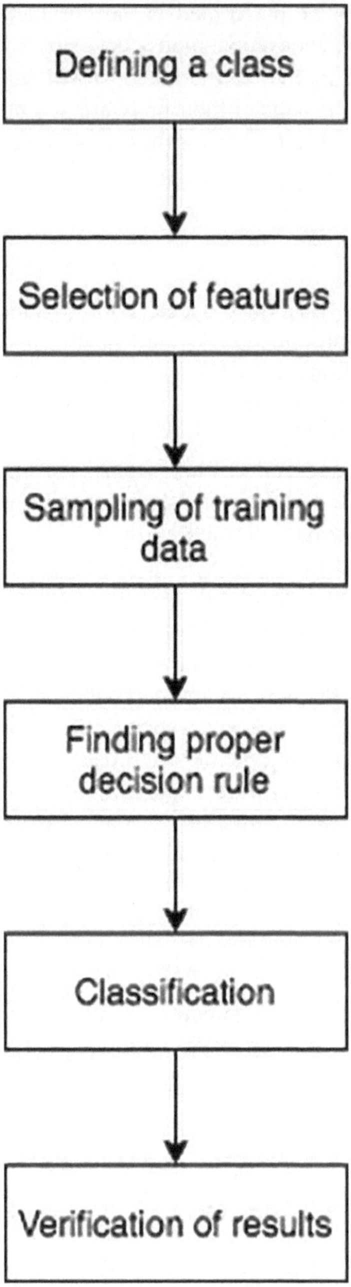

FIGURE 8.6 Steps followed in the classification of an image.

The current practices that are used for the classification of images follow the steps written above. These steps are explained below:

Step 1: Defining a class

The number of classes in which the dataset needs to be classified is defined based on the characteristics of the dataset and the requirement of the solution, the classes should be defined by taking the utmost care.

Step 2: Selection of features

The characteristics that will differentiate between the classes need to be described using multiple spectral and temporal characteristics such as color, textures, and size, i.e., only the useful characteristics are included because including unnecessary features can lead to problems like

- Overfitting
- Occam's razor: lose explainability
- Garbage in -> Garbage out

Step 3: Sampling of training data

It is one of the most important steps because sampling performs a vital role in formulating the various decision rules leading to the elimination of any biases.

Step 4: Finding proper decision rule

There exist many classification techniques, so after looking at the classes, the most relevant decision rule is picked for classification.

Step 5: Classification

Concerning the rule that is selected, all pixels are classified in a single class.

Step 6: Verification of results

Verification is important to cross-verify the results and thereby find the accuracy of the model. Good and reliable accuracy is essential especially to any models related to medical fields.

The major disadvantage that comes in view with this process is the feature matrix. There is a very high probability that there are some features that are not visible to the naked eye and hence appropriate features are not extracted, and this can result in a flawed classifier and the predictions made could be wrong too. To find the correct features, there were several trials made which turned out to be tiresome; additionally, the construction of algorithms when given a dataset is another monotonous and tedious job. All these issues were collectively solved by deep learning. Deep learning which is an extended branch of machine learning is inspired by the functioning of neurons in the brain, thereby forming a structure of its own using artificial neurons commonly known as ANNs (Artificial Neural Networks).

A few of the deep learning architectures are as follows:

1. Convolution Neural Network (CNN)
2. Recurrent Neural Network (RNN)

3. Auto-Encoders
4. Deep Belief Networks
5. Generative Adversarial Networks (GAN)

This chapter focuses on solving the issue at hand using the CNN.

8.10 CONVOLUTION NEURAL NETWORK

CNNs, which can also be termed as ConvNet, are regularized versions of a multilayer perceptron, which means that there are various layers, and each neuron present in one layer is connected to all the neurons in the neighboring layer. Since all the neurons in the preceding layer are connected to all the neurons in the next layer, this model is prone to overfitting. Convolutional Neural networks function in a similar way to the biological processes where the connection between various neurons resembles the formation of the animal visual cortex. The unique selling point of this algorithm is that the network learns from the filters as compared to the traditional algorithms that require a human to define the filters. This reduction of human efforts makes this one of the best reasons for the surge in usage of CNN nowadays [16] (Table 8.1).

CNN can further be classified into six major NETs:

1. LeNet(8)
2. AlexNet(7)
3. ZFNet
4. GoogleLeNet(19)
5. VGG Net(16)
6. ResNet(152)

These Nets can be differentiated from each other as shown in Figure 8.7.

As one can see from the graph above that ResNet has the lowest top-5 error rate among all and was the winner of the ImageNet challenge in 2015. The principal

TABLE 8.1
Different Types of CNN Nets and Their Difference [17]

Year	CNN	Developed by	Place	Top-5 Error Rate	No. of Parameters
1998	LeNet(8)	Yann LeCun et al.			60 thousand
2012	AlexNet(7)	Alex Krizhevsky, Geoffrey Hinton, llya Sutskever	1st	15.3%	60 million
2013	ZFNet()	Matthew Zeiler and Rob Fergus	1st	14.8%	
2014	GoogLeNet(19)	Google	1st	6.67%	4 million
2014	VGG Net(16)	Simonyan, Zisserman	2nd	7.3%	138 million
2015	ResNet(152)	Kaiming He	1st	3,.6%	

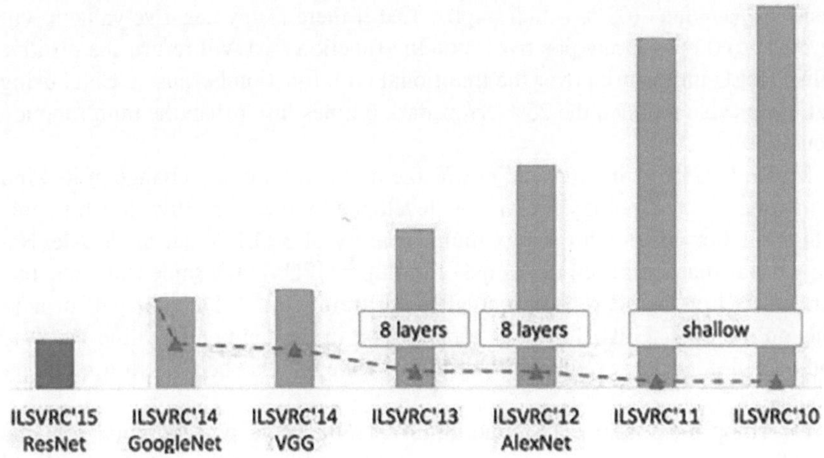

FIGURE 8.7 Error rates in different Nets [17].

TABLE 8.2
Number of Convolution Layers in Different Nets

ImageNet	No. of Convolution Layers
AlexNet	8
VGG	19
Inception	22
GoogleNet	22
ResNet	152

breakthrough with ResNet was that it allowed training an extremely deep neural network of 152 layers successfully. With the algorithms before ResNet, it was difficult to train models with a lot of layers because it resulted in gradients being lost (Table 8.2).

Even though ResNet has the lowest error rate, it is not recommended for small datasets as is the case here. So, the authors chose AlexNet because of its smaller architecture which is suitable for smaller datasets to avoid underfitting. AlexNet was the first deep CNN to have successfully outperformed the traditional object recognition techniques used at that time. This was a firebrand designed by Alex Krizhevsky, AlexNet became well known when it won the ImageNet challenge in 2012 [18].

The most important feature of AlexNet is not the number of layers it consists of but the following new features that changed people's approach towards Convolution Neural Networks:

ReLU nonlinearity: ReLU stands for Rectified Linear Unit and is considered to be the most applied activation function in deep learning. It can be understood easily as $f(x) = \max(0, x)$, which implies that if there is any negative value it will be made to 0 but for any positive value the function $f(x)$ will return the positive value. ReLU unit is used over the traditional tanh function because a CNN using ReLU was able to reach the 25% error mark 6 times faster than the tanh function would [19].

Multiple GPUs: Since AlexNet was the first revolutionary change of its kind in its time, the technology was not as developed by then, and this forced people to use smaller GPUs with a maximum memory of 3 GB. What made AlexNet unique was that it allowed the usage of multiple GPUs at the same time, i.e., one can put half of the neurons on one GPU for training and the other half of neurons on the other GPU. This was a revolution in the field of Machine learning and helped in not only reducing the training time but also help in training larger models.

Overlapping pooling: The traditional CNN used to pool outputs from the neighboring groups of neurons and prevented overlap, but later on they realized that overlapping led to the reduction of error by 0.5%, which is a very significant amount especially in larger datasets.

8.11 PROPOSED ARCHITECTURE

The architecture consists of eight layers among which five are convolutional and the other three are fully connected layers as shown in Figure 8. After the data is processed in these layers, the output from the last layer is fed to the Softmax layer where distribution is produced over 1000 different class labels [17] (Figure 8.8).

After looking at the architecture, it is necessary to understand what convolution layers are and what is max-pooling which we can see is applied as shown in the diagram above.

The simple process of multiplication of an input with a filter that results in the formation of activation is called convolution. Repeated multiplication of the same filter results in the formation of a map of activations termed as feature map, which in turn sums up the features that are detected from the input. The USP of CNN is its ability to learn a colossal number of filters in parallel, specific to a dataset while keeping in mind the constraints [21] (Figure 8.9).

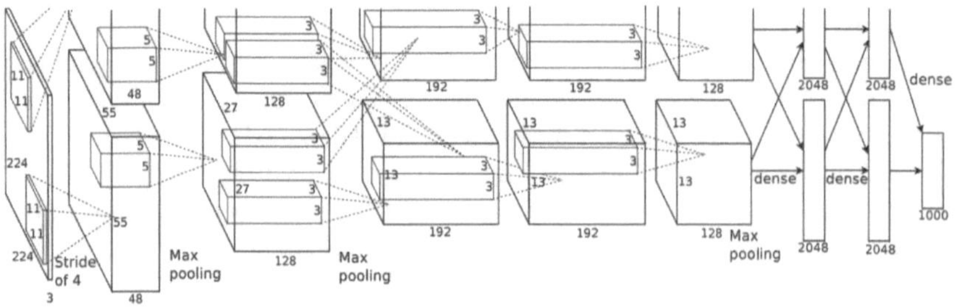

FIGURE 8.8 Architecture used (AlexNet) [20].

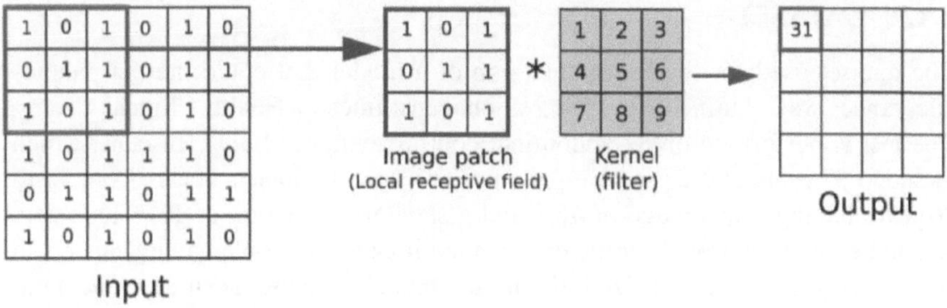

FIGURE 8.9 Methodology behind convolutions.

- Max Pooling
- Average Pooling

Input Matrix			
-4	20	55	100
-60	2	-10	39
19	-43	6	76
22	-47	57	119

Output Matrix			
	20	100	
	22	119	

FIGURE 8.10 Understanding max pooling.

8.11.1 Max Pooling

After passing the input through the convolution layer, a feature map is formed, which is very sensitive to the position of the features that are there in the input. So the function of the pooling layer is to reduce that sensitive information by summarizing the information that is found in the feature map. The main feature of this layer can be termed Dimensionality Reduction. There are two types of pooling [22,23] (Figure 8.10).

- Max Pooling
- Average Pooling

8.12 DATASET

The dataset used for the implementation of a model that could predict pancreatic cancer was obtained from the National Institutes of Health Clinical Center. The test is performed on 82 abdominal contrast-enhanced 3D CT scans, which included 53 male and 27 female patients. The age of subjects varies from 18 to 76 years having a mean age of 46.8 and a standard deviation of 16.7. This data was not enough to provide sufficient scanned images of people having cancer, so additional sources were utilized to train the model for better accuracy. The dataset is divided into two parts, i.e., people having cancer in the pancreas and people who do not have any cancer in the pancreas. Figure 8.11 on the left shows a tumor pointed with a bold arrow while the other image is of a person who does not have pancreatic cancer. The dataset consists of a total of 200 images.

8.13 PREPROCESSING OF DATA

Since the data we collected is from various sources, there is a need to preprocess the data to get some uniformity, but before that, we need to understand what is meant by the term preprocessing. When someone first hears the word data, the first thing that comes to mind is an Excel sheet but that is an archaic mindset which does not hold today. With the development of technology in the 21st century, data is found in various forms such as images, texts, audios, videos although, unfortunately, machines still understand only binary values, 1 & 0 so we convert all our data types into binary, so the computer can understand and work on it. This is termed **preprocessing**.

1. **Converting image to grayscale**: Since CT scans will not carry any information in RGB (Red, Green, Blue) or HSV (Hue, Saturation, Value) color spaces, the most logical first step will be to convert the image to grayscale as seen in Figure 8.12 where 8.12a shows the original image and 8.12b after converting to grayscale. This not only reduces the size of our dataset but also normalizes any anomaly in it.

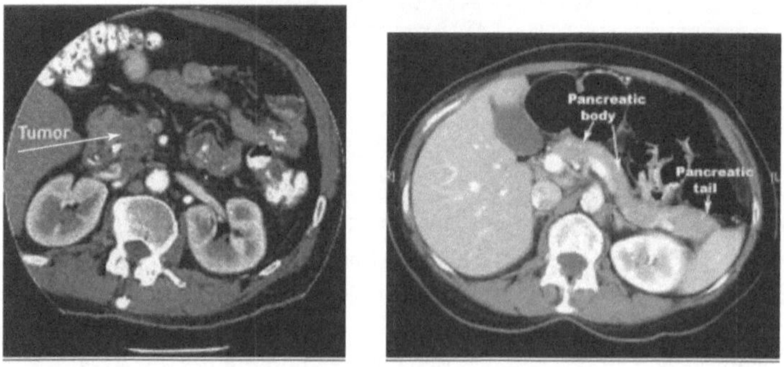

(a) (b)

FIGURE 8.11 (a) Pancreas with a tumor and (b) without tumor.

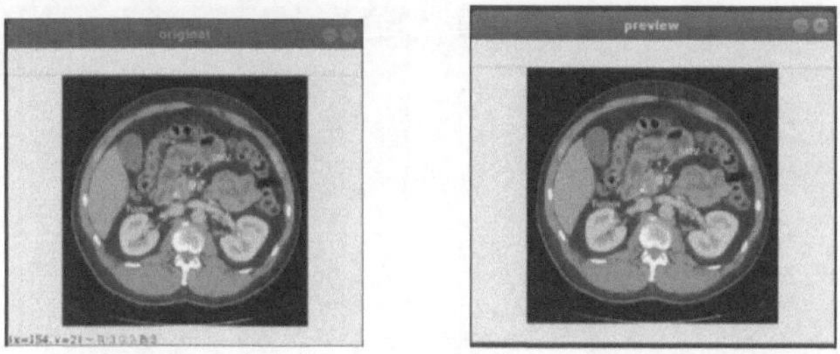

(a) (b)

FIGURE 8.12 (a and b) Grayscale image of the pancreas with a tumor and without tumor.

2. **Resizing**: A neural network can only take a specific size as input. This determines the next most intuitive step of resizing. The size selected here is to resize it to (256, 256).
3. **Sharpening**: Pancreatic tumor/cancer is a cyst formation in the pancreas. Any preprocessing always focuses on enhancing or making the anomaly stand out which makes it much easier for the model to learn the specific feature.

   A 2D matrix becomes key to perform such an operation. A sharpening kernel is a 3×3 matrix.

$$K = \begin{matrix} -1 & -1 & -1 \\ -1 & 9 & -1 \\ -1 & -1 & -1 \end{matrix}$$

   A kernel like this is slid across the window in a convolution a manner (Figure 8.13).
4. **Thresholding**: The above three steps allow us to enhance the image. Thresholding allows us to enhance more on the tumor. A safe threshold value to be selected, e.g., 127.

   But this number can be changed depending on the lighting conditions of the CT scan image (Figure 8.14).
5. **Pickle the data (only for python-based approach)**: Pickle files are read much faster by python rather than most of the common formats like NumPy arrays, JSON, etc. Hence, the data is converted to a NumPy array, and then pickling it will help in deserializing it.

8.14 DATA AUGMENTATION

8.14.1 Need of Augmentation

(a) **Generalization**: The model can generalize much better once the data is augmented.

(b) **Increases training data**: Since there are not many images available for pancreas-related CT scans and the tumor detection is not orientation

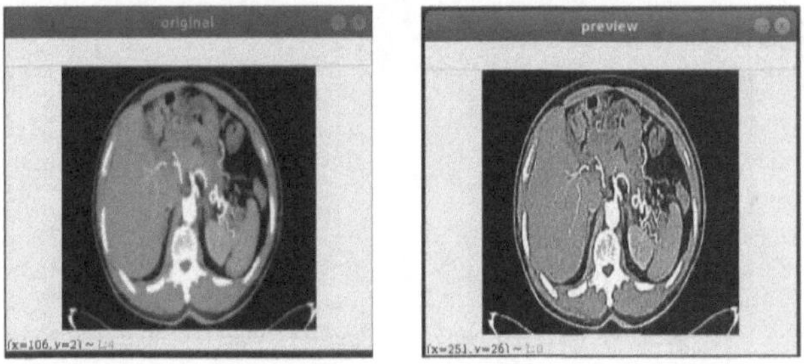

(a) (b)

FIGURE 8.13 (a) Original image and (b) sharpened image.

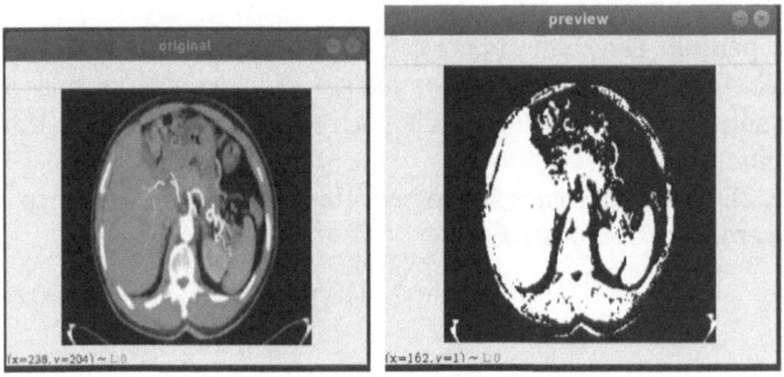

(a) (b)

FIGURE 8.14 (a) Original image and (b) the image after thresholding.

dependent, we can augment a certain amount of data for leveraging the performance of the model.

1. Rotation: As the name suggests, rotation involves rotating the image in a clockwise or anticlockwise direction, which helps the model generalize tumor detection.

   Usually, the user decides a rotation range about which the images have to be rotated (Figure 8.15).
2. Width shifting: Width shift is moving the image across the *X*-axis in such a way that some part of the image gets clipped either from the left or from the right side.

   The width shift range is usually a floating-point number between 0 and 1 (Figure 8.16).
3. Height shifting: Similar to width shift, height shifting refers to moving the image across the *Y*-axis, allowing some part of it to get clipped from above or below.

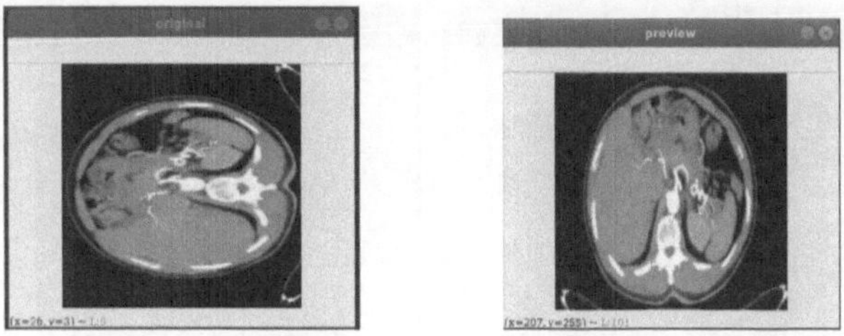

(a) (b)

FIGURE 8.15 (a) Original image and (b) the image after rotation.

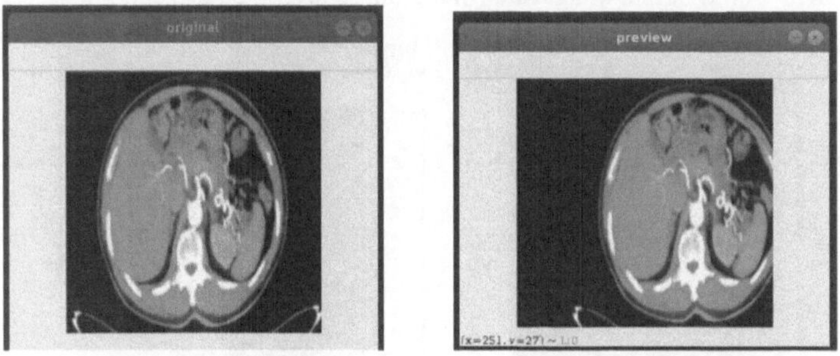

(a) (b)

FIGURE 8.16 (a) Original image and (b) width shifted image.

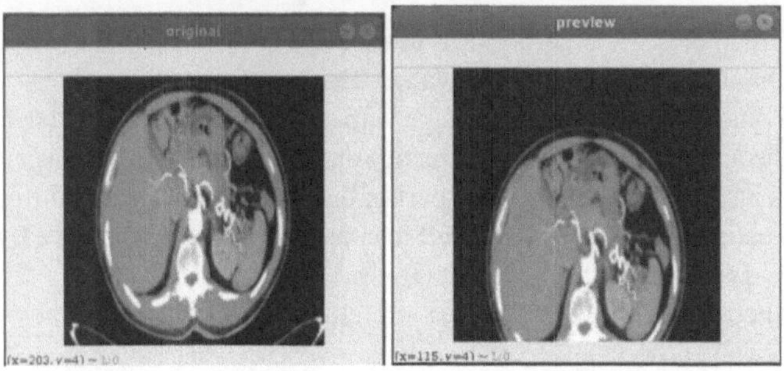

(a) (b)

FIGURE 8.17 (a) Original image and (b) the image after height shifting.

The height shift range is also a floating point between 0 and 1 (Figure 8.17).

4. Horizontal flip: The process of horizontal flip involves flipping the image across a vertical axis that passes through the midpoint of the horizontal axis of the image (Figure 8.18).

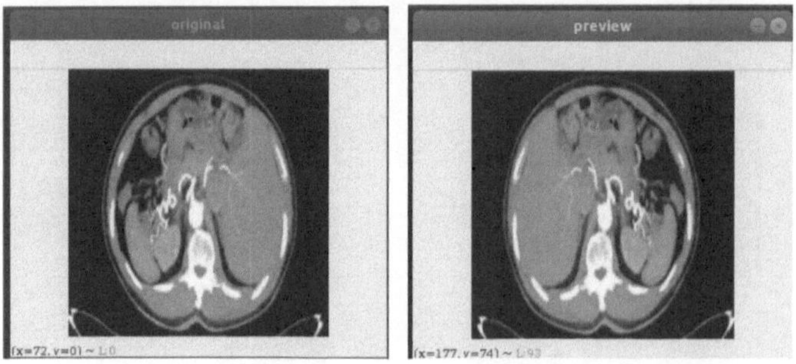

(a) (b)

FIGURE 8.18 (a) Original image and (b) the image after the horizontal flip.

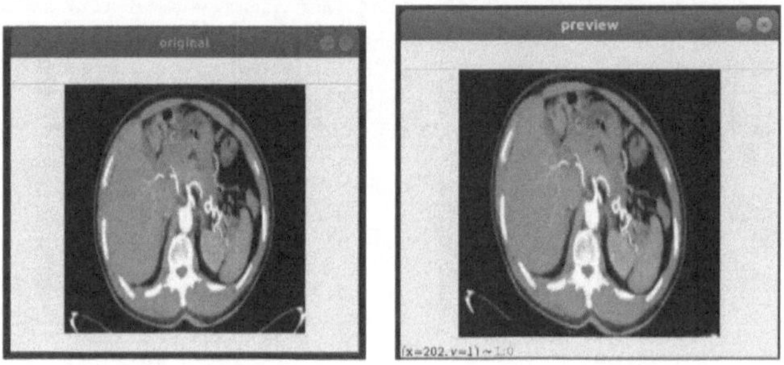

(a) (b)

FIGURE 8.19 (a) Original image and (b) the image after sheering.

5. Channel shift: Channel shift randomly changes the pixel value of the channel in the image with values close to the present values.
6. Zoom: Zooming into the image with a certain value assigned to control the amount of zoom is an effective data augmentation that even though is just focusing on a certain part of the image prevents overfitting.
7. Featurewise center: It is used to convert the image to have a 0 mean.

   Basically Image – Mean (Image).
8. Sheering: Sheering involves stretching the image out in any direction specified.

   This should be used carefully as the structure of the tumor should be preserved.

   Hence ideally only the edges of the images should be sheered (Figure 8.19).

8.15 MODEL USED

The lack of data renders most of the state-of-the-art architectures unreliable since most of these architectures are data-hungry and require a huge number of parameters to learn efficiently.

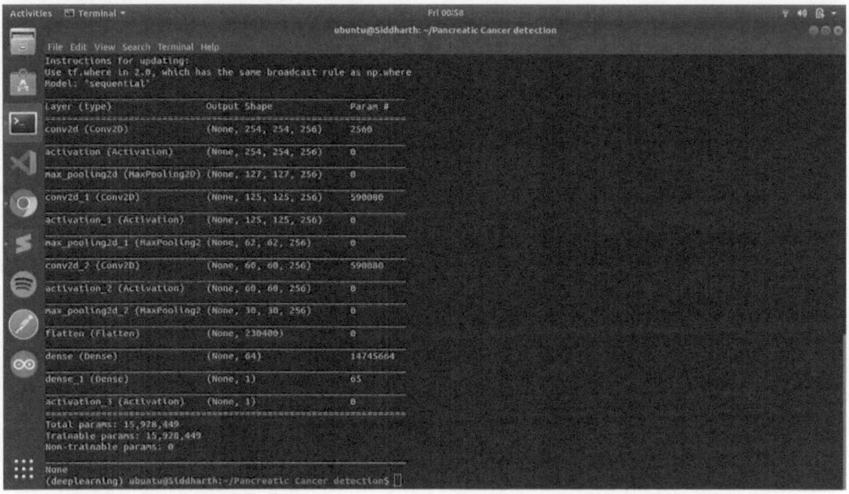

FIGURE 8.20 Layers in the architecture used.

Hence, the next logical step is to try classical architectures. A good architecture to begin with for tackling any type of vision-based task is AlexNet. AlexNet is a set of convolution and pooling layers followed by dense layers.

For the case in hand, given the small dataset, three sets of Convolution and Max Pool is ideal. The activation used will be Rectified Linear Unit or ReLU. Once we place the sets of convolution and pooling, the dense layers can be instantiated. For a small dataset like this one, two dense layers should suffice. The first one has 64 nodes, and the last one has one node which has a sigmoid activation. This sigmoid activation allows the model output to be squashed between 0 and 1.

Now a basic forward pass will include three sets of Convolution, Pooling, and Activation. To be specific, the number of filters in the convolution is 256 and the sliding window has a dimension of 3×3 and in pooling the size of the window is 2×2. Three sets of the above-mentioned will be followed by 2 dense layers with 64 and 1 nodes, respectively. Tuning of hyperparameters may result in better or worse performance (Figure 8.20).

Some Tips for Performance Improvement in CNN Architectures:

1. Normalizing input images:

   Normalizing generally improves the performance of the neural network as it confines the range of inputs to a user-specified range. In the above-mentioned case, dividing the input image array by 255 will help normalize the input in the range of [0, 1].

2. Decaying optimizer momentum:

   Decaying optimizer momentum allows the optimizer to converge much faster onto the minima.

3. Increasing the number of filers in deeper layers of the neural network will help the network learn finer details, but this step comes with added computation cost.
4. Regularization:

   Regularization is great in reducing overfitting and also facilitates in introducing some amount of randomness into the training process.

   For example, L1, L2 Regularization.
5. Dropouts:

   Dropouts are just the random removal of neurons from a network in the training process. This also has effects of regularization as it adds a certain amount of randomness to the training.

   Dropouts are usually applied by giving a probability score to the deep learning framework in use.
6. Using skip connections:

   Using skip connections may or may not help to depend on the type of data used in training.

8.16 RESULTS

After successful implementation of the above architecture, an accuracy of 75 was obtained. Due to the unavailability of a large dataset, the model suffered from an acute lack of parameters to learn from. There is scope for improvement if a larger dataset is used and if the dataset has more diverse cases of the tumor. A change in architecture can also be done upon the availability of a better dataset. State-of-the-art architectures usually require a lot of data to learn efficiently. However, classical deep learning architectures like AlexNet are made for most of the vanilla deep learning tasks. The training time is less than 3 minutes per epoch as the dataset only has 200 images, which is also due to the fewer number of layers being used as adding a layer requires at least a certain number of parameters to be added to the model. However, the model has learned well considering the dataset size and lack of diversity in it.

8.17 CONCLUSION

To conclude this chapter, it can be inferred that pancreatic cancer is a very fatal disease and it is difficult to predict with 100% accuracy. Several studies are going on for the early detection of this disease. The above algorithm is a stepping stone to predict the same, and it can be made better if the data is available in abundance. Till now, there is not sufficient information available to know the actual cause of disease; once enough parameters are available, the use of AI will be revolutionary and it will produce better outcomes.

REFERENCES

1. Krahl, Maurice E. "Endocrine function of the pancreas." *Annual Review of Physiology* 36, no. 1 (1974): 331–360.

2. Klein, Alison P., Kieran A. Brune, Gloria M. Petersen, Michael Goggins, Anne C. Tersmette, G. Johan A. Offerhaus, Constance Griffin et al. "Prospective risk of pancreatic cancer in familial pancreatic cancer kindreds." *Cancer Research* 64, no. 7 (2004): 2634–2638.
3. Siegel, Rebecca L., Kimberly D. Miller, and Ahmedin Jemal. "Cancer statistics, 2019." *CA: A Cancer Journal for Clinicians* 69, no. 1 (2019): 7–34.
4. Surveillance, Epidemiology, and End Results (SEER) Program. "SEER* Stat Database: North American Association of Central Cancer Registries (NAACCR) Incidence Data-CiNA Analytic File, 1995-2015, for Expanded Races, Custom File With County, ACS Facts and Figures Projection Project (Which Includes Data From CDC's National Program of Cancer Registries [NPCR], CCCR's Provincial and Territorial Registries, and the NCI's Surveillance, Epidemiology, and End Results [SEER] Registries)." (2018).
5. Torre, Lindsey A., Britton Trabert, Carol E. DeSantis, Kimberly D. Miller, Goli Samimi, Carolyn D. Runowicz, Mia M. Gaudet et al. "Ovarian cancer statistics, 2018." *CA: A Cancer Journal for Clinicians* 68, no. 4 (2018): 284–296.
6. Chari, Suresh T., Cynthia L. Leibson, Kari G. Rabe, Jeanine Ransom, Mariza De Andrade, and Gloria M. Petersen. "Probability of pancreatic cancer following diabetes: a population-based study." *Gastroenterology* 129, no. 2 (2005): 504–511.
7. Kaur, Sukhwinder, Michael J. Baine, Maneesh Jain, Aaron R. Sasson, and Surinder K. Batra. "Early diagnosis of pancreatic cancer: challenges and new developments." *Biomarkers in Medicine* 6, no. 5 (2012): 597–612.
8. McGuigan, Andrew, Paul Kelly, Richard C. Turkington, Claire Jones, Helen G. Coleman, and R. Stephen McCain. "Pancreatic cancer: a review of clinical diagnosis, epidemiology, treatment, and outcomes." *World Journal of Gastroenterology* 24, no. 43 (2018): 4846.
9. Islami, Farhad, Ann Goding Sauer, Kimberly D. Miller, Rebecca L. Siegel, Stacey A. Fedewa, Eric J. Jacobs, Marjorie L. McCullough et al. "Proportion and number of cancer cases and deaths attributable to potentially modifiable risk factors in the United States." *CA: A Cancer Journal for Clinicians* 68, no. 1 (2018): 31–54.
10. Qiu, Le, Lei Zhang, Vladimir Turzhitsky, Umar Khan, Yuri Zakharov, Douglas K. Pleskow, Mandeep Sawhney et al. "Multispectral endoscopy with light gating for early cancer detection." *IEEE Journal of Selected Topics in Quantum Electronics* 25, no. 1 (2018): 1–9.
11. Chen, Hong, Gary Y. Hou, Yang Han, Thomas Payen, Carmine F. Palermo, Kenneth P. Olive, and Elisa E. Konofagou. "Harmonic motion imaging for abdominal tumor detection and high-intensity focused ultrasound ablation monitoring: an in vivo feasibility study in a transgenic mouse model of pancreatic cancer." *IEEE Transactions on Ultrasonics, Ferroelectrics, and Frequency Control* 62, no. 9 (2015): 1662–1673.
12. Khokhlova, Tatiana D., and Joo Ha Hwang. "HIFU for palliative treatment of pancreatic cancer." *Journal of Gastrointestinal Oncology* 2, no. 3 (2011): 175.
13. Zhou, Yufeng. "High-intensity focused ultrasound treatment for advanced pancreatic cancer." *Gastroenterology Research and Practice* 2014 (2014). doi: 10.1155/2014/205325.
14. Klein, Alison P. "Identifying people at a high risk of developing pancreatic cancer." *Nature Reviews Cancer* 13, no. 1 (2013): 66–74.
15. Zhang, Qiubo, Linjuan Zeng, Yinting Chen, Guoda Lian, Chenchen Qian, Shaojie Chen, Jiajia Li, and Kaihong Huang. "Pancreatic cancer epidemiology, detection, and management." *Gastroenterology Research and Practice* 2016 (2016): 8962321.
16. Albawi, Saad, Tareq Abed Mohammed, and Saad Al-Zawi. "Understanding of a convolutional neural network." In *2017 International Conference on Engineering and Technology (ICET)*, pp. 1–6. IEEE, Antalya, 2017.

17. Kim, Yoon. "Convolutional neural networks for sentence classification." arXiv preprint arXiv: 1408.5882 (2014).
18. Grm, Klemen, Vitomir Štruc, Anais Artiges, Matthieu Caron, and Hazım K. Ekenel. "Strengths and weaknesses of deep learning models for face recognition against image degradations." *IET Biometrics* 7, no. 1 (2017): 81–89.
19. Agarap, Abien Fred. "Deep learning using rectified linear units (ReLU)." arXiv preprint arXiv: 1803.08375 (2018).
20. Krizhevsky, Alex, Ilya Sutskever, and Geoffrey E. Hinton. "ImageNet classification with deep convolutional neural networks." In *Advances in Neural Information Processing Systems*, pp. 1097–1105. New York, 2012.
21. O'Shea, Keiron, and Ryan Nash. "An introduction to convolutional neural networks." arXiv preprint arXiv: 1511.08458 (2015).
22. Wu, Haibing, and Xiaodong Gu. "Max-pooling dropout for regularization of convolutional neural networks." In *International Conference on Neural Information Processing*, pp. 46–54. Springer, Cham, 2015.
23. Iandola, Forrest N., Song Han, Matthew W. Moskewicz, Khalid Ashraf, William J. Dally, and Kurt Keutzer. "SqueezeNet: AlexNet-level accuracy with 50x fewer parameters and <0.5 MB model size." arXiv preprint arXiv: 1602.07360 (2016).

9 Applications of the Map-Reduce Programming Framework to Clinical Big Data Analysis

Current Landscape and Future Trends

Gagandeep Kaur and Satish Saini
RIMT University

Sachin Minocha
Galgotias University

CONTENTS

DOI: 10.1201/9781003050827-9

9.1 INTRODUCTION

This section covers the big data along with its characteristics, use-cases, and history. This section also covers the Hadoop and Map-Reduce framework with importance and features. These are the basic terminologies required to understand the challenges and trends of Map-Reduce programming for clinical big data.

9.1.1 Big Data

The technological term big data became famous in the early 2000s. "Big data" is being used for highly complex, fast, and huge amounts of data, which cannot be processed by traditional methods. Traditional methods include storing and accessing a large amount of information. The industry analyst Doug Laney included the concept of three Vs to elucidate the term "big data". Volume, Velocity, and Variety are three Vs of big data. While at SAS, two more Vs, i.e. variability and veracity are also considered making five Vs of big data as shown in Figure 9.1.

These 5 Vs shown in Figure 9.1 are described below.

Volume: In the past, storing large data that is collected by organizations from various sources like smart devices, social media, videos, business transactions, and more, was extremely difficult, but platforms like Hadoop reduced the burden.

Velocity: Due to fast-growing internet technologies, data flow into businesses at an unusual velocity should be controlled in a lively fashion by devices like smart meters and RFID tags.

Variety: There are two types of formats in which we get data, i.e. structured format and unstructured format. Big data mainly consists of text documents, emails, videos, audios, etc. that are a type of unstructured data that leads to variations in data.

Variability: Data flows are quite unpredictable as data varies often but businesses should understand the necessity to manage data loads daily or maybe seasonal and when something trends on social media.

Veracity: Veracity is a term used to show data quality. As data is collected from heterogeneous sources, it becomes difficult to connect, match, cleanse, and transform data in systems. To control large and complex data, businesses should link up and correlate data.

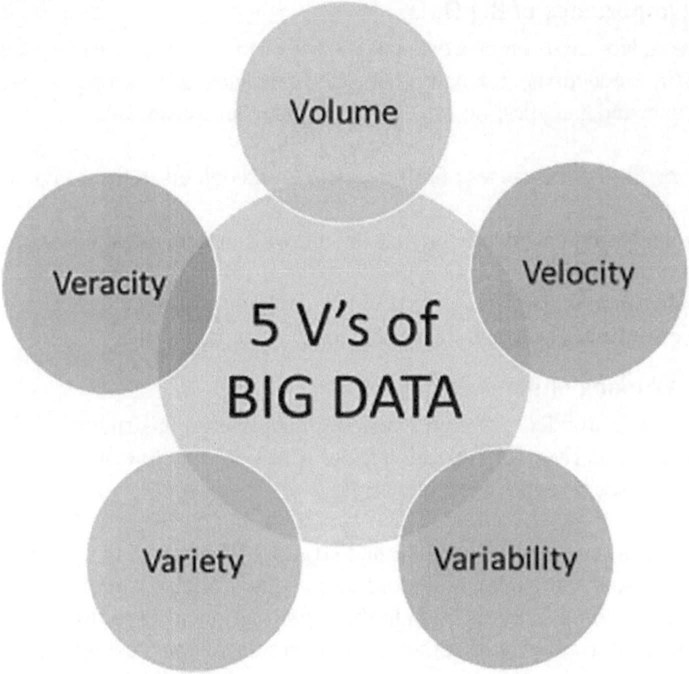

FIGURE 9.1 Five Vs of big data.

These 5 Vs of big data make the processing of big data a challenge (Dhavapriya and Yasodha, 2016). The next subsection will cover the history of big data to get an idea that how the term "big data" evolved.

9.1.1.1 History of Big Data

The term big data is comparatively new, but the concept of large data came into existence during the 1960s and 1970s with the development of a relational database. A large number of data users were created around the 2000s through Facebook, YouTube, etc. Hadoop and NoSQL were also developed during this year and gained popularity to store and analyze big data. Open-source frameworks like Hadoop (also famous by the name Spark) provided an edge in the development of big data by making it less complex and inexpensive. After the 2000s, the volume of big data has increased. Huge data is generated every day – but it's not just amounts of humans who are generating data. Due to the advancement in IoT, a number of devices can be easily attached to the internet connection and it is possible to converge data following customer's usage and also with product performance. The computing capabilities along with machine learning techniques have expanded possibilities to analyze big data.

9.1.1.2 Importance of Big Data

By analyzing large and complex data taken from various sources, we can get answers related to time reductions, cost, new product development, and smart decision-making. With big data and analytics, we can easily conclude the reasons for:

1. Determining the causes of failure, issues, and defects in things related to business.
2. By analyzing customer's buying habits, we can generate coupons at the point of sale.
3. Evaluating the whole risk portfolios quickly or with very little time.
4. Setermining false fraudulent behavior in an organization.

9.1.1.3 Working of Big Data

Big data flow is divided into five steps. Big data works on structured as well as unstructured data. Every organization should consider the flow of big data among sources, systems, owners, users, etc.

i. **Big data strategy**: A pre-planned strategy is required to enhance the way to share, manage, and store data. This strategy works within the organization as well as outside the organization. Due to this, big data should be treated with other crucial assets in the business and has helped to achieve future and existing business and Tech goods. So big data strategy sets a path for business sources among an ample sufficiency of data.
ii. **Know the sources**: Due to IoT, a lot of devices are connected to the internet and data flows into IT systems from medical devices, industrial appliances, smart cars, and more. By analyzing the "big data", organizations can make certain changes to meet future needs. Big data comes from Facebook, YouTube, Instagram, etc. in the form of images, videos, voice, etc. Mostly, such data is in an unstructured and semi-structured form that makes it very difficult to analyze.
iii. **Access, store, and manage**: To store, manage, and access huge amounts of datasets is not an easy task. But cloud solutions, data lakes, and Hadoop provide flexible and extremely inexpensive options for the storage and handling of big data. Modern computing methods are so flexible and provide quick access along with methods for data integration, ensuring data quality and data governance.
iv. **Analyze big data**: For analysis of big data, there are two methods; in the first method, the organization can use the whole big datasets for analysis and can use the results for their betterment, while in the other method, the analysis of the relevant amount of data, instead of analyzing the whole data, is done.
v. **Make intelligent, data-driven decisions**: For organizations to stay in competitive markets, they should analyze the trusted and well-managed data. Organizations should take trusted decisions on the evidence

presented by big data. Data-driven organizations perform exceptionally well and make more profits.

9.1.1.4 Big Data Use-Cases

We can use big data in every field. Some of the big data examples are explained in Table 9.1.

9.1.2 Hadoop

Apache Hadoop is an open-source framework used and helpful for storing and processing large and complex datasets (Olson, 2010). It includes the following components:

i. **Hadoop-distributed file system**: It is one of the bottom-layer components. It is used for storage purposes. First of all, it breaks up files into chunks and then distributes them across cluster nodes.
ii. **Yarn**: It is also one of the indispensable components. It is used for job scheduling. It is helpful in cluster resource management.

TABLE 9.1
Big Data Use-Cases

Field	Use-Case
Financial services	• The entire prospect of customer • Liability as well as scam supervision and management • Real-time transaction analysis
Healthcare/life sciences	• Syndrome examination analysis • Genomic statistics
Telecommunications	• Real-time calling details, CDR handling, as well as the processing • Customer portrait monetization and study • Real-time association component auditing • Real-time analysis of network faults • IVR calls
Digital media	• Real-time ad matching, inquiry, as well as focusing • Website data and reconstruction capturing
Retail	• Cross-channel commerce • Customer clustering as well as breaking • Click-stream scrutiny • Market Basket scrutiny • Real-time suggestion • Sentiment study
Law enforcement	• Multimodal inspection • Real-time cyber-security disclosure • Energy and utilities • Smart meter inquiry

iii. **Map-Reduce**: This component is used for parallel and synchronized processing.
iv. **Common libraries**: Hadoop subsystems used libraries while processing data.

9.1.2.1 Hadoop History

In the late 1990s and early 2000s, the World Wide Web (www) initially arose. To locate the data among text-based contents, search engines and indexes were designed. Automation was necessitated as within years web grew from few to millions of pages. Web crawlers and many university-led research projects also came into the picture. Nutch was an open-source web search engine and an example of such projects. Nuch was the brainchild of Poug cutting and Mike Cafarella; the whole project was based on multi-tasking to send web search results quickly. One of the other projects based on a similar concept for storing and processing data faster was named Google. Later in 2006, Yahoo took ideas from Google early work and moved forward with project Nutch, but now the idea was automating distributed large data storage and processing. Yahoo in 2008 releases Hadoop as an open-source project. Today, ASF is a global non-profit community of software developers that works on such projects.

9.1.2.2 Importance of Hadoop

It is useful in processing and storing large complex data quickly.

i. **Computing power**: Hadoop uses a distributed computing model that processes large and complex data fast. As the number of computing nodes increases, the processing power also increases.
ii. **Fault tolerance**: Data processing should be protected from hardware failure. In case if one node goes down, the other nodes come into force automatically so that distributed computing does not fail.
iii. **Flexibility**: Due to Hadoop, data does not need any preprocessing for storing. Even the storage and use of unstructured data is completely dependent on you.
iv. **Low cost**: It uses commodity hardware to store big datasets and an open-source framework is free of cost.
v. **Scalability**: The system is quite adjustable. In the case of more data, you can add nodes accordingly with little administration.

9.1.3 Map-Reduce

Map-Reduce is a system application that enables massive vulnerabilities across hundreds or thousands of Hadoop cluster servers. Map-Reduce is called the heart of Apache Hadoop to perform two distinct and separate tasks parallel. First, it maps the job, i.e. datasets are broken down into smaller datasets known as tuples/key/value pairs. Second, Reduce task comes into action (Dean and Ghemawat, 2004). It takes the map inputs and combines the tuples to form a set of tuples. As the name implies, the map step is always performed first.

9.1.3.1 Importance of Map-Reduce

Map-Reduce programming paradigm offers several characteristics that are explained as follows:

i. **Scalability**: Petabytes of stored big data can be easily processed within HDFS.
ii. **Flexibility**: Data from multiple sources and types can be accessed with more ease.
iii. **Speed**: It offers fast speed while processing huge amounts of data due to parallel processing and minimal data movement.
iv. **Simple**: There is a choice for the developer to write code in any language, e.g. Java, C++, and Python.

9.1.3.2 Hadoop Map-Reduce Working

Map-Reduce works in two steps, i.e. Map and Reduce, as shown in Figure 9.2.

Map step: The data of the master node gets divided into smaller parts known as sub-tasks. The result of the sub-task after processing gets stored in the local file form where the reducer can get access easily. All the sub-tasks using in the Map phase in form of key or value pairs are combined in a list.

Reduce step: The data after processing the Map phase gets processed and analyzed using the Reducer Phase. Job Tracker is responsible for the parallel processing of jobs in this phase.

The data is effectively divided and combined using Map-Reduce as shown in Figure 9.2. The processing speed is high as well. The Map-Reduce function is capable of processing and storing real-time data computation of complex tasks (Marozzo et al., 2012).

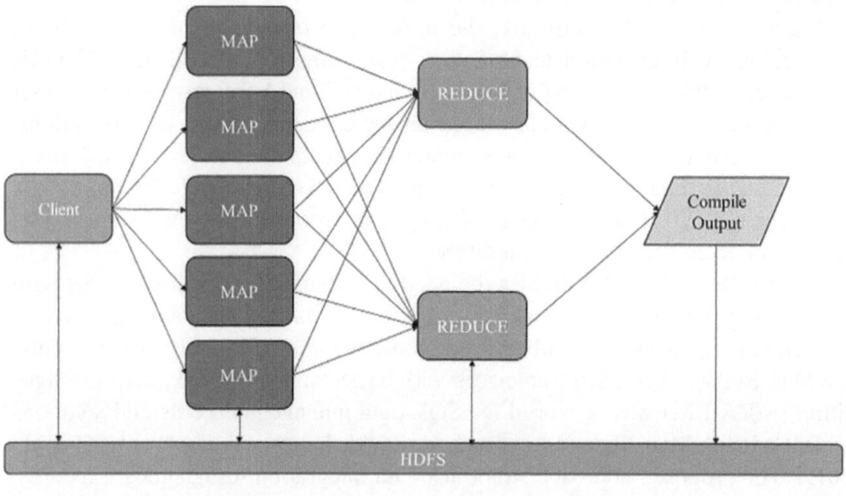

FIGURE 9.2 Map and Reduce functioning on the top of HDFS.

9.2 BIG DATA FRAMEWORKS FOR HEALTHCARE

Various healthcare frameworks that use the Hadoop Map-Reduce platform for handling a large amount of data accurately and with negotiable ambiguity have been on the top in recent researches (Nishadi, 2019). In this section, we would review the recent advancements and contributions in the healthcare sector.

Pradeep and Sundar (2020) proposed a QAOC, novel architecture. According to the authors, Hadoop while processing with big data degrades the performance and system. The two limitations, huge data management and query analysis, were focused on Hadoop. To support massive users, the fuzzy and neural network combination was used (Pradeep and Sundar, 2020). Clusters were used to manage data in Hadoop. On the other hand, a binary search tree was implemented for the data search mechanism. The novel architecture was implemented on Hadoop and better results were achieved in terms of memory consumption, speed, and execution time. The novel architecture provides better results as compared to traditional algorithms.

Harb et al. (2020) discuss that patient's health observation becomes a hard task as the number of patients is increasing and disease-related new symptoms come into existence. The authors proposed a robust framework for the observation of patients in real time (Harb et al., 2020). Sensor devices that work remotely with the use of big data technology were combined to healthcare systems; an efficient solution with low cost. The proposed framework relied on four layers of big data analytic techniques. Observation of patients in real-time, data storage, and decision-making in real time; symptom diagnosis; and classification of a patient, data visualization, and retrieval were the four layers used to implement the framework of big data using Hadoop in real-time data provided in healthcare systems. The false diagnosis was prevented by studying the vital sign correlation of patients between the same cluster. The effectiveness of the framework was measured using real-time healthcare systems.

Pramanik et al. (2020) discuss the importance of healthcare data analytics and healthcare information technologies. The authors proposed the HCIandA framework within the context of big data. The HCIandA framework was further classified into three technologies based on the concepts of system applications, concepts used to evaluate systems, underlying technologies, and forthcoming research fields. The three stages of HCIandA were HCIandA 1.0, HCIandA 2.0, and HCIandA 3.0 (Pramanik et al., 2020). The proposed framework study was carried out to contribute to the literature.

Dharavath et al. (2019) discuss the need for efficient healthcare systems with the growing population of Asia. The authors also discuss the data analysis and collection challenges in healthcare domains. For managing unstructured data, the Map-Reduce Approach framework with Expectation and Maximization algorithm (MRA-EM) was proposed. NoSQL data management tools and solutions were used with Map-Reduce Approach to provide better results (Dharavath et al., 2019). The proposed approach simplifies data integration for healthcare records. The authors concluded that the data mapping problem for healthcare records can be resolved using the proposed framework. The functionality of the proposed framework can still be enhanced.

Santur et al. (2019) discuss that people use web technologies for gathering information to meet their daily needs. The technologies with mobile version applications made it convenient. Blood-glucose monitoring, pulse oximeters, sleep and activity trackers, and blood pressure monitoring are some of the applications which are synchronized with mobile applications. By utilizing these mobile applications, patients can track their records themselves for diagnosis. The records can be stored or monitored via cloud applications. The authors proposed a cloud application for women's pregnancy monitoring (Santur et al., 2019). The application was accessible to only members. Period tracker, baby percentile, pregnancy calendar, and baby's schedule for vaccination were monitored efficiently using the proposed framework.

Shaikh and Ali (2019) discussed the exponential growth of data rate. The data number is increasing in an unstructured way, i.e. data belongs to different sources, and different sources generate different formats. Traditional systems are incapable of storing data generated from these numbers of sources. Big data solutions can be used to overcome this challenge. The authors proposed a big data framework based on Hadoop HDFS and Map-Reduce (Shaikh and Ali, 2019). The focus of the research was on the effectiveness of using big data solutions in Indian healthcare. The framework was implemented stepwise to make Indian healthcare systems smarter.

Ismail et al. (2019) proposed a Remote Patient Monitoring (RPM) model for monitoring patients while at their homes. The proposed model can work efficiently for old-age patients who are staying alone and for whom coming to the hospital for check-ups is a tedious task. The data generated using this framework can be analyzed and stored on the cloud using big data. To access data with higher speed, Spark with machine learning was used as a solution to big data (Ismail and Shehab, 2019). A case study of patients suffering from disorders related to blood pressure was monitored using the proposed framework and the results obtained were efficient.

Ismail and Shehab (2019) proposed a Message Queue Telemetry Transport (MQTT) technique based on Hadoop Map-Reduce. The proposed framework worked on a wearable device that monitors a patient's temperature, oxygen consumption, sleep rate, blood levels, and food ingestion. The wearable device is made from a combination of different sensors (Ismail and Shehab, 2019). The data from the patient was also collected individually or using a smartphone. The data integration technique was used to club both source data to generate efficient results. Hadoop Map reduction solutions were used as a middleware for accurate results. The prototype system was developed to reduce the burden of a doctor as a solution to cover all the patients in the era of high population.

Santhosh and Nayak (2019) discussed the need for big data as a solution to the challenges of traditional algorithms in terms of data analytics, data collection, and data amounts. The focus of the research was to analyze diabetic patient's sugar levels to prescribe an accurate amount of insulin. The authors proposed a data mining algorithm with Hadoop Map-Reduce as the platform for precisely analyzing the patient's clinical database. The proposed algorithm was performed on available datasets but to achieve accurate and precise results, more attributes will be required (Santhosh and Nayak, 2019).

El Aboudi and Benhlima (2018) discussed that big data techniques for quality healthcare systems have led to generating large amounts of data. The authors focused on the emergency prevention challenge due to a large amount of data. A novel architecture to prevent emergencies was proposed using a big data framework. Real-time alerts were made by using the reliability feature of batch computing and stream computing. The proposed system focused on accurate monitoring of patient's health (El Aboudi and Benhlima, 2018). The proposed architecture was tested on a prototype system that generates real-time alerts.

Sebaa et al. (2018) discussed the enormous generation of data due to an increase in medical devices. According to the authors, Hadoop comes as a solution for handling large amounts of data as traditional algorithms are not capable enough to process and handle large amounts of unstructured data (Sebaa et al., 2018). A review of the literature was provided regarding the challenges of traditional algorithms and Hadoop solutions.

Sivaparthipan et al. (2018) discussed the effectiveness of big data and Hadoop in the technological revolution. The authors proposed a statistical assessment model for the analysis of diabetes. The accuracy of the proposed model and *F*-measure were analyzed (Sivaparthipan et al., 2020). The results obtained were comparatively higher than the traditional models.

Din et al. (2019) discussed the gap between innovative tools like big data and users. The gap between technological advancement and user leads to inefficiencies in the systems. Constraints grown due to the use of IoT devices used in healthcare systems and user inefficiency were discussed. The novel architecture was proposed by authors to ease the real-time access of IoT devices so that real-time online, as well as offline data, should be handled efficiently by users (Din and Paul, 2019). Sensor datasets were tested efficiently on the proposed system.

Binu et al. (2017) discussed the popularity of using IoT in healthcare systems. A smart healthcare monitoring system was proposed for observing the health status of a child. The status results were analyzed through the results of games played by the child on android (Binu et al., 2017). The game was integrated with sensors through android mobiles.

Yuvaraj et al. (2019) discussed the need for smart healthcare systems to predict diabetes accurately. With a growing population, more cases of kidney failure, blindness, and heart attacks are diagnosed as diabetic at later stages. A novel algorithm was proposed to predict diabetes with smart healthcare systems (Yuvaraj and SriPreethaa, 2019). The working of the algorithm was tested using different national databases.

Kalyankar et al. (2017) proposed a predictive analysis method to predict the risks of diabetes in India. Past datasets were used to focus on the risks using predictive analysis. Machine learning advanced algorithms were used in the analysis

(Kalyankar et al., 2017). The predictive analysis was carried out effectively to find out the type of diabetes.

A framework was established in the form of a BDA, i.e. Big Data Analytics, by Chrimes et al. (2017). This framework used real big data volumes of healthcare data. Existing technologies like HPC architecture with HBase and Hadoop (HDFS) were utilized to generate results. Special adjustments were made to generate the results. HBase took a week to investigate 1 billion records, out of which it generates a result of approximately 10 TB of data and took only 3 seconds. Both queries take less than 2 seconds (Chrimes et al., 2017). All the queries, i.e. simple or complex produced, accurate figures. Significantly large volumes of data were captured via the BDA platform. Data retrieval times were reduced.

Pramanik et al. (2017) evaluated a study of smart systems using big data and carried out a critical analysis. A big data-enabled Smart Healthcare System Framework (BSHSF) was proposed. Ubiquitous healthcare solutions were provided by the proposed smart healthcare conceptual model. The cost was reduced with the increase in intelligence (Pramanik et al., 2017). It also provides opportunities for healthcare organizations to deliver intelligent smart system services. Machine Learning, Intelligent Agents, and Text Mining are the three branches upon which the framework was based.

Fang et al. (2016) surveyed extensively on the applications of big data in healthcare systems. A framework called the "Health informatics processing pipeline framework" was implemented. The extensive research work on the other hand took interest in the issues about data heterogeneity, the complexity that exists in the available healthcare data, discovered pattern visualization, and data privacy. A systematic data processing pipeline was offered by the proposed framework. Advancements in sensor technology were responsible for providing novel solutions to the healthcare domain. Data generated from a wide range of sensors, mobile devices, and wearable devices have been used extensively on demand for healthcare services (Fang et al., 2016).

Sakr and Elgammal (2016) discussed the major challenges of the healthcare systems which can be resolved using ICT technology advancements. The smart healthcare system was proposed which integrates big data analytics, the internet of things, and cloud computing. Improved efficiency and effectiveness of healthcare systems were made possible due to these emerging sensing technologies. The flaws and identified gaps were also discussed as well as the road map which can be used for future research advancements for developing smart healthcare systems. For improving a patient's profile analytics, the proposed architecture played a vital role (Jaseena and David, 2014). On the other hand, the framework lacks the ability to handle complex data sources.

Dencelin and Ramkumar (2016) focused on machine learning-based Multilayer Perceptron (MLP) algorithms. The algorithm was used to predict protein secondary structure. The algorithm was tested using a different set of inputs and the outcomes produced were accurate. The input datasets of protein were α-helix, β-sheet, and coils. Apache Spark framework was used to enhance the performance (Dencelin and Ramkumar, 2016). The algorithm was implemented for security, bandwidth, and fault-tolerant features in a distributed environment. It offers great opportunities in research if MLP and Apache Spark framework are used together.

Forkan (2015) proposed a knowledge discovery-based approach. The proposed approach was based on real-time healthcare services monitoring. Big data analysis was used to find the patient's specifics from the data lake. The two-step learning methodology was advocated to generate big data from various living systems of patients with the advent of cloud technology. A Map-Reduce algorithm was applied in the first step to finding out the correlation among vital parameters of patient data. Supervised learning algorithms were implemented in the second step. The advantages of machine learning, remote monitoring, and cloud data were discussed using the proposed framework. It performs efficiently in a real-time decision-making framework for healthcare services (Forkan et al., 2015). Developing novel solutions for protein structure prediction via cloud-based big data technologies has gained the attention of researchers. With the advancements in sensor technologies, it can work more effectively.

Lin et al. (2015) discussed the importance of self-caring devices for users for emergencies. They advocated that self-caring is possible with big data platforms because massive data can be stored securely and patients can self-diagnose by matching it with similar records. A self-caring service called "home diagnosis" was proposed. The framework was based on cloud services. A prototype system was designed (Lin et al., 2015) and was presented with examples to demonstrate efficiency and scalability.

Mohammed et al. (2014) discuss the applications of the Map-Reduce programming framework on clinical data. A review using different Map-Reduce programming frameworks and their implementations was carried out. The review was solely based on healthcare data and healthcare systems. The authors further advocate the benefits of using Hadoop Map-Reduce-based programming paradigms in clinical processing and also takes the insights to its implementation, challenges, and forthcoming claims (Mohammed et al., 2014). The potential for Map-Reduce implementation for clinical data was discussed in detail. They demonstrated the significance of implementing the Hadoop Map-Reduce framework.

Raghupathi and Raghupathi (2014) proposed a big data analytics healthcare framework. Specific operations were performed using Hadoop ecosystem tools and were carried out on HDFS via the Map-Reduce programming model. Many examples were shared and improved results were discussed thoroughly. There was no practical or experimental evaluation of results carried out (Raghupathi and Raghupathi, 2014). Several challenges were addressed. Cloud-based big data and Map-Reduce frameworks have the potential to cope up with the challenges of the healthcare systems if used in real-time applications.

Kim et al. (2014) proposed a big data framework for ubiquitous healthcare systems. The proposed framework worked on real-time services. Transmission and analysis of data were also a major part of the framework. The healthcare services were based on the vital information provided by the patient's record. The framework used the basic Hadoop structure to handle physiological signals (Kim et al.,2014). Standardized services were also provided to secure interoperability. Real-time analysis was carried out. The proposed algorithm was novel in extracting and processing vital signals.

Youssef (2014) proposed big data and mobile cloud-based Healthcare Information Systems (HISs). The framework provided a high level of integration, availability, interoperability, and was shared that healthcare data among medical practitioners, drug developers, and patients through cloud components. Security and access control protocols were applied to the proposed framework to guarantee security and privacy (Youssef, 2014). The proposed framework provides high-quality health services in the mobile cloud computing environment, but it fails to discuss security policies.

Jokonya (2014) proposed a consolidated framework based on big data. The proposed framework was suggested in the healthcare systems for the control and prevention of HIV/AIDS, Tuberculosis, and silicosis. The framework mainly focused on Mining Industry as silica dust affects the lungs and there are chances of silicosis and tuberculosis. The patient's body is at high risk of tuberculosis infections due to a weak immune system as an adverse effect of HIV/AIDS (Jokonya, 2014). To reduce the risks, the need for a predictive Epidemiology model for control and prevention was addressed. Data was captured considering various risk factors of tuberculosis. The big data component combines the data for analysis using the proposed framework by predictive and descriptive models. The proposed framework gained novelty as it performs effectively for identifying tuberculosis (TB) disease.

Chawla and Davis (2013) proposed a collaborative filtering-based approach to a personalized patient healthcare framework. The proposed approach worked on personalized disease risk profiles by capturing similarities for individuals. The collaborative filtering approach was designed to predict the patient's opinion based on known preferences (Chawla and Davis, 2013). The approach consists of a large group of user records. The healthcare history of the individual patient was compared with the record of other patients' in terms of some defined similarities. Symptoms, family history, lab results, and demographic data were defined as similarities. Prediction of disease was carried out by computing the pool of similar patients' records. The proposed framework provides a solution to the healthcare challenges by maintaining electronic healthcare records in the context of big data (Table 9.2).

The analysis done in this section enables us to determine the role of Map-Reduce in the analysis of clinical datasets described in the next section.

9.3 ROLE OF HADOOP MAP-REDUCE FRAMEWORK IN CLINICAL ANALYSIS

In recent times, the arithmetic manufacturing of data has established a modern era among IT called "big data". In medical contexts, such datasets are built utilizing significant laboratory information system (LIS) data, test itemization data, biological data, gene expression data, electronic medical record (EMR),

TABLE 9.2
A Summary of Big Data Frameworks

Authors	Study Year	Technology/Framework Used	Implications	Scope
Pradeep and Sundar	2020	QAOC architecture	Limitations of using Hadoop with big data resolved using the novel QAOC algorithm	Generalized systems
Harb et al.	2020	Robust big data analytics framework for real-time monitoring	Remote sensing devices were used in collaboration with big data and Hadoop technology to achieve effective results	Generalized healthcare systems
Pramanik et al.	2020	HCIandA framework within the context of big data	The study was carried out to contribute to the literature	Generalized healthcare systems
Dharavath et al.	2019	Map-Reduce Approach framework with Expectation and Maximization algorithm (MRA-EM)	Data integration for healthcare records was simplified using the proposed framework	Generalized healthcare systems
Santur et al.	2019	Cloud-based framework for pregnancy monitoring	Period tracker, baby percentile, pregnancy calendar, and baby's schedule for vaccination were monitored efficiently	Self-care diagnosis healthcare systems
Shaikh and Ali	2019	Hadoop HDFS and Map-Reduce	For handling data from different sources and with different formats, big data solutions were implemented stepwise to achieve smarter results	Generalized healthcare systems
Hassan et al.	2019	Remote Patient Monitoring (RPM) Model with Spark and machine learning-based on big data solutions	Case study of a patient was monitored remotely, and the results obtained were accurate	Smart healthcare systems
Ismail and Shehab	2019	Message Queue Telemetry Transport (MQTT) based on Hadoop Map-Reduce framework	The wearable sensor device prototype was made to collect and integrate data efficiently	Smart healthcare systems

(Continued)

TABLE 9.2 (*Continued*)
A Summary of Big Data Frameworks

Authors	Study Year	Technology/Framework Used	Implications	Scope
Kumar et al.	2019	Data mining algorithm with Hadoop Map-Reduce platform	The sugar level of a patient to prescribe the accurate level of insulin was analyzed using available datasets	Generalized healthcare systems
El Aboudi and Benhlima	2018	Big data architecture	Real-time alerts for emergency were tested on a prototype system to avoid the prevention of emergency issues	Emergency healthcare systems
Sebaa et al.	2018	Big data Hadoop	Provides a case study of implantation of data ware technology with big data Hadoop for generating efficient results	Generalized healthcare systems
Sivaparthipan et al.	2018	Big data Hadoop	Statistical assessment framework in context with big data Hadoop was efficiently used for diabetics analysis	Generalized healthcare systems
Din et al.	2019	Big data Hadoop	A lifetime of smart monitoring devices was enhanced using an energy harvesting algorithm	Smart healthcare systems
Binu et al.	2017	IoT within the context of Hadoop	Behavior and health records of children were monitored using sensor-based games	Smart healthcare systems
Yuvaraj et al.	2019	ML algorithms in context with Hadoop	Novel implementation for diabetes prediction was implemented effectively	Generalized healthcare systems
Kalyankar et al.	2017	Hadoop Map-Reduce based on machine learning algorithms	The predictive analysis was used to identify the type of diabetes to deal timely with future risks and to provide efficient treatment	Generalized healthcare systems

(*Continued*)

TABLE 9.2 (*Continued*)
A Summary of Big Data Frameworks

Authors	Study Year	Technology/Framework Used	Implications	Scope
Chrimes et al.	2017	BDA framework for real data volumes of healthcare systems	Data retrieval time was reduced and the queries were processed accurately with less than 2 seconds	Generalized healthcare systems
Pramanik et al.	2017	Smart system technology based on the conceptual model of big data	BSHSF provides ubiquitous healthcare solutions with reduced time and increased intelligence	Generalized healthcare systems
Fang et al.	2016	Pipeline framework for processing systematic healthcare data	Advancements in sensor technology were responsible for providing novel solutions to healthcare frameworks	Generalized healthcare systems
Mahmud et al.	2016	Data visualization and analytics framework	Minimum financial and human efforts will be required by using smart devices	Generalized healthcare systems
Sakr and Elgammal	2016	Integration of big data analytics, IoT, and cloud computing via smart healthcare systems	The efficiency and effectiveness of healthcare systems was improved but it lacks to handle complex data sources	Generalized healthcare systems
Dencelin and Ramkumar	2016	Knowledge discovery-based approach	A novel solution for protein structure prediction was developed effectively	Generalized healthcare systems
Forkan et al.	2015	Self-care/self-diagnosis with big data platform based on cloud computing	A prototype system for self-health care was designed and was tested for efficiency and scalability	Self-care diagnosis healthcare systems
Lin et al.	2015	A cloud-based framework for home diagnosis service	Retrieval of medical records based on symptoms based on a user inquiry	Self-care diagnosis healthcare systems
Mohammed et al.	2014	Hadoop Map-Reduce-based programming paradigm	The potential of using Hadoop Map-Reduce technology in the healthcare environment was discussed in detail	Generalized healthcare systems

(*Continued*)

TABLE 9.2 (*Continued*)
A Summary of Big Data Frameworks

Authors	Study Year	Technology/Framework Used	Implications	Scope
Raghupathi and Raghupathi	2014	Big data analytics healthcare framework	Various tools and platforms are outlined for healthcare big data for improving results	Generalized healthcare systems
Kim et al.	2014	Real-time services-based ubiquitous healthcare system using big data framework	A novel algorithm was developed in extracting and processing vital signs	Generalized healthcare systems
Youssef	2014	Healthcare Information Systems (HISs) based on a cloud computing environment	Provides high-quality healthcare services in a mobile cloud computing environment	Generalized healthcare systems
Jokonya	2014	An integrated framework based on big data	The novel algorithm developed to identify tuberculosis disease	Generalized healthcare systems
Chawla and Davis	2013	Collaborative filtering-based approach for personalized patient healthcare framework	Provides a solution to the healthcare challenges with the context of big data	Generalized healthcare systems

biometrics data, and data from other domains (Palanisamy and Thirunavukarasu, 2019). Grand datasets are extremely difficult to examine and query using traditional procedures, especially when the queries themselves are complicated. A Map-Reduce method, in essence, divides the query and datasets into core pieces (Kavitha, 2015). The query's defined components can be managed simultaneously or reduced to quickly provide results (Lyu et al., 2016).

By working with the Map-Reduce structure on top of the Hadoop-distributed file system, huge datasets have been handled effectively. Generally, the impact of big data analysis in healthcare framework and bioinformatics (Dai et al., 2012) has been represented. On the other hand, our reconsideration broadens the breadth of the practices of the Map-Reduce structure. Its open-source utilization leads to a wide range of applications in a variety of fields.

1. **Openly accessible scientific datasets**: It involves the online data published in the reports of the United States Food and Drug Administration (FDA).
2. **Biometrics datasets**: It involves the feature of measurable data concerning the characteristics of a human being. This data is used to access control or for identification.
3. **Bioinformatics datasets**: It involves the patient's biological statistics (for instance, DNA sequence, protein structure, and many more).
4. **Biomedical signal datasets**: It involves the data recorded from the patient's vital signs (for instance, electroencephalography (EEG), electrocardiography (ECG), and many more).
5. **Biomedical image datasets**: It involves the statistics that appear in the form of images by examination of the patient's body (for instance, magnetic resonance imaging (MRI), ultrasound imaging, histology images, and many more).

The role of the Hadoop Map-Reduce framework is huge in the clinical analysis, and this can be extended further by understanding the mapping between clinical data and Map-Reduce context in big data described in the next section.

9.3.1 Mapping of Clinical Data with Hadoop Map-Reduce in Context with Big Data

With the detailed study, it is clear that Hadoop Map-Reduce offers solutions to the clinical data challenges. The mapping is explained in Table 9.3.

Table 9.3 describes the challenges faced by clinical datasets and corresponding solutions offered by the Map-Reduce programming framework. These solutions are summarized in Figure 9.3.

The solutions offered by the Map-Reduce programming framework for the clinical challenges are shown in Figure 9.3, while the challenges faced by the Map-Reduce framework to analyze clinical data are discussed in the next section.

TABLE 9.3
Mapping of Clinical Data with Hadoop Map-Reduce in Context with Big Data

Clinical Data Challenge	Solution with Hadoop Map-Reduce in Context with Big Data
Huge data amount for processing	It can handle a large amount of data
Data sources and formats are different	It can handle data from various sources and of various formats
Needs more storage capacity	Storage capacity is increased using big data cloud
Real-time data management is needed	Real-time data can be managed efficiently
Needs less time to compute data	Hadoop Map-Reduce increases computation time
Smart systems needed	Smart systems can work effectively with Hadoop Map-Reduce

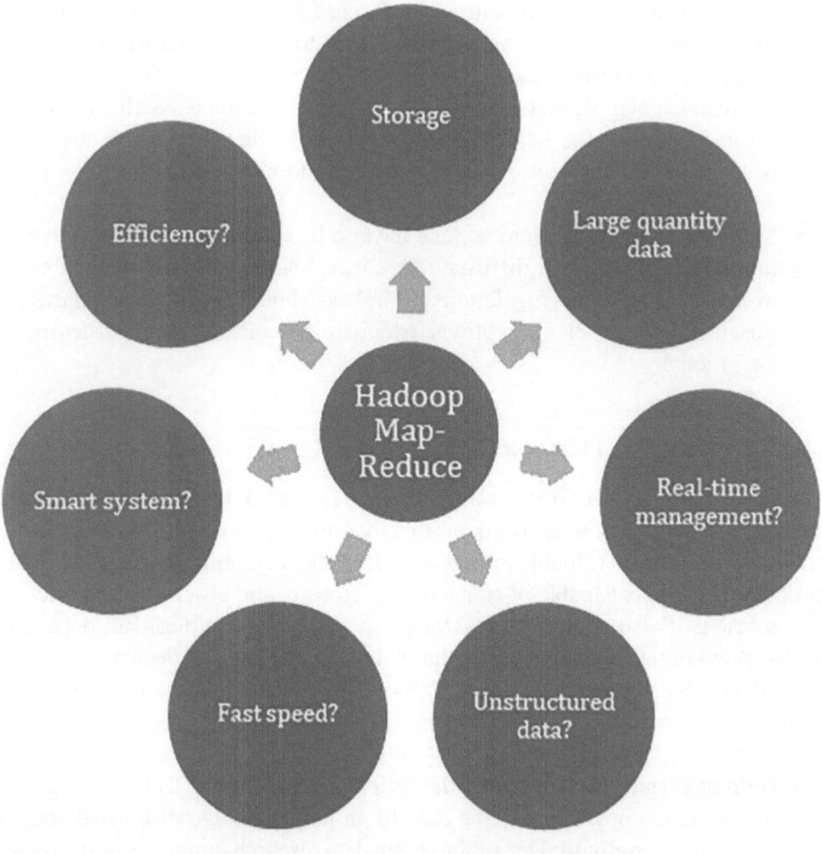

FIGURE 9.3 Hadoop Map-Reduce solution to clinical challenges.

9.4 CHALLENGES OF HADOOP MAP-REDUCE FRAMEWORK IN CLINICAL ANALYSIS

Generally, healthcare systems go through temporary outlay and loss of data utilization. Accordingly, there is an essential demand to find key solutions that could diminish needless outlay (Schadt, 2012). Overture in health quality results and control in outlay measures disclose on using the potential of large-scale integrated databases to featured arrangements and visions. How this scientific data could be clustered, managed, unveiled, and adopted though there is less certitude is one of the major challenges. In healthcare systems, the main problem is a lack of statistical data that would be used to make critical decisions. This framework represents the subsequent challenges to big data solutions in medical facilities:

i. Healthcare departments are not approving the redesigns of healthcare systems which will prompt efficiency.
ii. Medical data comes into existence from several sources (for instance, labs, providers, financial, data vendors, etc.). It prompts the actual need for data synthesis and managing structure to hold the statistics into a formative data warehouse.
iii. Medical data if shared among scholars and researchers will provide great benefits to the healthcare systems. These benefits are strangled because of privacy concerns and laws that do not grant permission to access that data.
iv. Solutions to big data architectures have to be formative and supportive to maintain the various diffused references. Solutions should be flexible to maintain rule and regulations as well as should meet the standards (for instance, novel encryptions which might need system structure alterations).

9.4.1 Clinical Big Data and Forthcoming Claims

Big data casually advise direct profits, on the other hand, analytics based on big data could affirm many sensitive decisions that might prove useful in a scientific domain. Extraction of valuable information from big data might be considered as a processing channel capable of transmitting, storing, and observing analytics of large datasets. It also comprises several setup steps to accomplish full data possession and maintenance, with particular claims challenged and decreased at each level (Martin-Sanchez and Verspoor, 2014). The forthcoming claims of clinical big data can be as follows:

i. **Data aggregation**: At present, the simple method to heap a largr amount of data is to copy/transmit the data to an ample storage disk drive and trek it to the configured terminal(s). Big data research initiatives typically include numerous firms, diverse geographic locations, large numbers of research or IT teams, and the establishment of a separate or duplicate database from the immediate database that is used in production. This type of statistics/data transfer among various groups and databases is very arduous to synchronize and manage. Also, another way is the usage of a network to transmit the statistics. On the other hand, transmitting a huge amount of data into or out of a data repository poses an essential networking claim. Another approach is that data be copied from many sources and created again across objects, similar to how Hadoop replicates file blocks and stores them in a rack-aware manner across numerous data nodes.

ii. **Data maintenance**: Since big data is associated with big accumulation of scientific/medical datasets, it is arduous to store efficiently as well as to manage the statistics on a single disk drive by using conventional data management structures. Furthermore, there is a heavy IT load for small organizations or various departments to maintain a big number of statistics. Also, at a glance, the content of health big data, the data must be continuously managed its portrayal of real patient's statistics of a hospital system in volume, its metadata, and data profiles as well; else way, the analytics is rendered useless.

iii. **Data integration**: The integration process comprises combining and transmitting statistics into a suitable format for statistics analysis. Since big data in healthcare are incredibly big in size, distributed, unorganized, and different, data integration is very challenging and very arduous to maintain over the years. Architecture for broad multi-media statistics storage and care utilizing Hadoop via the BDA platform for healthcare necessitates the establishment of patient-centric statistics in the database and especially integration with HDFS to minimize mistakes and faults and to ensure data integrity. Furthermore, not only is disorganized data integration a significant danger to BDA but so is structured EHR data integration.

iv. **Data analysis**: For analytic algorithms, the estimating interval raises enormously even with a tiny upswing in statistics. For instance, for doing casting knowledge in computational biology as well as bioinformatics, the Bayesian network is a trendy algorithm. Despite the calculative intricacy of the Bayesian network, the calculative time to search out a better network likewise soars exponentially. It might take several days even for the simple statistical analysis, even months, to attain the outcome when facts are extensive in scale. For some intense calculative problems, the data can be effectively analyzed using parallelism, i.e. to distribute the tasks into sub-tasks throughout various computers. However, if we could not correlate the coherent algorithm, it would be more demanding to tame the power of those massive parallel processing (MPP) hardware as well as tools that novel technologies are proposing for big data.

v. **Pattern interpretation for application**: In healthcare, security, confidentiality, off-the-record, and privacy of patient's data is requisitioned by rules and regulations and privacy commissioners at government levels, which could be large-scale obstacles for BDA implementation. This is all accessible to expose purposeful worth with the extensive datasets. Privacy problems might be reduced by using unique key-value store technologies in databases, but extensive setup, as well as technical skill, is required. For instance, Pattuk et al. (2013) proposed an architecture for secure big data management comprising an HBase database known as Big Secretive that helps in secure outsourcing and processing of encrypted facts throughout public key-value stores. Therefore, our understanding to preserve specific reclusion in the era of big data is technologically unoccupied (Chawla and Davis, 2013). Data validation is another important term related to big data claims. Most of the populace reckon that for decision-making, bigger statistics regularly provide the best information. Dismally, big data tools to track knowledge from rapid statistics science might not be able to save us from various deceptions, lacking information, and wrong theory. Knowledge representation is yet another claim. Big data is limited in its value for having the ability to analyze if agreement makers could not perceive the discovered structures. Unluckily, the presentation of the outcomes, statistics fantasies, and its interpretation by nontechnical professionals are a bigger threat due to the hard essence of analytics in healthcare. Data privacy protection is of utmost importance while talking about claims. Assuring patient's statistics privacy is taking into consideration the processes as well as storage on the BDA platform, and the access to the statistics and platform for potential public disclosure. Indexes created from HBase could securely encrypt key-value stores assuring that the index domain of patient's statistics generated could not be pervaded, for instance, like that by Pattuk et al. (2013). Likewise, HBase 0.98 and newer versions incorporate transparent encryption, as well as compression, to save the statistics even more.

9.5 FUTURE TRENDS OF HADOOP MAP-REDUCE FRAMEWORK IN CLINICAL ANALYSIS

Big data can unzip the gross healthcare system's value chain. Big data analysis does the transition in the conventional context of healthcare structure from searching novel medicines to patient-centric healthcare for superior medical reports with increased efficiency. The big data utilization in healthcare organizations can enhance and stimulate interactions among administrators, scientists, logistic managers, researchers, and lab directors by recovering outlay, implementing preferable efficiencies which are based on outcome testing, diminishing hazard, and reforming personal care.

Researches should focus on clinical integration of processes like test procedures, specimen collection as well as their handling, etc. so that an integrated system should be ready to cope with all the challenges related to clinical data.

Relevant open-source software can be synchronized with cost-effective hardware to ease the process integration. The synchronized system can develop a smart laboratory that will deal with increased clinical data at the highest performance rate.

9.6 CONCLUSIONS

Smart integrated system solutions will eliminate the issues faced by traditional systems because they lack technological advancements. The efficiency of clinical systems plays a vital role while dealing with the patient's healthcare records. Hadoop Map-Reduce in conjunction with big data demonstrates its significance while processing healthcare records. Large-scale datasets for clinical data processing are provided as big data solutions. Data in terabytes/petabytes can be accessed or stored using big data solutions. Hadoop is an open-source software that makes the task easier and accessible. Map-Reduce task parallelization solution makes big-tasks processing faster and efficient. Hadoop Map-Reduce framework acts as a backbone due to its efficiency, cost-effectiveness, and capability to managing unstructured and large amounts of data. As Hadoop Map-Reduce can work with any programming language, it is the topic of interest for researchers as well. Many pieces of research are going on to make clinical systems more efficient using Hadoop Map-Reduce Paradigm.

REFERENCES

Binu, P. K., Akhil, V., and Mohan, V. (2017). Smart and secure IoT-based child behavior and health monitoring system using Hadoop. *2017 International Conference on Advances in Computing, Communications, and Informatics, ICACCI 2017*, 2017 January, 418–423. doi:10.1109/ICACCI.2017.8125876

Chawla, N. V., and Davis, D. A. (2013). Bringing Big Data to personalized healthcare: A patient-centered framework. *Journal of General Internal Medicine*, *28*(Suppl. 3), 660–665. doi:10.1007/s11606-013-2455-8

Chrimes, D., Kuo, M. H., Moa, B., and Hu, W. (2017). Towards a real-time Big Data analytics platform for health applications. *International Journal of Big Data Intelligence*, *4*(2), 61. doi:10.1504/ijbdi.2017.083131

Dai, L., Gao, X., Guo, Y., Xiao, J., and Zhang, Z. (2012). Bioinformatics clouds for Big Data manipulation. *Biology Direct*, *7*, 1–7. doi:10.1186/1745-6150-7-43

Dean, J., and Ghemawat, S. (2004). Map-Reduce: Simplified data processing on large clusters. *OSDI 2004–6th Symposium on Operating Systems Design and Implementation*, *51*(1), 137–149. doi:10.21276/ijre.2018.5.5.4

Dencelin, L. X., and Ramkumar, T. (2016). Analysis of multilayer perceptron machine learning approach in classifying protein secondary structures. *Biomedical Research (India)*, *2016*(Special Issue 2), S166–S173.

Dharavath, R., Nyakotey, S., and Edla, D. R. (2019). Map-Reduce based integration of health hubs: A healthcare design approach. *Health and Technology*, *9*(5), 737–750. doi:10.1007/s12553-019-00321-8

Dhavapriya, M., and Yasodha, N. (2016). Big Data analytics: Challenges and solutions using Hadoop, Map Reduce, and Big Table. *International Journal of Computer Science Trends and Technology (IJCST)*, *4*(1), 5–14. Retrieved from http://www.ijcstjournal.org/volume-4/issue-1/IJCST-V4I1P2.pdf

Din, S., and Paul, A. (2019). Erratum to "Smart health monitoring and management system: Toward autonomous wearable sensing for Internet of Things using Big Data analytics" (Future Generation Computer Systems (2019) 91 (611–619), (S0167739X17315078), (10.1016/j.future.2017.12.059)). *Future Generation Computer Systems, 91*, 611–619. doi:10.1016/j.future.2019.06.035

El Aboudi, N., and Benhlima, L. (2018). Big Data management for healthcare systems: Architecture, requirements, and implementation. *Advances in Bioinformatics, 2018*. doi:10.1155/2018/4059018

Fang, R., Pouyanfar, S., Yang, Y., Chen, S. C., and Iyengar, S. S. (2016). Computational health informatics in the Big Data age: A survey. *ACM Computing Surveys, 49*(1). doi:10.1145/2932707

Forkan, A. R. M., Khalil, I., Ibaida, A., and Tari, Z. (2015). BDCaM: Big Data for context-aware monitoring – A personalized knowledge discovery framework for assisted healthcare. *IEEE Transactions on Cloud Computing, 5*(4), 628–641. doi:10.1109/tcc.2015.2440269

Harb, H., Mroue, H., Mansour, A., Nasser, A., and Cruz, E. M. (2020). A Hadoop-based platform for patient classification and disease diagnosis in healthcare applications. *Sensors (Switzerland), 20*(7), 1–20. doi:10.3390/s20071931

Hassan, M. K., El Desouky, A. I., Elghamrawy, S. M., and Sarhan, A. M. (2019). A Hybrid Real-time remote monitoring framework with NB-WOA algorithm for patients with chronic diseases. Future Generation Computer Systems, 93, 77–95.

Ismail, A., and Shehab, A. (2019). *Security in Smart Cities: Models, Applications, and Challenges. Future Generation Computer Systems* (Vol. 9). Springer International Publishing, Cham. doi:10.1007/978-3-030-01560-2

Jaseena, K. U., and David, J. M. (2014). Issues, challenges, and solutions: Big Data mining, 131–140. doi.:10.5121/csit.2014.41311

Jokonya, O. (2014). Towards a Big Data framework for the prevention and control of HIV/AIDS, TB, and silicosis in the mining industry. *Procedia Technology, 16*, 1533–1541. doi:10.1016/j.protcy.2014.10.175

Kalyankar, G. D., Poojara, S. R., and Dharwadkar, N. V. (2017). Predictive analysis of diabetic patient data using machine learning and Hadoop. *Proceedings of the International Conference on IoT in Social, Mobile, Analytics, and Cloud, I-SMAC 2017*, (Dm), 619–624. doi.:10.1109/I-SMAC.2017.8058253

Kavitha, G. (2015). Clinical data analytics in Big Data using Hadoop. *Journal of Scientific & Engineering Research, 6*(5), 24–27.

Kim, T. W., Park, K. H., Yi, S. H., and Kim, H. C. (2014). A Big Data framework for u-Healthcare systems utilizing vital signs. *Proceedings – 2014 International Symposium on Computer, Consumer and Control, IS3C 2014*, 494–497. doi:10.1109/IS3C.2014.135

Kumar, S. and Mohbey, K. K. (2019). A review on big data based parallel and distributed approaches of pattern mining. Journal of King Saud University-Computer and Information Sciences.

Lin, W., Dou, W., Zhou, Z., and Liu, C. (2015). A cloud-based framework for home-diagnosis service over big medical data. *Journal of Systems and Software, 102*, 192–206. doi:10.1016/j.jss.2014.05.068

Lyu, D. M., Tian, Y., Wang, Y., Tong, D. Y., Yin, W. W., and Li, J. S. (2016). Design and implementation of clinical data integration and management system based on the Hadoop platform. *Proceedings – 2015 7th International Conference on Information Technology in Medicine and Education, ITME 2015*, 76–79. doi:10.1109/ITME.2015.86

Mahmud, S., Iqbal, R., and Doctor, F. (2016). Cloud-enabled data analytics and visualization framework for health-shocks prediction. *Future Generation Computer Systems, 65*, 169–181. doi:10.1016/j.future.2015.10.014

Marozzo, F., Talia, D., and Trunfio, P. (2012). P2P-Map-Reduce: Parallel data processing in dynamic Cloud environments. *Journal of Computer and System Sciences*, *78*(5), 1382–1402. doi:10.1016/j.jcss.2011.12.021

Martin-Sanchez, F., and Verspoor, K. (2014). Big Data in medicine is driving big changes. *Yearbook of Medical Informatics*, *9*, 14–20. doi:10.15265/IY-2014-0020

Mohammed, E. A., Far, B. H., and Naugler, C. (2014). Applications of the Map-Reduce programming framework to clinical Big Data analysis: Current landscape and future trends. *BioData Mining*, *7*(1). doi:10.1186/1756-0381-7-22

Nishadi, T. A. S. (2019). Healthcare Big Data analysis using Hadoop Map-Reduce. *International Journal of Scientific and Research Publications (IJSRP)*, *9*(3), 87104. doi:10.29322/ijsrp.9.03.2019.p87104

Olson, M. (2010). Hadoop: Scalable, flexible data storage, and analysis. *IQT Quarterly*, *1*(3), 14–18.

Palanisamy, V., and Thirunavukarasu, R. (2019). Implications of Big Data analytics in developing healthcare frameworks – A review. *Journal of King Saud University – Computer and Information Sciences*, *31*(4), 415–425. doi:10.1016/j.jksuci.2017.12.007

Pattuk, E., Kantarcioglu, M., Khadilkar, V., Ulusoy, H., and Mehrotra, S. 2013, June. BigSecret: A secure data management framework for key-value stores. In *2013 IEEE Sixth International Conference on Cloud Computing* (pp. 147–154). IEEE.

Pradeep, D., and Sundar, C. (2020). QAOC: Novel query analysis and ontology-based clustering for data management in Hadoop. *Future Generation Computer Systems*, *108*, 849–860. doi:10.1016/j.future.2020.03.010

Pramanik, M. I., Lau, R. Y. K., Azad, M. A. K., Hossain, M. S., Chowdhury, M. K. H., and Karmaker, B. K. (2020). Healthcare informatics and analytics in Big Data. *Expert Systems with Applications*, *152*, 113388. doi:10.1016/j.eswa.2020.113388

Pramanik, M. I., Lau, R. Y. K., Demirkan, H., and Azad, M. A. K. (2017). Smart health: Big Data enabled health paradigm within smart cities. *Expert Systems with Applications*, *87*, 370–383. doi:10.1016/j.eswa.2017.06.027

Raghupathi, W., and Raghupathi, V. (2014). Big Data analytics in healthcare: Promise and potential. *Health Information Science and Systems*, *2*(1), 1–10. doi:10.1186/2047-2501-2-3

Sakr, S., and Elgammal, A. (2016). Towards a comprehensive data analytics framework for smart healthcare services. *Big Data Research*, 4, 44–58.

Santhosh, K. V., and Nayak, S. (2019). *Engineering Vibration, Communication, and Information Processing* (Vol. 478). Springer, Singapore. doi:10.1007/978-981-13-1642-5

Santur, Y., Santur, S. G., and Karaköse, M. (2020). Architecture and implementation of a smart-pregnancy monitoring system using web-based application. *Expert Systems*, *37*(1), e12379.

Schadt, E. E. (2012). The changing privacy landscape in the era of Big Data. *Molecular Systems Biology*, *8*(612), 1–3. doi:10.1038/msb.2012.47

Sebaa, A., Chikh, F., Nouicer, A., and Tari, A. K. (2018). Medical Big Data warehouse: Architecture and system design, a case study: Improving healthcare resources distribution. *Journal of Medical Systems*, *42*(4), 1–16. doi:10.1007/s10916-018-0894-9

Shaikh, T. A., and Ali, R. (2019). Big Data for better Indian healthcare. *International Journal of Information Technology*, *11*(4), 735–741. doi:10.1007/s41870-019-00342-6

Sivaparthipan, C. B., Karthikeyan, N., and Karthik, S. (2020). Designing statistical assessment healthcare information system for diabetics analysis using Big Data. *Multimedia Tools and Applications*, *79*(13–14), 8431–8444. doi:10.1007/s11042-018-6648-3

Youssef, A. E. (2014). A framework for secure healthcare systems based on Big Data analytics in mobile cloud computing environments. *The International Journal of Ambient Systems and Applications*, *2*(2), 1–11. doi:10.5121/ijasa.2014.2201

Yuvaraj, N., and SriPreethaa, K. R. (2019). Diabetes prediction in healthcare systems using machine learning algorithms on Hadoop cluster. *Cluster Computing, 22*. doi:10.1007/s10586-017-1532-x

WEB REFERENCES

https://data-flair.training/blogs/hadoop-Map-Reduce-tutorial/
https://mapr.com/products/apache-hadoop/
https://opensource.com/life/14/8/intro-apache-hadoop-big-data
https://www.oracle.com/big-data/guide/what-is-big-data.html
https://www.sas.com/en_in/insights/big-data/hadoop.html
https://www.sas.com/en_in/insights/big-data/what-is-big-data.html

10 An Investigation of Different Machine Learning Approaches for Healthcare Analytics

Kayal Padmanandam
BVRITH College of Engineering for Women

CONTENTS

DOI: 10.1201/9781003050827-10

There are infinite opportunities for machine learning (ML) in healthcare analytics, and here we will discuss a few shown in Figure 10.1.

10.1 MACHINE LEARNING FOR AILMENT IDENTIFICATION AND DIAGNOSIS USING ML MODELS

Diagnosis is a process of identifying the ailment the person has, using differential diagnostics. Diagnosis is a challenge for a medical practitioner because the symptom can be nonspecific, since the same symptom can be a sign of many disorders. So, differential diagnostics play a vital role in accurate ailment identification by performing multiple tests that are compared, correlated, and finally classified. This process can be improved with maximum prediction accuracy using various ML techniques. The diagnosis involves a procedure. The procedure has multiple processes. The first and foremost is a physical examination of the patient, getting to know about the patient's medical history. At this point, the practitioner would set a hypothesis for the diagnosis and start to perform differential diagnosis tests. This test will help the practitioner to eliminate the candidate ailments that have the same sign of symptoms but are not associated with the patient's illness, hence approaching the ailment identification. From this point, a practitioner can very well diagnose the ailment using the theory of

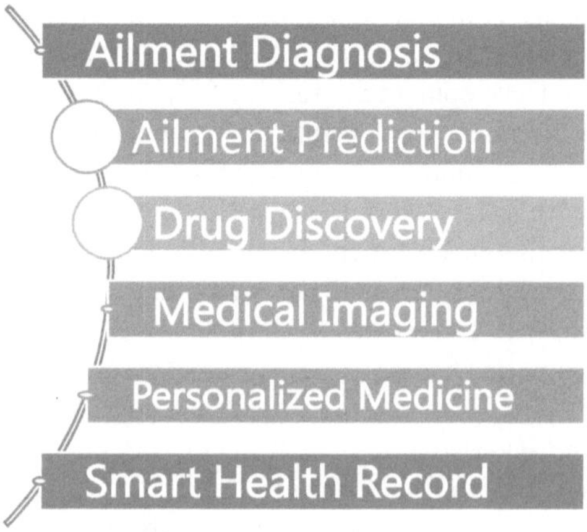

FIGURE 10.1 Machine learning opportunities in healthcare.

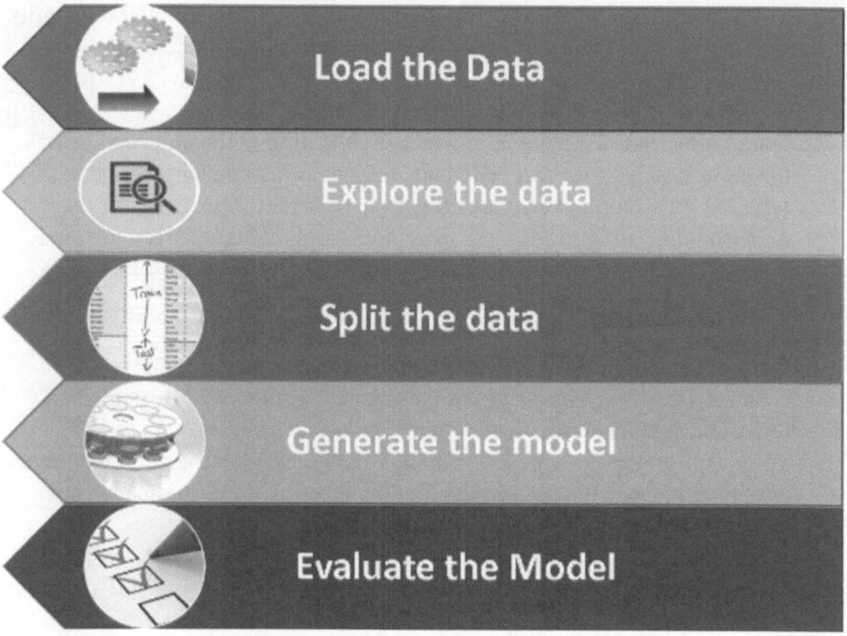

FIGURE 10.2 Machine learning model.

exclusion. If any disruption happens in this process, the practitioner would rule out the initial hypothesis and set a new one. This is called the diagnostic delay, which is common for unknown cause diseases or asymptomatic diseases.

Now we will see how a ML algorithm is helpful for a clinician. ML is a subset of artificial intelligence (AI), making machine intelligent. Intelligence requires learning. The machines are made to learn (training) and then predict (testing). The learning phase is called the training phase where a dataset is provided to the model. About 70% of the dataset will be used for learning. The algorithm learns the data to the maximum. Once the learning is done, the model enters the testing phase with the remaining 30% of the data. The prediction of each case by the model on every unseen data is compared with the actual case. This is will be analyzed using various measures and finally articulated as model accuracy. As the machines will learn from the data that is given, it has to be reliable and error-free data. The accuracy of the diagnosis (prediction) is directly dependent on the training data. Setting up a ML model follows various steps, which are explained in Figure 10.2.

10.1.1 Data Loading

ML algorithms necessitate data to be formatted in a very precise way. Loading data to the ML algorithm follows two preliminary steps, Data Collection and Data Preparation. Together they can be called the data preprocessing technique, which plays a vital role in the algorithmic prediction process. The quality and quantity of the data will dictate the accuracy of the proposed ML prediction/diagnosis.

10.1.1.1 Data Preprocessing

a. **Data collection**: This is the process of collecting the health records of the patients. This data will be in enormous quantity and with varied data types as we would have collected it from different sources. Also, these data would be structured, semi-structured, or unstructured. So, the data at hand is raw and cannot be directly used for any analysis. These data should be processed and made decipherable so that useful insights can be retrieved.
b. **Data preparation**: Raw Data will be in wrangled form which should be cleaned. The data cleaning challenges include dealing with missing data, removing duplicate data, dealing with inconsistent values, data type conversions, pruning outliers or anomalies, data normalization, and feature enrichment. Once the data is prepared by meeting these challenges, it is now ready for data loading into the ML algorithm.

10.2 EXPLORE YOUR DATA

Building a ML model is preceded by many imperative steps. After data preprocessing, the user should understand the data that is loaded for an ML model. This step ensures the birth of a better model in terms of accuracy. It is achieved to make overall observations about the data, abridge it, discover basic trends, or uncover hidden relations between attributes. We can explore our data with metrics like the number of samples used, the number of independent variables (attributes), number of dependent variables (target), number of categories in the dependent

FIGURE 10.3 Data cleaning.

FIGURE 10.4 Data loading.

	age	sex	cp	trestbps	chol	fbs	restecg	thalach	exang	oldpeak	slope	ca	thal	target
0	63	1	3	145	233	1	0	150	0	2.3	0	0	1	1
1	37	1	2	130	250	0	1	187	0	3.5	0	0	2	1
2	41	0	1	130	204	0	0	172	0	1.4	2	0	2	1
3	56	1	1	120	236	0	1	178	0	0.8	2	0	2	1
4	57	0	0	120	354	0	1	163	1	0.6	2	0	2	1
...	...	...	...	...	...	...	...	...	...	...	...	...	...	...
298	57	0	0	140	241	0	1	123	1	0.2	1	0	3	0
299	45	1	3	110	264	0	1	132	0	1.2	1	0	3	0
300	68	1	0	144	193	1	1	141	0	3.4	1	2	3	0
301	57	1	0	130	131	0	1	115	1	1.2	1	1	3	0
302	57	0	1	130	236	0	0	174	0	0.0	1	1	2	0

303 rows × 14 columns

FIGURE 10.5 Heart_disease dataset from UCI repository.

variables, the frequency distribution of the samples across the target variables, mean, median, etc. This Exploratory Data Analysis (EDA) familiarizes us with the new dataset.

For instance, the dataset and the summary of the dataset (https://archive.ics.uci.edu/ml/datasets/Heart+Disease) heart_disease are shown in Figures 10.5 and 10.6, respectively. Here there are various independent variables (age, gender, etc.,) and one target variable (has_heart_disease) present. There are 303 records, in which 97 are male and 206 are female and the dataset also says 164 patients do

```
summary(heart_disease)
     age            gender    chest_pain resting_blood_pressure serum_cholestoral fasting_blood_sugar resting_electro
Min.   :29.00   female: 97   1: 23      Min.   : 94.0          Min.   :126.0     0:258               0:151
1st Qu.:48.00   male  :206   2: 50      1st Qu.:120.0          1st Qu.:211.0     1: 45               1:  4
Median :56.00                3: 86      Median :130.0          Median :241.0                         2:148
Mean   :54.44                4:144      Mean   :131.7          Mean   :246.7
3rd Qu.:61.00                           3rd Qu.:140.0          3rd Qu.:275.0
Max.   :77.00                           Max.   :200.0          Max.   :564.0

max_heart_rate   exer_angina        oldpeak         slope       num_vessels_flour   thal     heart_disease_severity exter_angina
Min.   : 71.0   Min.   :0.0000   Min.   :0.00   Min.   :1.000   Min.   :0.0000   3   :166   Min.   :0.0000          0:204
1st Qu.:133.5   1st Qu.:0.0000   1st Qu.:0.00   1st Qu.:1.000   1st Qu.:0.0000   6   : 18   1st Qu.:0.0000          1: 99
Median :153.0   Median :0.0000   Median :0.80   Median :2.000   Median :0.0000   7   :117   Median :0.0000
Mean   :149.6   Mean   :0.3267   Mean   :1.04   Mean   :1.601   Mean   :0.6722   NA's:  2   Mean   :0.9373
3rd Qu.:166.0   3rd Qu.:1.0000   3rd Qu.:1.60   3rd Qu.:2.000   3rd Qu.:1.0000              3rd Qu.:2.0000
Max.   :202.0   Max.   :1.0000   Max.   :6.20   Max.   :3.000   Max.   :3.0000              Max.   :4.0000
                                                                NA's   :4

has_heart_disease
no :164
yes:139
```

FIGURE 10.6 EDA of Heart_disease dataset from UCI repository (only 14 variables used out of 75).

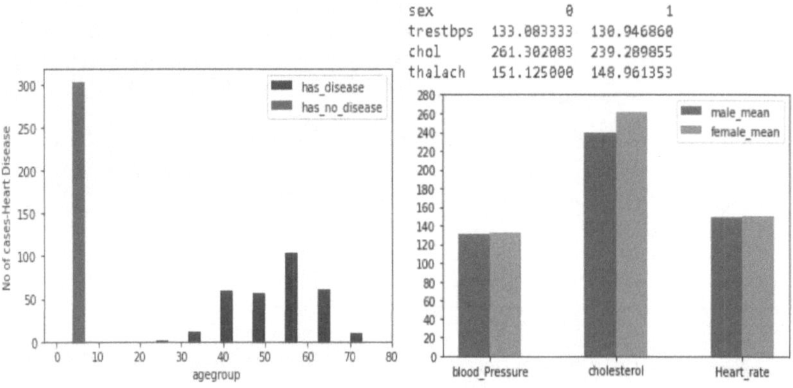

(a) (b)

FIGURE 10.7 (a and b) Visualization of EDA of Heart_disease dataset using Python matplotlib.

not have heart disease(target-0) and 139 patients have heart disease (target 1). There are exclusive data exploratory visualization tools that make the process more data-driven. These visualized insights give the utmost understanding of our data and enable targeted actions. Such amalgamation of data and targets plays a vital role in building a robust ML model. Data visualization of the same dataset is shown in Figure 10.7.

10.3 SPLITTING DATA

For any ML model, data is the source of the learning process. When we have a pre-processed data ready for algorithmic learning, the next question is about the split of data. Any ML algorithm should go through two phases called training and testing. During the training phases, the model understands the training data or labeled (with target) instances, and during the testing phase, it works with testing data or unlabeled (without target) instances. The aim of any ML model is to generalize data for the new instances beyond the training samples used in the learning process. But the new instances are unknown, and accuracy cannot be tested using it. So, we need to use some of the existing data as unknown /testing data (by removing the target) instances. When this is possible, the accuracy of the model can be evaluated by inspecting the prediction of (testing) data instances across the actual target values (also called ground-truth value) of the data instances used. This is the reason for the data split in ML.

Generalization refers to the extent of the ML model's ability to adapt the learning in the training phase, and how well it applies them to new, previously unseen data accurately. It is a kind of inductive learning that is executed by the model.

10.3.1 Efficient Split

The efficient split of training and testing data again plays a major role in the ML model building. The ratio of data used for training and testing should be considered with utmost care. A common approach is to take all labeled data and split them into training and testing subsets with the ratio of 7:3 or 8:2, respectively. If the ratio is not used appropriately, the model may suffer from overfitting or underfitting glitches. Always it is good to ensure that the test set meets two important conditions. First, it should be large enough to yield statistical insights and, secondly, the test set must represent the dataset as a whole.

Overfitting: When a model models the training data extremely well, the problem of overfitting comes in. Because the model has learned the details as well as the noise in the training data to an extent, it negatively impacts the prediction accuracy of the model on new samples. Overfitting befalls if the model exhibits low bias but high variance. Overfitting often occurs in an excessively intricate model but can be thwarted by placing multiple models and comparing their predictive accuracies on test data, using cross-validation techniques.

Underfitting: When a model, models the training data extremely poorly, the problem of underfitting comes in. Because the model has not learned the underlying trends of the training data, it again impacts the prediction accuracy of the model on new samples. Underfitting befalls if the model exhibits low variance but high bias. Underfitting is an outcome of an extremely naive model. The overfitting and underfitting are shown in Figure 10.8.

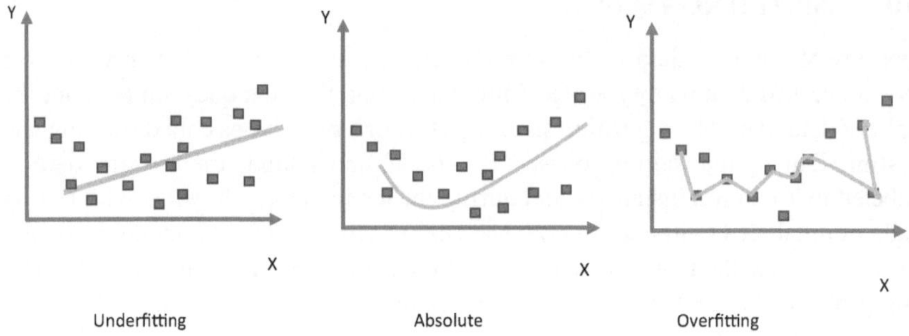

FIGURE 10.8 Example of overfitting Vs underfitting.

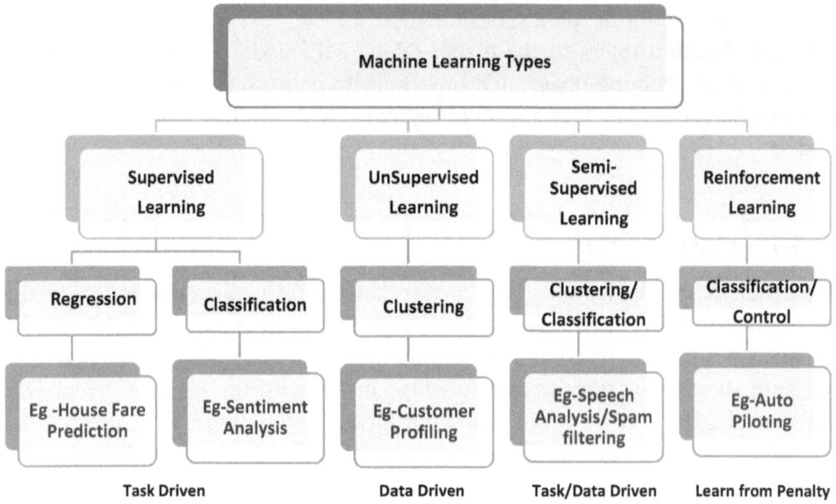

FIGURE 10.9 Types of machine learning.

10.4 GENERATE THE MODEL

ML is classified into four types – Supervised Learning, Unsupervised Learning, Semi-supervised Learning, and Reinforcement Learning – as shown in Figure 10.9. Each category plays a vital role in different scenarios of biomedical research. It has many algorithms based on various tasks. For instance, we have tasks like classification, clustering, regression, dimensionality reduction, and many more. Each of these tasks has many algorithms revolving around it.

10.4.1 Supervised Learning

Supervised learning is a task-driven approach where the algorithm learns from the dataset where all the items are labeled (Supervisor). There is a need to split the data into a training set and testing set. The algorithm is trained with a training set and learns all the possible insights from the data. For instance, many cat faces are given as a dataset, the algorithm can learn them. Now the algorithm is ready to predict an unseen face and can classify (label) it as "cat" or "not cat". The test dataset is unseen data for the algorithm, so the prediction can be made for all the test data. As it is supervised learning, its accuracy can be compared against the actual labels of the test data and the predicted label of the algorithm. This comparison gives the accuracy of the classifier algorithm. This type of learning can answer problems like classification and regression.

10.4.2 Unsupervised Learning

In unsupervised learning, the dataset is unlabeled. The algorithm does not have a supervisor to check its accuracy; rather, it has to understand the data and learn maximum insights and patterns on its own, hence called the data-driven approach. The training dataset does not have any outcome (label) variable. For instance, a dataset that has many animal faces that are not labeled (for instance, cats-1 and dogs-2). The algorithm tries to identify the pattern and extract the features of all the images and decides whether they are similar or not. With this insight, the algorithm will group (cluster) them as group 1 and group 2. These clusters can be named (cluster profiling) meaningfully related to the input variable. Unsupervised algorithms can solve problems like clustering, dimensionality reduction, anomaly detection, etc.

10.4.3 Semi-Supervised Learning

It is a mixture of supervised and unsupervised algorithms. It takes a minimum of labeled data and uses the insight on the maximum of unlabeled data. This algorithm partially learns from the labeled dataset and can customize the construction of categories. This algorithm can be used, where acquiring the fully labeled dataset is complex.

10.4.4 Reinforcement Learning

Reinforcement learning trains algorithms based on rewards and penalty systems. There is an AI agent in the environment that will learn the accurate move for the next step. If the agents move correctly, it is rewarded, or else it is penalized. The algorithm will learn the optimal way for future moves. It learns from feedback and it is iterative, so more the feedback, better the learning of the AI agent. This strategy is used in robotics, gaming, and autonomous system building.

The first and foremost job before generating a model is to decide the task and choose the right algorithm. When this step is done, the next is to train the model. The aim of training a model is to make predictions on unseen data accurately. Let us consider a classification algorithm for the dataset (shown in Figure 10.5) discussed already. The challenge here is choosing the right algorithm for the

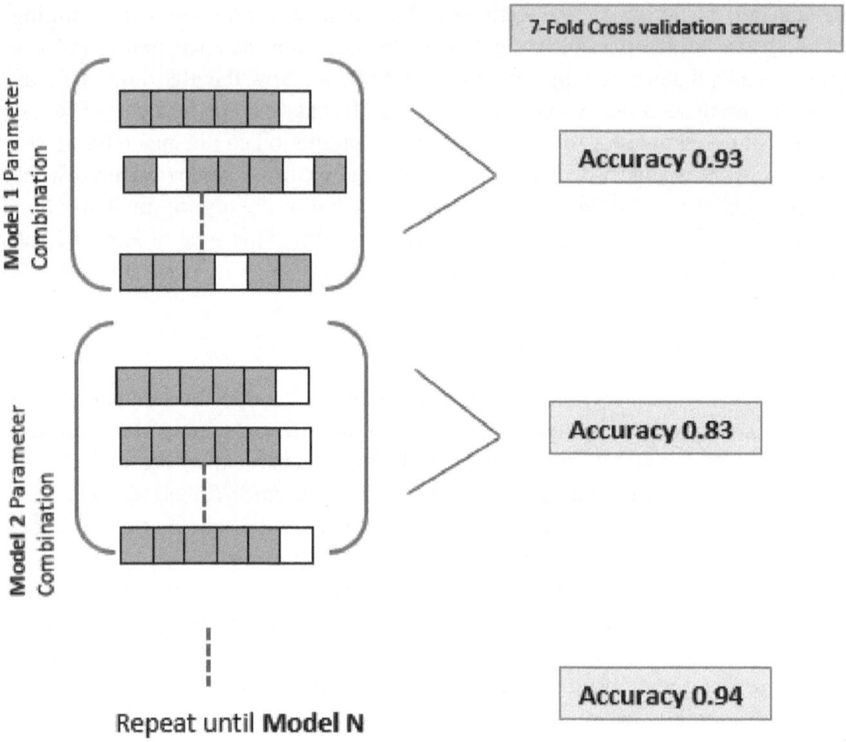

FIGURE 10.10 Hyperparameter optimization process.

right data, also called "choosing the right estimator". Over experience, this is possible by a ML expert on looking into the data and knowing the stakeholder's requirement. Different estimators fit different models based on the data and problems as the saying says "One size does not fil all". Python scikit-learn [1] library has so many algorithms for developers approaching a problem based on the type and amount of dataset. But it is always advised to run the data on multiple similar algorithms and compare the accuracy results of those algorithms to have a flawless predictor model. To improve the performance of an algorithm, we have another technique called hyperparameter optimization. "Before the learning process begins, the hyperparameter values are initialized". Every ML algorithm has a wide variety of parameters that can control the learning process. Tuning these parameters [3] depending on the dataset can increase the performance of the model built. The process of identifying the algorithm's best set of parameters is called hyperparameter optimization, shown in Figure 10.10.

10.5 EVALUATE MODEL

Model evaluation of an algorithm is a performance measurement technique [4] that is an integral part of any ML project. It always aims at the accuracy of the

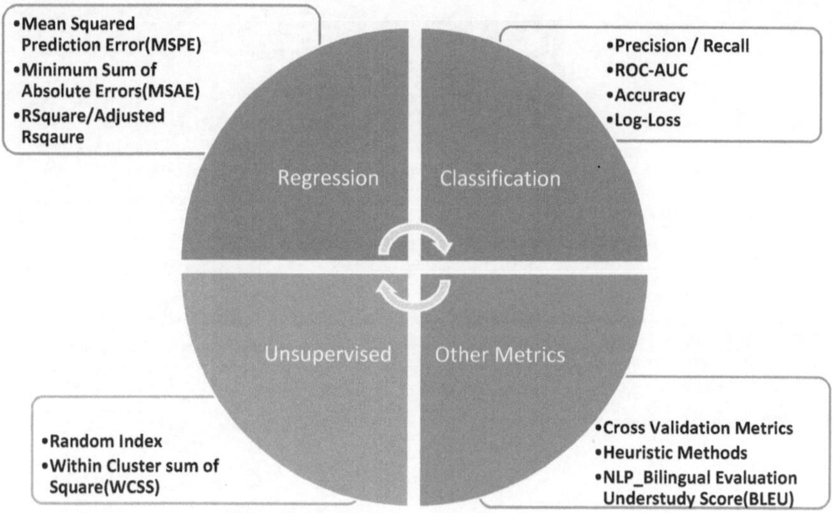

FIGURE 10.11 Task-based evaluation metrics.

generalization of the model on unseen data. We have methods like holdout and cross-validation and metrics like classification accuracy, confusion matrix, area under the curve (AUC), mean absolute error, and many more depending on the learning task, shown in Figure 10.11. It is always good to remember that at each stage of ML the amount of clarity and accuracy spent towards building the model will play a big role in improving the model's efficiency. This can be observed as accuracy in the evaluation period.

10.5.1 Drug Discovery

The drug discovery and development cycle is a very long complex process. It has multiple phases (0–4) (shown in Figure 10.12) that it take decades to be an innate drug. Here we will look into the multiple challenging phases in drug development and how ML plays a vital role in improving the discovery and minimizing the development time taken. ML can be applied to all the stages of drug development with clear insight and high-quality data [5]. A diversified dataset that may include a variety of data from chemical, transcriptomic, or genomic level will be used. Transforming such biological data into an interpretable model is a challenge, but with ML approaches this has been efficiently used in drug development [6].

10.5.2 Drug Development Phases

Phase 0: The basic research behind drug development concentrates on preclinical challenges. For instance, understanding the accountable targets (target identification) for the disease like DNA mutations, misfolded proteins, and miscellaneous biomarkers. The primary footstep is to establish an association between the target

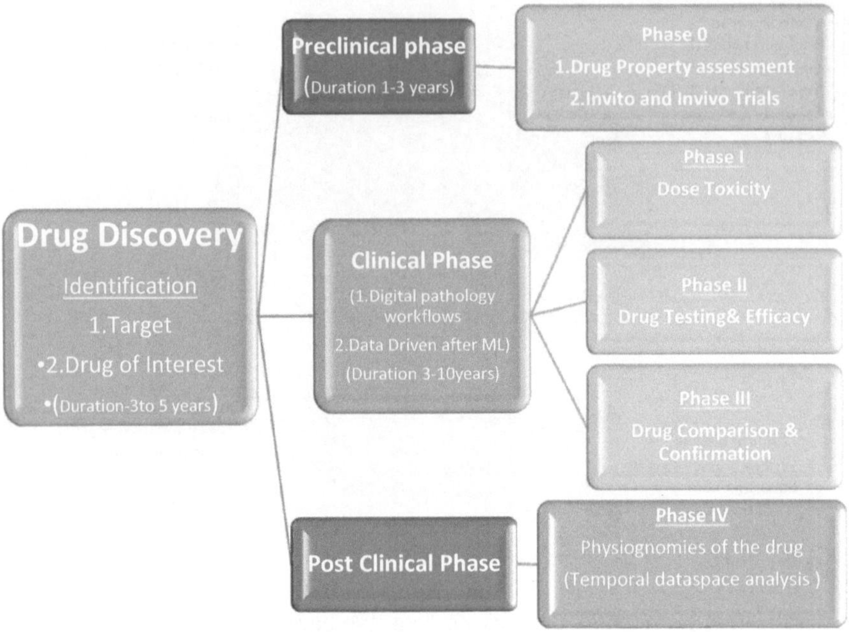

FIGURE 10.12 Drug development phases.

and disease. This is a highly challenging and risky process as it screens thousands of chemical compounds that interfere with target variables. ML plays a magnificent role in the automation and analysis of high-dimensional, noisy biological data, and its attributes association.

The goal of this process is to suggestively narrow down the number of target components (attributes) considered and to select the ones that are directly responsible for the disease. This is a longer phase where trial and error take place and is prone to development failure. Usually, the failure rate is higher than the success rate. So, the development may go on regressive plans. These target selection and target validation can be applied with the ML feature selection process that selects relevant attributes (features), tests them sequentially, and rates their usefulness towards the target variable. Once done, it can be optimized using hyperparameter tuning. This will effectively reduce the amount of time, cost, and failure rate experienced in the target identification and validation process. Likewise, ML plays multiple roles in the development of drugs at all stages.

Phases 1–3: The clinical research period concentrates on the dose toxicity, dose side effects, the performance of the drug, pre- and post-comparison of the molecule, and their interaction with disease targets. Though it is all clinically done, there is a lot of data space to be collected, analyzed, and interpreted. This path has now been made data-driven after the invention of ML as it does the job efficiently in very little time. Machine learning can be used in computational pathology by developing digital pathology workflows. It is recognized as a strong candidate for drug diagnosis and biomarker analytics [7].

The application of Artificial Neural Network (ANN), Deep Neural Network (DNN), Convolutional Neural Network (CNN), and Recurrent Neural Network (RNN) in the biological computational process of drug discovery benefits the cost and time spent on introducing the drugs to the market. Technically speaking, there are numerous ML libraries like PyTorch, Keras, and scikit that learn to execute these complex computations in a fraction of time. The most renowned one is the TensorFlow library by Google for differential programming across ML applications.

There is a need for the high-end requirement for hardware too, as the size and complexity of these data are exponentially high. A wide variety of platforms like Graphical Processing Unit (GPU) and Tensor Processing Unit (TPU) are now available on the cloud for free of cost. GPU is a dedicated electronic circuit premeditated to swiftly manipulate and adjust memory to accelerate image data generated very commonly in bioresearch using high-throughput scanning devices. TPU is an "AI accelerator application-specific integrated circuit" (ASIC) designed by Google used for neural network ML tasks, particularly for the TensorFlow library.

Phase 4: Post-drug research is motivated by promotional deliberations, conducted after the drug has been approved to find its toxicity. This phase is also called post-drug research to analyze the physiognomies of the drug regarding safety, effectiveness, drug interaction with various combinations, treatments, and side effects. This phase takes a lot of temporal dataspace analysis of the drug on humans after periodic consumption. Efficient ML techniques like correlation and regression analysis will help analyze the drug reaction with various drug combinations.

The use of AI and ML will drastically benefit the modeling and prediction of biological processes and their effects involved in drug discovery in quite less time. It plays a key part in discovering an accurate cure for incurable and unknown cause diseases.

Medical Imaging is the process of visualizing and digitizing images of various parts of the human body system. These digitized images are used for the diagnosis and treatment of any disease. These digitized images are commonly called radiological images. Some of the techniques used are X-ray, Magnetic Resonance Imaging (MRI), ultrasonography or ultrasound, endoscopy, and many more.

The medical imaging technique is considered as one of the most complex techniques, yet it is a powerful source of information about the patients' health. Image patterns can be observed and can be compared with normalcy for any diagnostic decision. But, identifying patterns is a challenge for human beings as the image has so many features and details. This is the arena where ML helps. ML helps pathologists to make rapid and more accurate diagnoses.

As discussed in Section 10.1, the ML strategy is to make predictions and support the decision-making system. With a balanced dataset and enhanced training, ML can learn well and the model can be generalized. This strategy can be applied

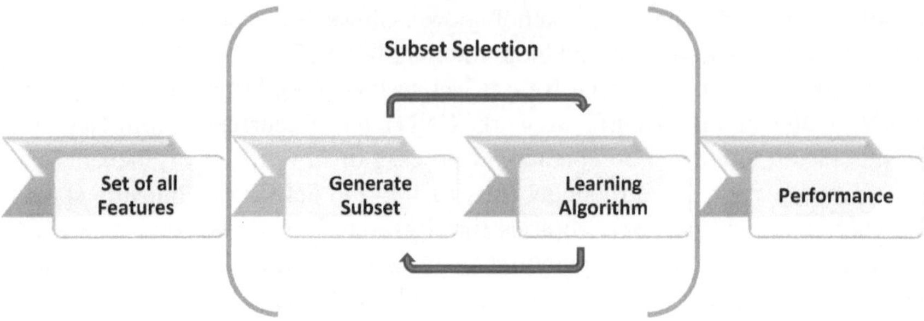

FIGURE 10.13 Feature selection cycle.

in medical imaging too to identify/recognize patterns in the image. As there will be a huge number of features in the image, identifying the relevant or right features is called feature extraction, which is a significant part of medical imaging.

It is also called a dimensionality reduction technique which will identify only the relevant profile details of a pattern through procedures called ML algorithm.

Now, the question is how to compute the relevancy? A feature that is "believed to be of importance" in the diagnostic process is called relevancy. An image can have an n-number of features. Selecting all the features will lead to overfitting and selecting a few will lead to underfitting as discussed in Section 10.3.1. Finding the subset of relevant features and their combinational relevance will help the system to classify the image or analyze metric for the image Region of Interest (ROI) is called feature selection. Though it is a very powerful technique, it can be hazardous sometimes as extracting features is complex without knowing their interdependency fraction. For instance, the features' height and weight have a high correlation to decide on a class label obese, and isolating them will lead to inaccurate prediction [8]. Once the subset is decided, it is learned with the ML model. Depending on the accuracy, the feature selection process will be continued as shown in Figure 10.13. The accuracy also depends on how well it is harnessed with the ML application.

10.5.3 Machine Learning Technique for Image Classification

Imagine, we have to classify cells as malignant (cancerous) and benign (non-cancerous). A practitioner after analyzing the internal risk factors (few listed) age, gender, genetic risk factors, menstrual cycle, dense tissue, obesity, and external factors (few listed) alcohols, taking hormonal therapies, birth control measures [9] will further proceed with a technical prediction support system where many algorithms are in existence for the prediction like k-nearest neighbors, SVM, decision trees, neural network, and deep neural networks. Feature selection applies weight (importance) to each attribute based on the level of importance. This weight application is very sensitive and vital. It is executed by assumption, made on the data, and weight adjustment based on the accuracy level obtained. Here we will discuss how Support Vector Machine (SVM) and Deep Learning algorithms undertake medical imaging.

10.5.3.1 Support Vector Machine (SVM)

The SVM is a basic yet powerful algorithm that plays a major role in biomedical image analysis. It works with the principle of separating feature space into two by drawing a hyperplane, which helps classification and prediction problems.

It is a supervised learning algorithm where it has the training and testing phases as discussed in Section 10.4.1. Here, we plot each data item (support vector) as a point in n-dimensional space, where n is the number of features you have, with the value of each feature being the value of a particular coordinate. Then, we perform classification by finding the hyperplane (machine) that best differentiates the two classes. It has many rules of thumb to differentiate the data into class labels.

The extremely complex high-dimensional image data when trained with SVM can learn the image features, pattern in the image based on labels, etc. It can easily classify separable data. But a real-time imaging data will be mostly inseparable. The support vector gained its fame because of the transformation it could do with inseparable data. It can convert data in a two-dimensional plane into an n-dimensional plane by applying an objective function and classifying the inseparable data of labels. This is made possible by an intelligent kernel trick. Kernel trick has different ways (linear, polynomial, radial, sigmoid) depending on the complexity of the data.

In medical imaging, for instance, data in original space will be as collected (CT attenuation) and is transformed to a higher dimensional space where it is represented with newly constructed co-ordinates (cosine of CT attenuation) followed by constructing a hyperplane that can separate data into labeled classes [8]. Different vector machine formulation and improved techniques have been developed to establish a SVM which can unravel classification/prediction problem.

10.5.4 The Neural Network to Deep Learning

ML Neural Networks is a process where the computer becomes capable of performing a task by learning called "training" and, for example, called "training set". The image recognition system is fed with numerous images labeled as different objects like houses, trees, cars, etc. In our case, the medical images are fed which are already classified as class labels. These visual data are learned by the system and the algorithm becomes capable to identify a pattern in the image. For instance, 100 images of cancerous cells (malignant) and 100 images of non-cancerous (benign) cells are fed as data. The system learns these data by penetrating feature correlation, the significance of features, the pattern of being malignant and benign labeled, and many more until it could be generalized. The neural network works like a human brain where there are millions of nodes interconnected and layered as an input layer, hidden layer, and output layer. Each layer has a set of nodes and they work in a "feedforward" fashion. The output of layerl (L1) is fed as input for the next layer (L2). Each node in a layer will have incoming and outgoing nodes to receive and send information. In this interconnected network, each incoming node receives data and a weight (feature importance) from its intra-connections. The sum of the weighted inputs (sum of the product of data and weight from each connection) is compared against a threshold; on success, it will proceed to send data to the outgoing layers. Random values are set for threshold and weight and the training is continued until the classification error is reduced to the minimum. Increasing the number of hidden layers in the network increases the learning as "more the depth more the learning is". It should be supported by hardware too which has led to the rise of the GPU whose architecture is similar to neural networks. GPU is capable of having millions of processing units (nodes) on a single chip which can have *n*-number of layers where *n* was 1, 2, 3 initially but with GPU it can be 10, 20, 50, and even more [10]. This increased depth of learning is called Deep Learning, which makes complex image recognition, feature selection, and feature selection based on ROI possible, and thus plays a vital role in medical imaging too.

10.6 MACHINE LEARNING IN TREATMENT STUDIES

As we have discussed ML plays a big role in diagnosis, it is no wonder it can be extended for the treatment of the diagnosed disease. There are multiple diseases where the support of ML is necessary for automated treatment.

10.7 MACHINE LEARNING FOR PERSONALIZED MEDICINE

Biologists throughout the globe are actively involved in the research of personalized medicine with the advent of ML. Personalized medicine is attractive to many as it is giving customized medical treatment to individual patients based on their identified features like genetic pattern, inheritance, and lifestyle. It again proves the theory "one size does not fit all". Researchers believe that AI and ML can bang the cipher on personalized medicine. With exponentially growing medical data, characterizing and categorizing the patient based on their health response, personalized medicine is looked at as a viable alternative to conventional treatments. Research says it is now used across various complex diseases like cancer, cardiac diseases, diabetes, rheumatology, psychotic disorders, and many more that are highly dependent on genetic patterns, inheritance, and lifestyle [11].

The ML algorithm can learn and predict. The increased data on the patients' medical record and response across various attributes to a disease can be periodically monitored and given as the training set to the learning algorithm. These data are enormous in size and varied in the structure where a human doctor alone cannot correlate or predict the outcome. The ML algorithm is capable to uncover the patterns and correlation between medical attributes after learning the model accurately. This system not only finds a solution but updates the treatment strategy applied for varied patients. This enhances further learning of the system with an increased dataset as a dataset is a fuel for ML. The process involved in personalized medicine is shown in Figure 10.14.

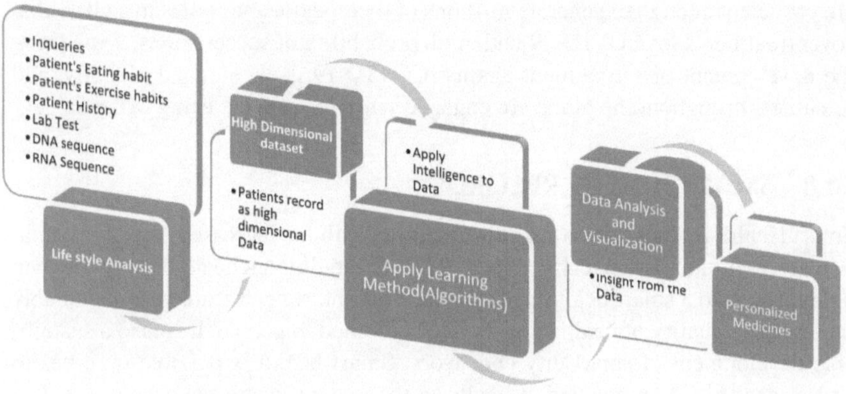

FIGURE 10.14 Processes of personalized medicine.

10.7.1 Radiation -Oncology

It is used in radiation treatment planning, says a case point in RayStation, which is an enhanced treatment planning system with "machine learning, multi-criteria optimization, ultrafast computation" for radiation therapy developed by the RaySearch laboratories. They develop much software with ML for cancer treatment in collaboration with global leading cancer centers. It is also said that, for better treatment of cancer, oncology software has to be extremely fast, easy, and agile [12].

10.7.2 Rheumatic Disease

Rheumatic disease affects the human body's joints, ligaments, muscles, and bones. Some kinds of rheumatic disease have no treatment yet. But now the field of rheumatology is seeing massive growth in the findings and interpretation of the immunological process, which paves the way for target treatments. A defined treatment is not applicable for this disease as it is heterogeneous in nature which gets differentiated in disease progression, response to treatment, concomitant development on patients. The emergence of AI and ML will be a trendsetter in the treatment of rheumatic disease. Research also says that there is a vast increase in rheumatology studies after the advent of AI and ML.

10.7.3 COVID-19

COVID-19, the threat of today's world, has no cure yet. Researchers around the globe are actively involved in finding drugs as a cure for COVID-19. Usually, the drug process takes decades to evolve, but ML algorithms can significantly reduce the time and cost as this process drastically reduces the failure rate in the drug production process. With this poise, researchers are utilizing ML algorithms that are proficient enough to generate millions of therapeutic antibodies to swiftly discover treatments for COVID-19 with high probability of success rates. Expediting the development of a treatment against COVID-19 is encouraged globally, and scientists throughout the globe are engaged with this process using ML support.

10.8 SMART HEALTH RECORDS

Smart Health is a new direction of healthcare. With the increased use of information and communication technologies (ICT), the traditional healthcare system can be improved to a smart healthcare system. This smart healthcare will undeniably increase the quality of healthcare services provided to people. It is also a sustainable development of smart city endeavors. Smart health is possible by e-health and m-health system enabled. e-health is the system where the usage of ICT is used for any kind of health procedure in terms of early diagnosis, accurate treatment, extensive care for the patients, educating the health workers, keeps track of diseases, monitoring public health and a vital role in medical researches too [13]. The m-health is the medical health, supported by mobile gadgets like a mobile phone or patient monitoring device supported by 4G telecommunication services for instant updates and action required for the patient under care.

10.8.1 Smart Health Data Process

With the convergence of IoT, Big Data, and ML techniques, the healthcare society is experiencing a revolution. Smart Health is for the people by the process through data and things. For any smart health service to be intact, there is a pipeline of a process that comprises the following.

i. Data acquisition
ii. Data exploration
iii. Diagnosis/prediction
iv. Decision-making/recommendation.

IoT and big data play a major role in data acquisition. With the increased population and growing health hazards among people, data acquisition becomes a big challenge. Because health monitoring gadgets are generating data that are exponentially growing. To manage and explore such huge real-time data, the support of IoT and Big data is essential. Diagnosis and decision-making are the phases where ML plays its role.

10.8.2 Deep Learning Algorithms in Healthcare

The advancement in computer algorithms has moved to DL (Deep Learning), a part of the ML arena. As the healthcare data are complex in structure, DL will suit many of the decisive applications. Important DL techniques like DNN, ANN, and CNN will highly be used in health predictive analysis.

10.8.3 Artificial Neural Network

ANNs are systems that imitate the human brain with multilayered interconnected nodes. These systems will learn to perform tasks by experience without any explicit programming requirements. Unlike supervised learning, they generate distinct characteristics differences to classify the dataset they are processing.

An ANN with nodes (neurons) and connections (edges) is shown in Figure 10.15. Each neuron will receive signals from the edges and the output is processed using nonlinear object functions. The strength of the signal is decided by the associated weight that gets adjusted throughout the learning as a measure of accurate tuning. Signal travels through the input layer, hidden layer, and the output layers that performs diverse transformation in the learning process.

10.8.4 Deep Neural Network

A DNN [14] is an extensive ANN with several layers of hidden layer between the input and output layer. It can be used to model intricate associations. It is a variant model with rudimentary ANN approaches. It exactly finds the logical operation to turn input into an output. They are called feedforward networks as data flow in one direction from the input layer to the output layer. It constructs a map of visualizable neurons with random weights, where the product is returned as output in the range from 0 to 1. For example, the original input image is transformed to a higher level of representation like edges, shapes, parts, and categories of objects. These are highly used in the image identification and feature selection process.

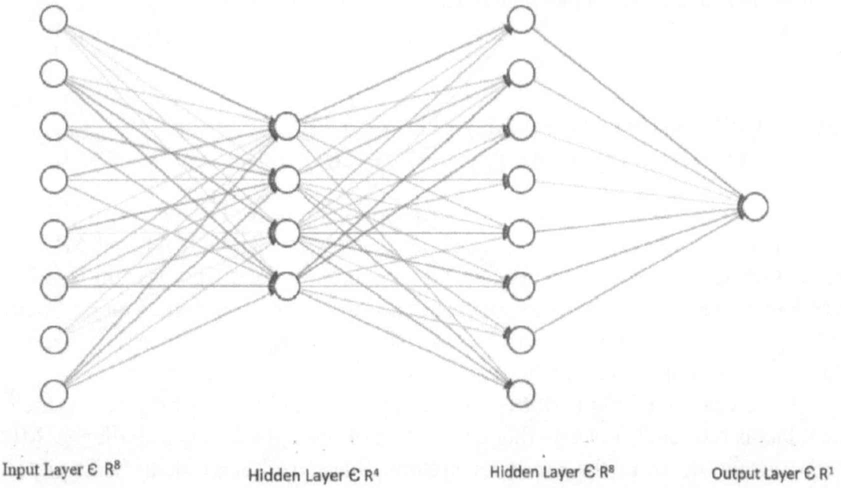

FIGURE 10.15 Structure of ANN (This figure generated using the tool NN-SVG Github (http://alexlenail.me/NN-SVG)).

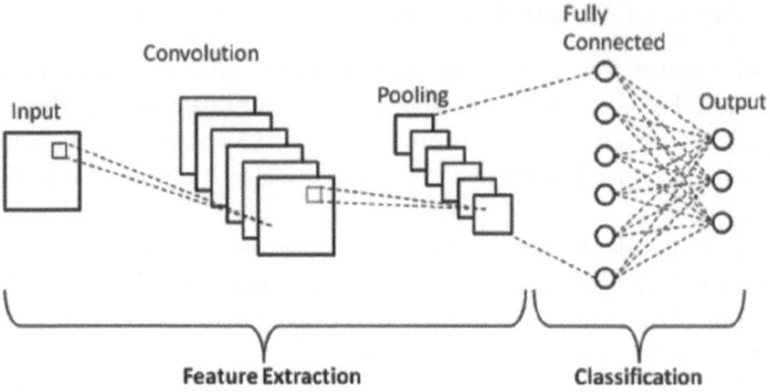

FIGURE 10.16 Structure of ANN.

10.8.5 Convolutional Neural Network

A CNN is a division of deep neural networks that helps to analyze visual images. They are applied in recognition systems, recommender systems, medical imaging, and time series challenges. It is a multilayered perceptron where each neuron is connected to every other neuron of the next layer. It is used mainly in image recognition, classification, and detection. It takes an input image and looks at it as an array of pixels. CNN models using training and testing phase on each input image which passes through a series of convolutional layers and pooling which applies complex mathematical object function to classify objects. Figure 10.16 explains the process of CNN which classifies images built on values [15].

> Just as machines made human muscles a thousand times stronger, machines will make the human brain a thousand times more powerful.
>
> –*Computer Scientist Sebastian Thrum*

10.9 IMPLEMENTATION CHALLENGES OF MACHINE LEARNING IN HEALTHCARE

The human system is intelligent. Applying intelligence to such an intelligent system is complex and requires a cautious approach. Though AI and ML are revolutionizing the healthcare sector, the application of the right system is still a challenge. We have to consider various measures before it becomes an on-board intelligence system. Here we will discuss a few challenges to apply ML in healthcare which is a promising technology to improve healthcare analytics.

The necessity of right data: As the saying "Machine learning is a vehicle and data is the fuel", it is true that acquisition of the right data is a challenge. ML requires data to train the predictor system. Since healthcare deals with physiological data that are interrelated with further numerous other components, it is a highly tedious process. Even after acquiring data, the challenge lies with understanding and structuring the data in a way it can benefit the intelligent system.

10.9.1 High Level of Data Customization

Any ML implementation requires levels of customization. These levels are exceptionally high in healthcare data because of the complexity of the human body, accessibility of data sources, and integration into the existing system. These deterrents can be answered when there is an efficient way to acquire the right data on time.

10.9.2 Accessibility of Expertise and Talents

Though the ML and healthcare system blends, the need for the right person to deal with, is a challenge again. We have very few professionals who are experts in ML and AI with healthcare domain knowledge. The assortment of deep understanding of ML techniques and healthcare knowledge would solve this challenge.

10.9.3 Venture Budget

It is a long-term decision to construct efficient and trustworthy AI and ML system practices for any healthcare organization. It is also true that every healthcare organization could not afford the cost and time to build such systems. Also, the support system required for analytics hubs like insurance companies, technology partners, will take a big budget. Though these challenges are prevailing, the organizations started realizing the future of the ML healthcare system and its role in the smart health era and will get equipped for transformation.

10.10 CONCLUSION

ML in healthcare is undeniably an innovative paradigm that can extend a promising clinical decision support system and radically bring down the decisive time and diagnostics errors if it is harnessed properly. AI and related cognitive technologies have already revolutionized the healthcare system, and it is witnessed by many innovations made possible which were looked impossible decades ago.

ML can increase the accessibility of treatments when there are scanty specialists, or considering the safety of the specialist in treating certain rare and dreadful diseases. It adds predictive power and customized treatments for the patients. Despite the implementational challenges and menaces, the healthcare professionals apprehend the changes required to be in the competitive world and accelerate them as the saying goes "when disruption is ubiquitous, transformation is everyone's job". It is no doubt that ML and cognitive studies have refined healthcare systems globally, and it is also an expected truth that it becomes a victim of its success.

REFERENCES

1. F. Pedregosa, et al, "Scikit-learn: Machine learning in Python," *J. Machine Learn. Res.*, vol. 12, pp. 2825–2830, 2011.
2. "About us — scikit-learn 0.22.2 documentation." [Online]. Available: https://scikit-learn.org/stable/about.html#citing-scikit-learn. [Accessed: 08-May-2020].

3. "3.2. Tuning the hyper-parameters of an estimator — scikit-learn 0.22.2 documentation." [Online]. Available: https://scikit-learn.org/stable/modules/grid_search.html. [Accessed: 08-May-2020].
4. P. Flach, "Performance Evaluation in Machine Learning: The Good, The Bad, The Ugly and The Way Forward."
5. J. Vamathevan, et al., "Applications of machine learning in drug discovery and development," *Nat. Rev. Drug Discov.*, vol. 18, no. 6, pp. 463–477, 2019, doi: 10.1038/s41573-019-0024-5.
6. C. Réda, E. Kaufmann, and A. Delahaye-Duriez, "Machine learning applications in drug development," *Comput. Struct. Biotechnol. J.*, vol. 18, pp. 241–252, 2020, doi: 10.1016/j.csbj.2019.12.006.
7. A. Serag, et al., "Translational AI and deep learning in diagnostic pathology," *Front. Med.*, vol. 6, p. 185, Oct. 2019, doi: 10.3389/fmed.2019.00185.
8. B. J. Erickson, P. Korfiatis, Z. Akkus, and T. L. Kline, "Machine learning for medical imaging," *Radiographics*, vol. 37, no. 2, pp. 505–515, Mar. 2017, doi: 10.1148/rg.2017160130.
9. V. Chaurasia and S. Pal, "A Novel Approach for Breast Cancer Detection using Data Mining Techniques," *Int. J. Innov. Res. Comput. Commun. Eng.*, vol. 2, no. 1, pp. 2456–2465, 2014.
10. "Explained: Neural networks | MIT News." [Online]. Available: http://news.mit.edu/2017/explained-neural-networks-deep-learning-0414. [Accessed: 27-May-2020].
11. S. Zhang, S. M. H. Bamakan, Q. Qu, and S. Li, "Learning for personalized medicine: A comprehensive review from a deep learning perspective," *IEEE Rev. Biomed. Eng.*, vol. 12, no. XX, pp. 194–208, 2018, doi: 10.1109/RBME.2018.2864254.
12. "Treatment Planning System - RayStation | RaySearch Laboratories." [Online]. Available: https://www.raysearchlabs.com/raystation/. [Accessed: 28-May-2020].
13. "What is Smart Health and How do People Benefit? | ActiveAdvice." [Online]. Available: https://www.activeadvice.eu/news/concept-projects/what-is-smart-health-and-how-do-people-benefit/. [Accessed: 31-May-2020].
14. J. Schmidhuber, "Deep Learning in neural networks: An overview," *Neural Netw.*, vol. 61, pp. 85–117, 2015, doi: 10.1016/j.neunet.2014.09.003.
15. "Understanding of Convolutional Neural Network (CNN) — Deep Learning." [Online]. Available: https://medium.com/@RaghavPrabhu/understanding-of-convolutional-neural-network-cnn-deep-learning-99760835f148. [Accessed: 03-Jun-2020].

11 The Potential of Machine Learning for Clinical Predictive Analytics

Kunal Pant, Nikhil Sati, and Divyansh Agrawal
Galgotias University

Deepa Dangwal
Uttranchal University

CONTENTS

11.1 INTRODUCTION

Challenges of Big Data with primitive data analysis

Primitive data analysis is the process of analyzing the data with human expertise and experience. For instance, whenever a doctor goes through a CT or MRI scan, he analyses the scan with all the skills and knowledge he gained through studies and experience, and then the doctor suggests the plan ahead. This is called the primitive data analysis for the data, i.e., the human expert is manually analyzing data and making out inferences on it based on his skills and past experiences.

DOI: 10.1201/9781003050827-11

11.1.1 Big Data

Over the past few years, data has become the 'gold' of the digital world. Every company requires data to get insights into how their businesses are running. This is a direct outcome of the increasing competition in the market. After 2000, there came a boom, or we call it the 'digital transformation' which enabled people to take up businesses to a whole new level, to the internet! Since then, the internet has become an ocean of floating data. Now the companies knew what their customers need and whatnot. This trend was readily picked up by every business as soon as they could do a zenith in that domain in the market. And right after that, the data started to increase at an unprecedented rate, a rate which couldn't be handled by the technologies businesses had at that time. As of now, every business either wants the data for its customers or a solution to manage this big data. It has indirectly caused a lot of other businesses to boom, which is primarily focusing on these domains, providing these services as a third party to the companies [1]. As you can see from the graph below, the rate at which the data generated has increased over the years and would continue to do so, now it's time to explain what big data is (Figure 11.1).

Consider a situation where there is such extensive data to process all of a sudden? The big data… Which doesn't have any fixed shape or size or pattern? This is, perhaps, where our human expertise can be outmaneuvered.

Big Data can generally be understood with the help of five Vs, which acts as the pillars of big data [2]. There has been much interpretation of Vs over time, with experts suggesting the need to change this definition from time to time, some even suggesting including the sixth V to make the description more robust but the most popular interpretation of the Vs are shown here (Figure 11.2).

Volume: A lot of healthcare analysis is based on analyzing the images. The data can be as big as could be thought. Hence, the volume of data is too large to be handled manually.

Velocity: With the ever-increasing Earth population, it is natural that the health data generated is also growing at an unprecedented rate. The speed at which the data is generated is noted under velocity.

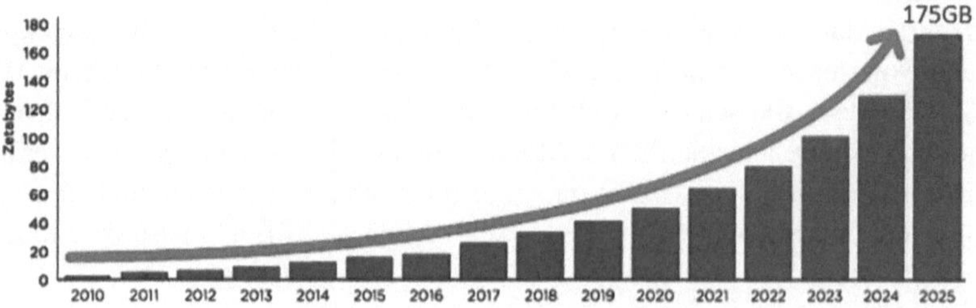

FIGURE 11.1 The increasing trend of data.

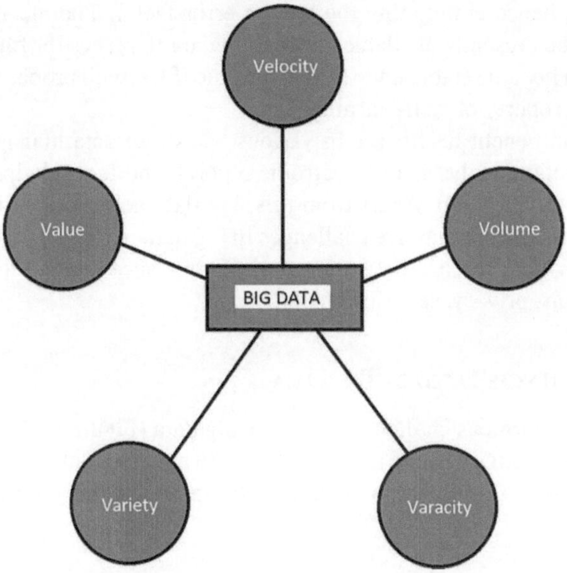

FIGURE 11.2 Vs of big data.

Value: Just think how many lives are at stake if that much data is being generated every day? How valuable this data is? If only the data can generate enough value!

Veracity: It is acceptable that the data accumulated is too much to be consumed. But is the data generated from a trusted source? Verifying and validating the data is considered under this section. If all the data from the personal sources were to be neglected, even then there's too much data being generated by trusted organizations like the WHO or World Bank.

Variety: Healthcare isn't a field with one or two subfields that would follow a particular norm to generate data. There are a variety of data generated from the healthcare sector in general, like image data from the MRI or CT scan images, speech data from the autism test, etc. This data cannot be generalized in a tabular form in general; there are multiple sources of this data with multiple formats.

With the increasing pace of lifestyle, people are getting more prone to diseases than ever before. Hence, a large number of populations need to be taken care of. In a world where we have 1.5 general physicians per 1000 people (Source: World Bank), it is really difficult to take notice of every human being. Plus, there come exceptional cases where initial symptoms are not visible but the person might be severely affected or the doctor might oversee some symptoms like malignant tuberculosis, which can be classified as benign by human mistake. These are very rare to be seen but still a possibility that cannot be overseen. Hence, the need for reducing human error is there. The overall problems laid by the ever-increasing data and the possibility of human error have led to the use of machine learning (ML) in the field of healthcare. Machines, unlike, humans tend to do work without tiring, hence eliminating the human error factor. The machine is made to learn with the presently available data and make inferences for future cases. A ML model can become stable up to 99% over time if fed with proper and accurate data with the property of 'self-learning' [3].

Big data can benefit healthcare in various ways like data management, data analysis, prediction analysis, and electronic reports, and it also helps to find and recognize the target group. Apart from this, big data helps to identify frauds in healthcare. But there are massive challenges in big data in terms of data retention, data protection, data storage, data management, data accessibility, the sensitivity of care decisions, privacy, security, etc. [4].

11.1.2 Challenges Faced by Big Data

Here are some enormous challenges faced by big data (Figure 11.3).

Privacy and security [5]: Privacy and security are essential in every business. Similarly, it is very significant in healthcare businesses as big data analysis uses

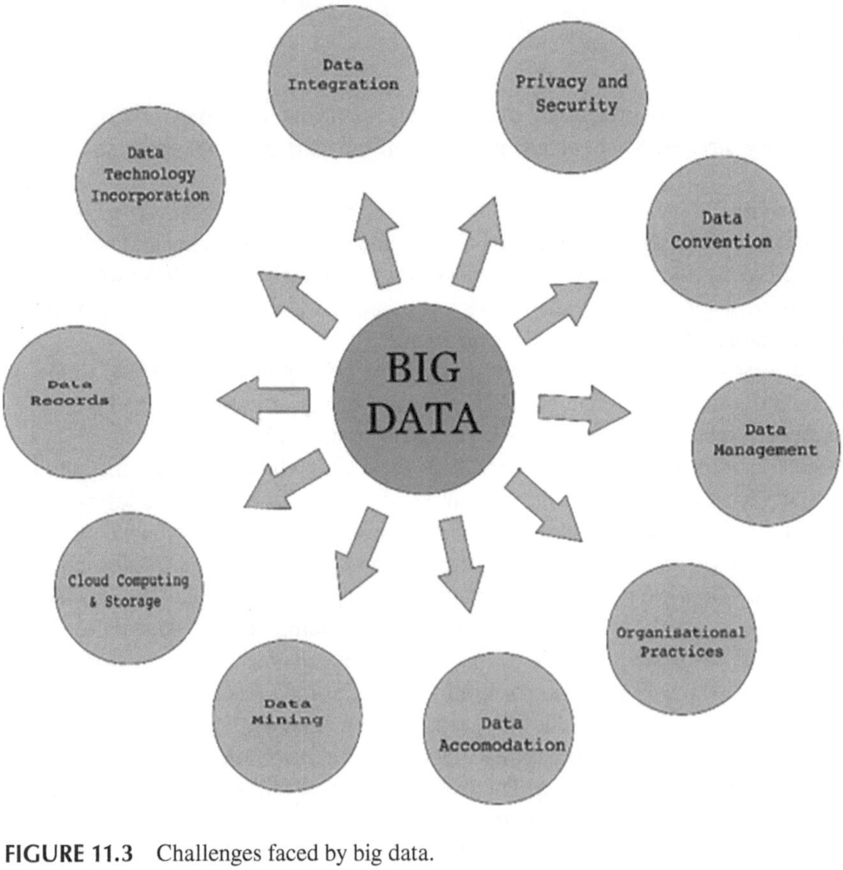

FIGURE 11.3 Challenges faced by big data.

the sophisticated tools and technologies that can modify the information archive and make decisions. Successful attacks on healthcare data can be exceedingly profitable for offenders and intensely harmful to the people and healthcare institutions. And sometimes, individuals whose data is stolen may suffer a financial crisis since health records contain personal data ranging from bank details to details about laboratory tests, which can lead not only to monetary but physical and mental harm as well.

Big Data increases the hazards to patient data for two reasons. The first is the risk posed by the information itself. Under big data settings, information is often replicated and kept without regard for space or time limits, and this trait is characterized by high risk and long-term risk. The second concern is the peril of big data technologies. Under big data technology conditions, whether a massive database uses anonymous personal encrypted data or not, there is still a user identity that will be re-identified by residual risk, and private integrity is frequently re-determined by circuit technology because big data uses pseudonymized personal confidential data that are anonymized but retain a user identity[6].

Data convention: Few data conventions in healthcare obstruct the global distribution of data analytics. Presently, the healthcare data are divided among different categories such that making a responsive and granulated database is not feasible. These standards appropriately increment the price of availing data, even when all the helpful information has been saved. Moreover, even well-structured data are often not available to doctors or researchers who could use them in meaningful ways. The health industries not providing the information comes in part from fear of violating terms and conditions [7].

Data management: Health industries face significant data challenges that will affect patient safety. All data that health industries collect must be formatted, free of redundancy and checked for accuracy, and made obtainable for different uses – medical, billing, administrative – and also the data storage and data processing of such vast data makes this task complicated. Nowadays, some hospitals hire patient safety experts who are having medical expertise, but people in these roles must understand how data management practices can be improved [8].

Organizational practices: All operations within the delivery of healthcare also create several barriers to the complete acquisition of data analytics. One understandable demonstration of the challenge is in each of the most promising areas of information analytics: clinical decision support. While data analytics could significantly improve the clinical decision-making process, the event of decision support tools do not pay sufficient attention to how decisions are made, and thus the related workflows are needed to support those decisions. The devices often assume that putting the proper information on one person's dashboard can induce them to form the appropriate choice, but really, most difficult clinical decisions involve many actors and sometimes follow organizational guidelines designed by committees. Data tools that do not fit into operational work and decision-making structures add load to physicians and are much less effective. As an example, many efforts to bring data analytics or other information technology into healthcare have created a tremendous data entry burden for physicians. It has led to high-profile inaccuracy, physician burnout, and general disappointment with the tools. As a consequence, most of the foremost reasons physicians cite for their resistance to the adoption of the leading recent data tools are related to workflow disruption. For data analytics to transform care, the designers of devices must be conscious of the context their tools are utilized in, and healthcare organizations must be willing to identify some elements of their practice [9].

Data accommodation: The current challenges in data storage are primarily due to excessive costs. One of the reasons for storage expenses is the vast volume of medical data. With the advent of medical information, the healthcare business has generated an enormous quantity of knowledge, ranging from medical diagnostic imaging to pathological map analysis. Also, the amount of data continues to grow exponentially. A patient's data typically must be retained for approximately 50 years, and it includes information like diagnosis and medicine recommendations, medical images, and other information. Because of this, the amount of medical data necessarily increases, and also the cost and difficulty of storing data increase [10]. Not only this, one extensive simplified system is required to make sure that the users are ready to retrieve the data with no hassle. Therefore, data accommodation and processing are among the most critical challenges in big data analytics.

Data mining: Although big data is excellent for modeling and simulation, there is a necessity to spot, structure, and pool the correct relevant data. Clinical big data comprises a disproportionate quantity of unstructured data, such as speech or other handwritten data, whose integration, analysis, and storage pose unique challenges. The sharing of unstructured data among organizations is difficult to realize [11]. Without properly structured data, it's challenging to investigate and visualize the output and to extract specific information or data.

Cloud computing and storage: Cloud computing now opens up new avenues for medical big data analytics and sharing. [12]. Although cloud computing offers a simple and flexible approach to mine resources, it also raises the danger of privacy exposure because a big volume of data is frequently required to be uploaded or downloaded to the cloud. Thus, the cloud will need to have sufficient space for storage and adequate speed for data upload at the same time. The cloud storage must be able to store images and graphics like CT, MRI, and X-ray, and must be capable of giving graphical presentations from the given set of data so the clinicians can easily visualize and understand quickly and make desired decisions accordingly.

Data records: Health data must stay available for a minimum of four years. That means corporations must take a long-term approach to data stewardship and keep a record of when the information gets accessed, by whom, and for what purpose. Medical data administration software allows users to determine access rights and processes, like those who give temporary data viewing capabilities to representatives in several departments during a hospital [13]. These products can show data and notes, and keep track whenever data inserts the system.

Organizations must put processes in place to systematically sort through the information to delete it when suitable or adjust and anonymize it to use it in new ways.

Data technology incorporation: Big Data in healthcare exceeded 150 Exabytes after 2011. The amount and complexity of data are growing exponentially, and other sophisticated features are becoming increasingly significant. Traditional computing and information technology face a significant difficulty as a result of this complexity. It is difficult to accommodate the availability, consistency of data, and to solve the healthcare data collection, processing real-time, lack of prior knowledge, and other troublesome issues [14]. Not only this, but the lack of information to support decision-making, policy planning, or strategy is also one of the problems in big data, and this can impact healthcare, care delivery, and research.

Data integration: The integration of data will require not only data within the healthcare system but also external data. Although it provides possible advantages, it is also challenging in terms of privacy, safety, and legal concerns. The healthcare data typically consists of patients that are striving for treatment in hospitals or dispensaries [15]. The addition of healthy individuals in the database will help to render a better knowledge of the nature of the disease and mediation as the data updates, so it is necessary that the information is passed to the users immediately for clinical decision-making and to improve the health consequences.

11.2 MACHINE LEARNING ALGORITHMS FOR ANALYTICS

In the previous section, the shortcomings of primitive data analytics in the healthcare sector due to large data being generated every day were discussed. So, how to tackle this? ML algorithms can come in pretty handy when a lot of data is to be analyzed without human intervention, if not the least after training the machine. ML algorithms have proved to be of vital importance in almost all walks of life. Especially, ML in healthcare can do wonders that were only dreamt about once. A lot of research is still going on and will be on how we can apply these algorithms to predict particular kinds of diseases well before it's too late to cure. But first, let's define ML.

ML is a subset of Artificial Intelligence that provides the systems the ability to learn and improve over time from the experience without being precisely programmed. It primarily focuses on the development of self-learning programs that can access the data and use it to learn [16]. Many statistical models are used extensively together with many algorithms to design machine learning models. In data analytics, the role of ML algorithms is to predict, classify, categorize, etc. data from a huge dataset with minimal errors.

11.2.1 Types of ML Algorithms

Based on the types of learning, ML algorithms can be classified into the following categories: supervised learning (labeling dataset), unsupervised learning (unlabeled dataset), semi-supervised learning (labeled and unlabeled), and reinforcement learning (learning from the past experiences).

11.2.1.1 Supervised Learning

Supervised learning refers to training with the **labeled dataset** [17].

Classification algorithms for analytics

- Decision Tree
- Random Forest
- Naïve Bayes
- K-Nearest Neighbor or KNN
- SVM

11.2.1.1.1 Decision Tree

A decision tree is used to represent decisions and decision-making visually because it uses a tree-like model of decisions. However, besides being a frequently applied tool in data mining for concluding a strategy to accomplish a selective goal, it is also popularly used in ML. A decision tree is a top-down tree having its root at the head. The importance of the features is evident, and relationships are apparent instantly. This system of the method is also known as a training decision tree. A decision tree is a method of leading from actual samples to general models, where the aim is to examine how to organize objects by analyzing a collection of instances [18] (Figure 11.4).

Learning input consists of a group of such vectors, each referring to a recognized section, and the output consists of a mapping of attribute values to classes. This mapping should carefully analyze both the given instances and other unseen instances. Decision Tree algorithms are committed to Classification and Regression Trees as well.

The universal purpose of preparing a decision tree is to constitute a training model that can prognosticate the number of target variables by studying decisions. The decision tree algorithm addresses the problem by illustrating it with a

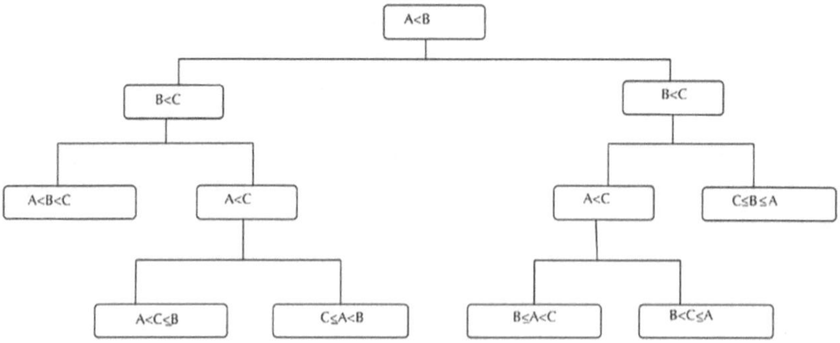

FIGURE 11.4 Training decision tree.

tree. Each internal node of the tree corresponds to an attribute, and each leaf node corresponds to a class label.

The decision of a solution with the help of decision trees begins by generating a set of explained cases. The entire collection is labeled as a training set and a testing set. First, all attributes which define the entre each case are described. Among them, individual character is taken that will explain a decision for the consigned problem. Concerning every input attribute, a specific class is defined. When we decide on an unsolved case, we generally start from the root of the decision tree and move along attribute nodes where values of the relevant attributes match the attribute values.

Let's take the example of a simple decision tree to detect common cold (Figure 11.5).

The decision tree is straightforward to understand. For example, from the tree displayed in the figure, we can conclude the following couple of rules:

1. If the person is suffering from a cold and has a blocked nose, then he/she needs medicine.
2. If the person is suffering from a cold and has a running nose, then he/she needs medicine.

Advantages of Decision Trees

1. Easy to know and interpret visual illustrations.
2. Easy to get rules.
3. Nonlinear correlations within parameters don't change tree appearance.
4. Can manage both analytical and categorical data. It can even handle multi-output queries.

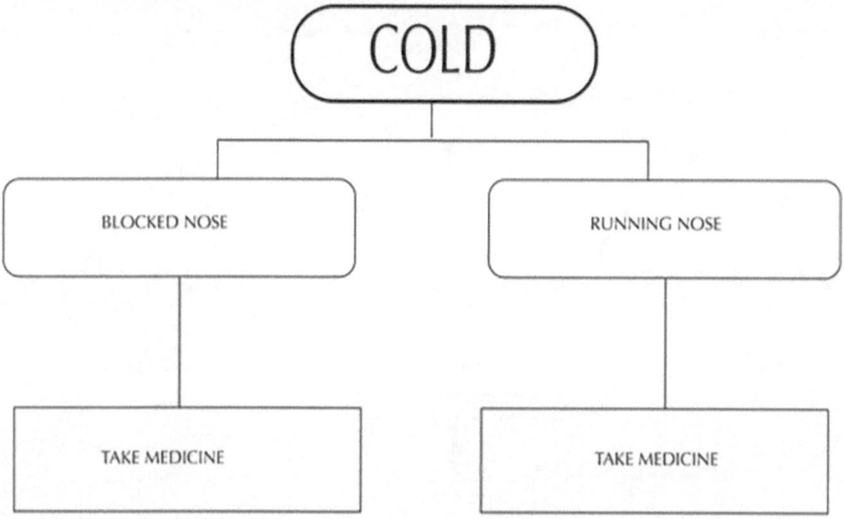

FIGURE 11.5 Decision tree to detect common cold.

Disadvantages of Decision Trees –

1. May suffer overfitting.
2. Information gain in a decision tree with categorical variables gives a biased response for attributes with more excellent no. of categories.
3. Calculations can become complicated when there are many class labels.

11.2.1.1.2 Random Forest

Random forest is a bootstrapping algorithm that is produced by multiple trees with different initial variables. It comprises a large number of individual decision trees.

Every single tree in the random forest presents a state prediction, and the state, including the highest majorities, becomes the model prognostication.

Random forest is a collective learning approach that works by building a collection of decisions during the training period. The key concept behind random forest is the knowledge of groups [19] (Figure 11.6).

Random forests are an instance of the ensemble idea in a decision tree. Ensembles are the most broadly accepted and powerful method in ML that includes building learning models. An ensemble demands multiple unique learning models and combines them to produce an aggregate model that is more powerful than any of its learning models. Ensembles are sufficient to assume we have different learning models, although each of them might perform well independently; they will definitely tend to make several kinds of errors on the dataset. Typically, this results because each model might over-fit with a distinct part of the data. By combining different individual models to an ensemble, we can average out their mistakes to reduce the risk of overfitting while maintaining strong prediction performance.

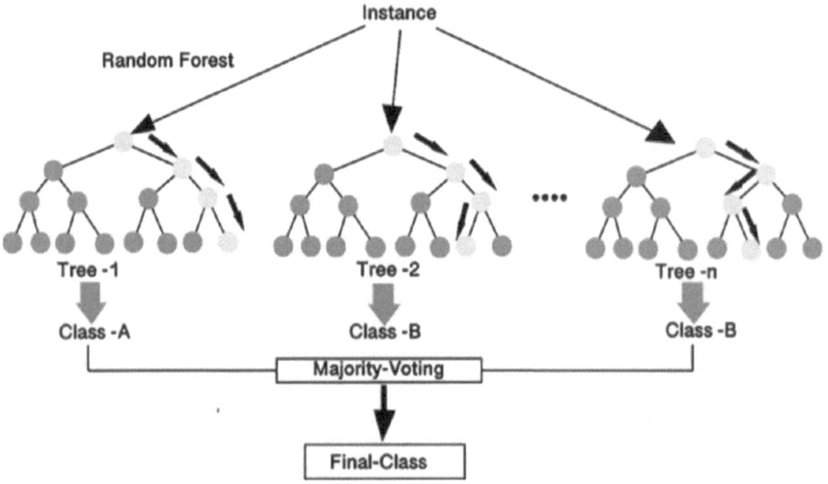

FIGURE 11.6 Random forest.

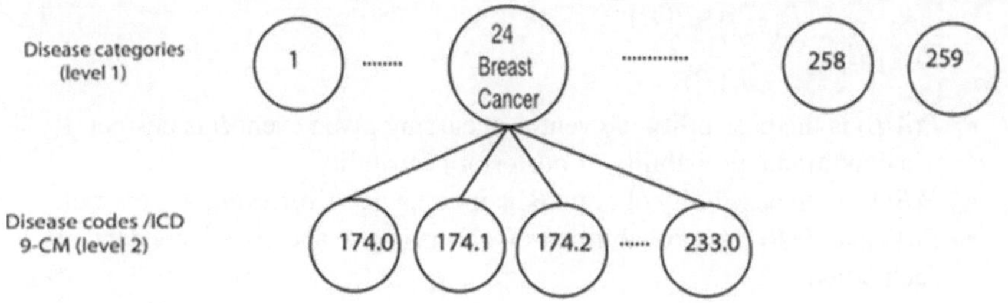

FIGURE 11.7 Disease codes and categories hierarchical relationship.

As its title would suggest, a random forest generates quantities of different decision trees on a training set, frequently in the order of tens or numbers of trees. The random forest algorithm has the most excellent accuracy when compared to other algorithms. The highest precision means the highest classification rate, which means it has the highest prediction rate. We can use a random forest algorithm for any data classification in healthcare and medical growth to find and predict any disease (Figure 11.7).

Let us understand the random forest algorithm by considering an example in healthcare.

Advantages of Random Forest

1. Excellent with high dimensionality.
2. It gives a quick prediction.
3. Manages unbalanced data.
4. It is low-biased and also has moderate variance.

Disadvantages of Random Forest

1. The complexity is very high.
2. It takes more time to train a model as compared to decision trees.

11.2.1.1.3 Naïve Bayes

Naïve Bayes is a classification technique based on Bayes' theorem by an assumption of independence among predictors or known as Bayes' theorem. In simple terms, a Naïve Bayes' classifier believes that the presence of a particular feature in a class is separate from the presence of any other element. Naïve Bayes is a robust algorithm for the predictive modeling field. Naïve Bayes on text classification usually performs well, even though word events are reliant on each other.

The Naïve Bayes algorithm depends on the Bayes theorem. Bayes' theorem (alternatively Bayes' law or Bayes' rule) represents the likelihood of an event, based upon previous information of conditions that might be related to the event [20].

Let us understand the Bayes theorem:

- $P(A/B) = \dfrac{P\left(\dfrac{B}{A}\right) * P(A)}{P(B)}$
- *P*(*A*/*B*) is the possibility of event *A* occurring given event *B* is correct. It is a conditional probability or posterior probability.
- *P*(*B*/*A*) is the possibility of event *B* happening has given event *A* is correct.
- *P*(*A*) and *P*(*B*) are probabilities of observing *A* and *B* independent of each other.

So, here the 'Naïve Bayes Classifier' combines the 'Bayes' model with the specific decision-making rules with the most feasible outcomes, respectively. It is an uncomplicated model that is pretty quick to execute and demands more insufficient training data.

Let's understand the Naïve Bayes algorithm by taking an example. The health prognostication system will have two cases: **a sick person** and **a healthy person** (Figure 11.8).

For the unhealthy person, the application will be loaded to a separate module and interact with the user, respectively. For them, the system first asks for the symptoms and then predicts the disease they may be undergoing. On the other hand, for a healthful person, the system will show the wholesome routines and diet they have to follow daily. The one looking for better health and endurance has to present inputs in terms of age, height, weight, and their corresponding medical archives.

Here, the required inputs are examined as symptoms. Bayes' classifier acknowledges each of the parameters to participate individually and hence not explore the relationship between parameters. Consequently, users can get their medical diagnosis. Bayes' algorithm is a supervised ML algorithm and presents correct distribution and forecast for the indications.

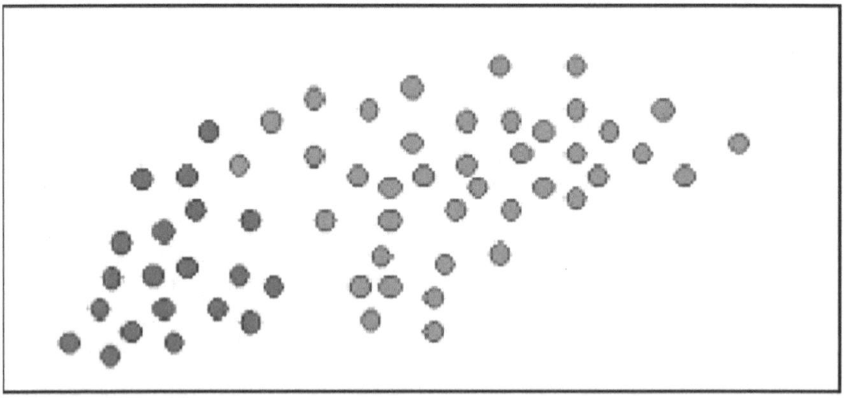

FIGURE 11.8 Naïve Bayes' classifier to classify sick and normal persons.

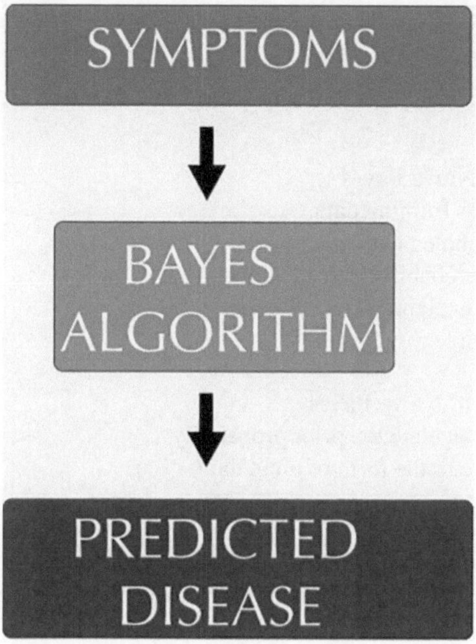

FIGURE 11.9 Bayes' algorithm to predict the disease.

In this, the prior probability will conclude based on the training dataset. The previous chance is the probability of a disease that can arise based on its symptoms.

$P(X=\text{CORONAVIRUS DISEASE})$ = (fever, tiredness, dry cough) / (overall symptoms)

The above is the conditional probability and will be used as a prior probability in the expected prediction (Figure 11.9).

Therefore, the Bayes algorithm will foretell the problem based on the greater value. Suppose user A wants to find whether he is suffering from TYPHOID, so the given prior probability is as:

$P(X=\text{TYPHOID})$ = (headache, fever, muscle weakness, loss of appetite, weight loss) / (overall symptoms)

Now a person is having fever and headache then:

$P(Y=\text{fever and headache})$, then provided that the probability of having TYPHOID (from training dataset), then what are the possibilities that this user *A* is suffering from TYPHOID, so

$$P(A|B) = P(B|A) * P(A) / P(B).$$

$P(B) \neq 0$ (zero)

Or

$P(A|B) = P(A \cap B)/P(B)$.

$P(B) \neq 0$ (zero).

Advantages of Naïve Bayes

1. It requires less training data.
2. It is easy to implement.
3. It can make probabilistic predictions.
4. Naïve Bayes explains the results easily.
5. It is quite simple.

Disadvantages of Naïve Bayes

1. We need to calculate the prior probability.
2. Very sensitive to the form of input data.

11.2.1.1.4 K-Nearest Neighbor (KNN)

KNN algorithm is a supervised ML algorithm that is quite simple and very powerful, and can be used to classify labeled data points to predict new data (Figure 11.10).

The KNN is the basic and most straightforward classification technique when there is practically no prior knowledge about the distribution of the data [21].

Applications of KNN in healthcare:

- KNN anticipated that whether the patient conceded because of respiratory failure, will have a subsequent coronary failure. This expectation is done depending on the segment, diet, and clinical estimations for that tolerance.
- Gauge the measure of glucose in the blood of a diabetic individual, from the infrared retention range of that individual's blood.

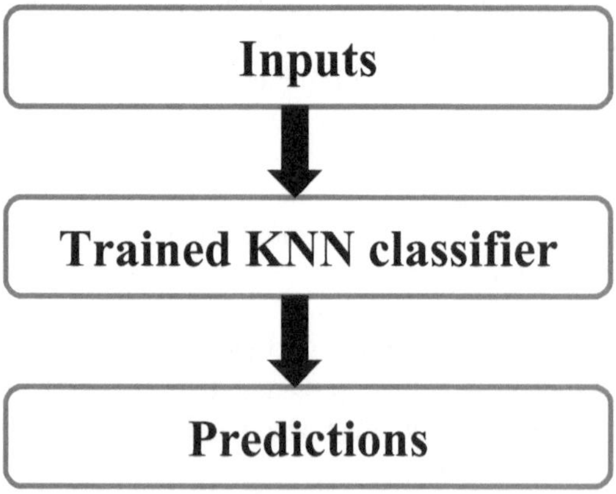

FIGURE 11.10 K-nearest neighbor algorithm.

- Distinguish the hazard factors for prostate malignant growth, because of clinical and segment factors.
- The KNN calculation has been likewise applied for dissecting smaller scale cluster quality articulation information, where the KNN calculation has been combined with hereditary calculations, which are utilized as an inquiry instrument.

KNN for regression:

When using KNN for regression issues, the expectation is based on the mean or median of the K-most similar cases.

KNN for classification:

When the KNN is used for classification, the output may be defined as the class with the highest frequency from the K-most similar examples. Each instance effectively votes for their class, and the class with the most votes is chosen as the prediction.

Advantages of KNN:

1. KNN is quite simple and unchallenging to implement.
2. SKNN algorithm does not require training before making predictions.
3. KNN can be used in classification as well as regression.

Disadvantages of KNN:

1. KNN does not work well with a huge amount of datasets.
2. KNN also does not work well with higher dimensions of datasets.
3. KNN seems to be sensitive to noisy data and missing values.
4. The computation cost is a little high.

11.2.1.1.5 Support Vector Machine

Support Vector Machine (also support vector networks) uses the concept of supervised learning. It's a learning algorithm mainly used for labeling the data into various classes, and the algorithm outputs an optimal hyper-plane that categorizes new examples.

An SVM can learn to acknowledge fraudulent credit card activity by examining thousands of fraudulent and non-fraudulent credit card activity reports. Alternatively, an SVM can learn to acknowledge handwritten digits by analyzing an oversized collection of scanned images of handwritten zeros, ones, and so forth.

SVM is used to predict unknown data as shown in Figure 11.11.

The Kernel Function:

The kernel function transforms the lower-dimensional feature into a higher-dimensional feature. Different Support Vector Machine algorithms use differing kinds of kernel functions [22].

At some moment, we see that our input file is in such large dimensional space and takes lots of computational power. At that instant, we use the kernel trick,

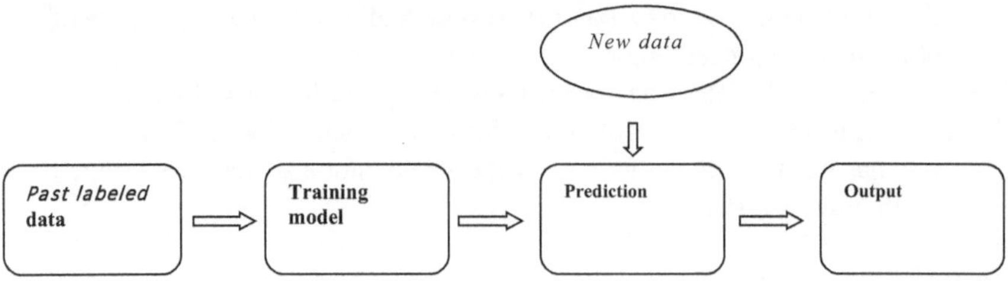

FIGURE 11.11 Support vector machine.

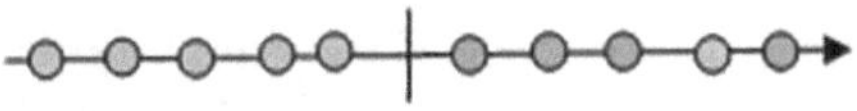

FIGURE 11.12 Kernel explanation with color balls with 1D diagram.

which allows us to compute the data within the original dimension without computing the coordinates of the data at a higher dimension.

The kernel function can be determined with the help of a fixed set of kernels by using cross-validation. But this approach is time-consuming and even doesn't guarantee that a kernel function that wasn't considered within the cross-validation procedure wouldn't perform better.

Why do we need kernels?

Let's understand this by taking an example.

In this example, we have to separate the yellow and green balls. Can you try to solve the given diagram in 1-D? (Figure 11.12)

NO!!

The yellow and green balls cannot be separated by a linear view. So to separate them, we have to go to a higher level because it is easier at a higher level (Figure 11.13).

In 2D, it looks a bit easier than 1D. Let's try this in 3D (Figure 11.14).

Now the problem is easily solved. For a good result, we need to introduce a kernel function here which will take the 2D input data and convert it into 3D output data.

Advantages of Support Vector Machine

1. They are accurate in high-dimensional spaces.
2. They are memory-efficient also because they use a subset of training points in the decision function.
3. SVM can be used for both classification and regression problems.

Disadvantages of Support Vector Machine

1. SVM algorithm is prone to overfitting if the number of features is much greater than the number of samples.

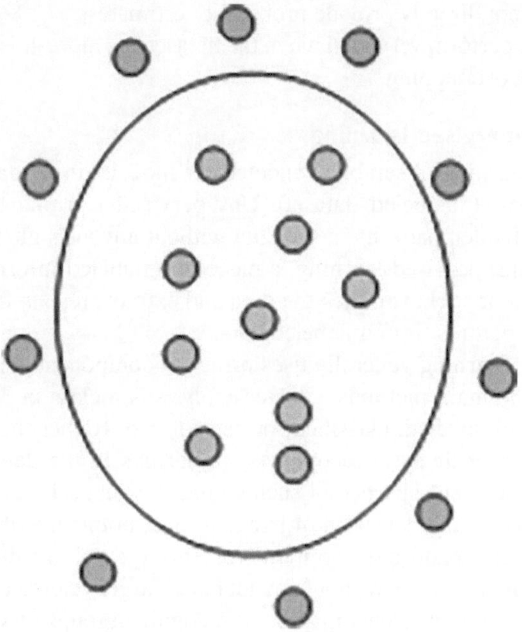

FIGURE 11.13 Kernel explanation with color balls with 2D diagram.

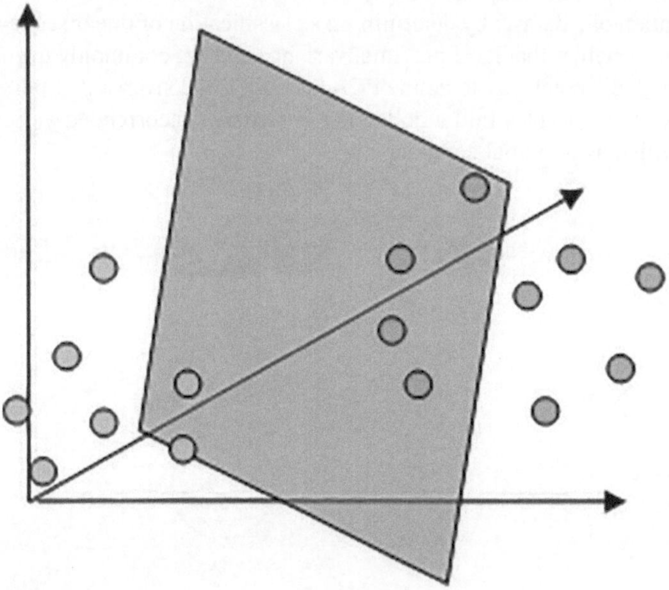

FIGURE 11.14 Kernel explanation with color balls with 3D diagram.

2. SVMs do not directly provide probability estimates.
3. It does not perform very well when the dataset has more noise, i.e., target classes are overlapping.

11.2.1.2 Unsupervised Learning

Unsupervised learning resembles undetected models in a dataset with no pre-existing labels (unlabeled dataset). Unsupervised learning means that the machine learns hidden patterns in the data without anyone's guidance or superintendence. In unsupervised learning, a piece of unlabeled information is given to the model. The model examines the data and extracts results from datasets to define hidden structures from unlabeled data.

Unsupervised learning generally uses principal component analysis and cluster analysis as the main methods. Cluster analysis is a class of ML that groups the data that is not labeled, classified, or categorized. Rather than acknowledging feedback, cluster analysis recognizes similarities in the data and responds based on the presence or absence of such similarities in each new piece of data. This method benefits the detection of irregular data points that do not belong to either group. In some pattern recognition problems, the training data consists of a combination of input vectors without any identical target values. The goal in such unsupervised learning problems may be to recognize groups of similar patterns within the data, where it is called clustering (Figure 11.15).

The two most commonly used techniques in unsupervised learning are as follows:

1. **PCA**: Principal component analysis produces a low-dimensional representation of a dataset by determining a classification of linear sequences of the variables that have maximal variance and are commonly uncorrelated. A different way to define PCA is that it is a conversion of possibly correlated variables into a collection of linearly uncorrelated variables identified as principal components.

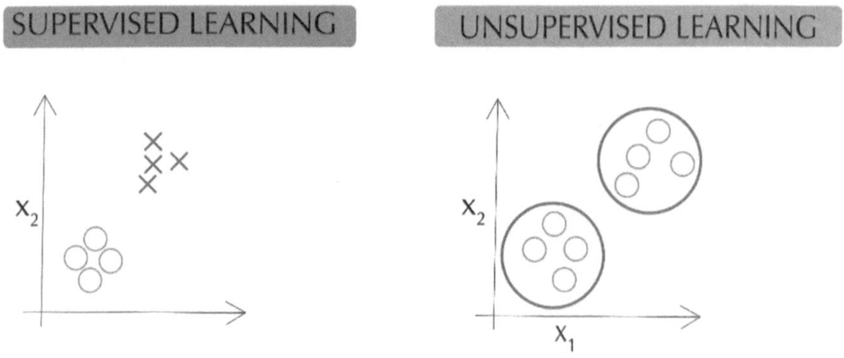

FIGURE 11.15 Cluster analysis.

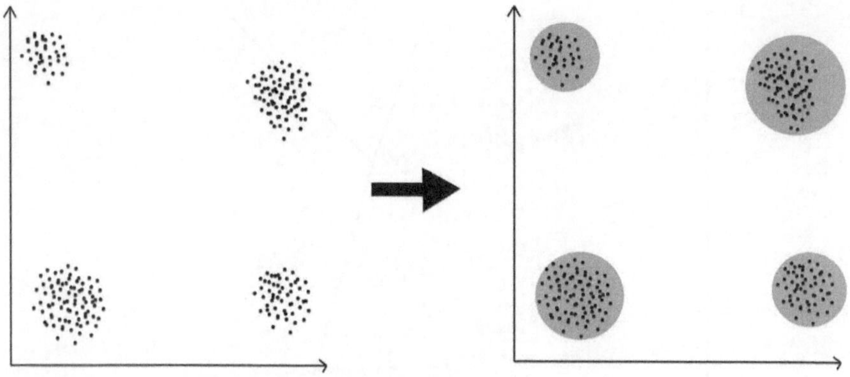

FIGURE 11.16 Clustered into four clusters.

2. **Clustering**: Clustering is associated with a set of methods and algorithms practiced to find groups (subgroups) in a dataset and includes partitioning the data into collections of similar conclusions (Figure 11.16).

11.2.1.2.1 K-Means Clustering

K-means is one of the most uncomplicated unsupervised learning algorithms that determine the well-known clustering problem. It is an iterative algorithm that tries to partition the dataset into K predefined different non-overlapping subgroups (clusters) where each data point relates to only one group. The method follows a simplistic and straightforward way to analyze a given dataset by a certain number of clusters. It manages to make the inter-cluster data points related while maintaining the clusters as different from each other [22].

It indicates data points to a cluster such that the sum of the squared distance among the data points and the cluster's centroid is the smallest. The more limited variation we have within clusters, the more comparable the data points are within the same cluster (Figure 11.17).

The main aim is to specify K points for each cluster so that these points must be placed in separate locations to get different results. So, the better choice is to put them far away from each other. The following step is to get each point belonging to a given dataset and correlate it to the nearest centroid.

The K-means algorithm works are as follows:

Step 1: Specify the number of clusters K.
Step 2: Initialize centroids and rearrange the given dataset and then choose K data points from the given centroid randomly without any replacement.
Step 3: Continue iterating until there is no change to the centroids, i.e., the distribution of data points to clusters is not changing.

Then, estimate the sum of the squared distance connecting data points and all centroids.

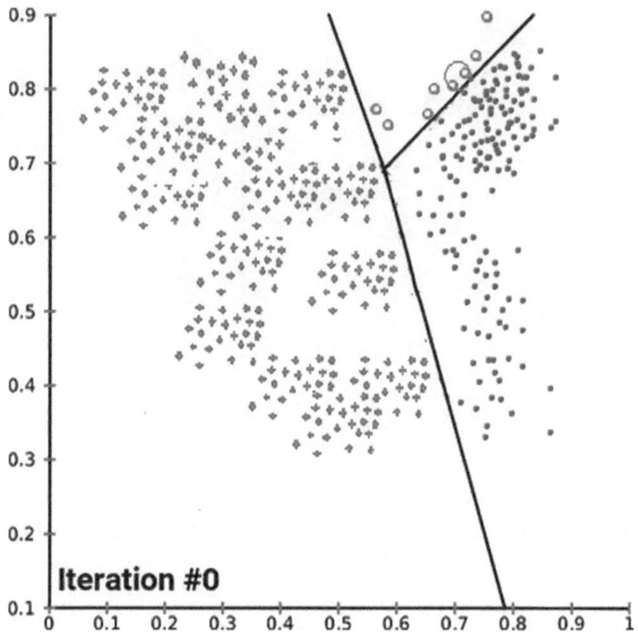

FIGURE 11.17 K-means clustering.

Assign each data point to the most neighboring cluster and calculate the centroids for the clusters by taking the average of all data points.

11.3 CONCLUSION

In this chapter, the following things were discussed:

- What is big data?
- How businesses rely on big data for their survival in the market?
- How is data increasing at an unimaginable rate and the related problems?
- Types of ML algorithms that can be used in the healthcare sector.
  - Supervised Learning
  - Unsupervised Learning
- Supervised learning algorithms can be implemented where we have labeled data, i.e., we know what our target is.
- Unsupervised learning is a ML technique where the model learns the latent patterns in the data itself and we do not need to label anything for the model to learn from. The model itself finds patterns and produces relevant results.
- Decision trees algorithm, a type of supervised learning algorithm, is a tree-based method to reach the general goal by the importance of features. A decision tree is a method of leading from actual samples to general models, where the aim is to examine how to organize objects by analyzing a collection of instances.

- Random forest, another supervised learning algorithm, is a bootstrapping algorithm that is produced by multiple trees with different initial variables. Every single tree in the random forest presents a state prediction, and the state, including the highest majorities, becomes the model prognostication.
- The probability of an event A happening given event B has already happened, which acts as a base for the Naïve Bayes algorithm can be calculated as follows:

$$P(A/B) = \frac{P\left(\frac{B}{A}\right) * P(A)}{P(B)}$$

- KNN algorithm is a simple and powerful supervised ML algorithm that can be used to classify labeled data points to predict new data. It is generally used when there is practically no prior knowledge about the distribution of the data.
- KNN can be used for both classification and regression purposes.
- Support Vector Machine, also a supervised ML algorithm, is a learning algorithm mainly used for labeling the data into various classes; the algorithm outputs an optimal hyper-plane that categorizes new examples.
- It also uses a kernel function that transforms the lower-dimensional feature into a higher-dimensional feature.
- PCA is an unsupervised learning technique that is used to reduce the dimensions in the data. The model aggregates the features internally and creates a single feature of multiple features keeping the essence of the actual data preserved.
- The data perseverance is up to the user, whether the data to be preserved should be what percentage similar to the data. The model then accordingly reduces the features.
- Clustering is another unsupervised learning technique that associates with a set of methods and algorithms practiced to find groups (subgroups) in a dataset and includes partitioning the data into collections of similar conclusions.
- One algorithm which is most commonly used for clustering is K-means clustering. It is an iterative algorithm that tries to partition the dataset into K predefined different non-overlapping subgroups (clusters) where each data point relates to only one group.
- KNN indicates data points to a cluster such that the sum of the squared distance among the data points and the cluster's centroid is the smallest.

11.4 COMMON TERMINOLOGIES

- **Accuracy**: The ratio of the number of correct prediction to the toal number of prediction.

- **Attribute**: An attribute refers to a particular type of data in the dataset. For example, the attribute of a person can be his age, weight, height, etc. It is one particular column of our data.
- **Bias metric**: It is the average difference between the prediction value and the correct value. Generally, it is a state of either **high bias** or **low bias**. The lower the value of bias, the more accurate the predictions are, and vice versa. Lower the bias, better the model.
- **Convergence**: A state while training the model is called converging if the loss changes scarcely in each iteration.
- **Dataset**: A dataset is a collection of data that is given as input to a ML model. A dataset is further broken down into three categories:
  - i. **Training Dataset**
  - ii. **Testing Dataset**
  - iii. **Validation Dataset**
- **Dimension**: Dimension in ML refers to several attributes of the dataset.
- **Feature**: A feature refers to an attribute. It is an input variable for an ML model. These terms are mostly used interchangeably.
- **Feature vector**: Feature vector refers to a set of values for all the features or attributes. For a relational database, it can be referred to as a row.
- **Feature selection**: The process of selecting relevant attributes from a dataset to be used for creating a ML model.
- **Hyperparameters**: It refers to the predefined parameters that are usually set before the starting of the training.
- **Label**: A label refers to the outcome of the model function that we are attempting to predict.
- **Labeled dataset**: It refers to the data which have been marked with label tags identifying certain characteristics.
- **Model**: Model is well-defined computation formed as a result of an algorithm that will predict some output for a particular given input.
- **Noise**: It refers to the distortion in data.
- **Overfitting**: Overfitting refers to a situation when our statistical model starts learning with too much from the training datasets.
- **Parameters**: It refers to a value that is selected by the algorithm in the process of learning.
- **Precision**: Precision is defined as the ratio of true positive concerning the total number of data points predicted as positive by the model:

$$\text{Precision} = \text{TP} / (\text{TP} / \text{FP})$$

- **Recall**: Recall is defined as the ratio of true positive concerning the total correct positive observation:

$$\text{Recall} = \text{TP} / (\text{TP} + \text{FN})$$

- **Regularization**: It refers to the adding of a penalty term to the objective function and control the model complexity using that penalty term.

- **Training**: Training is labeled with certain techniques to make it recognizable to the machine which helps them to understand the objects. It includes both input data and its corresponding expected output.
- **Under-fitting**: Under-fitting refers to a situation when our statistical model is performing poorly concerning the training data.
- **Validation**: Validation is used to evaluate the model and estimate an unbiased generalization performance.
- **Variance**: It refers to the amount by which the target function changes while being trained on the data.

REFERENCES

1. Baesens, B., Bapna, R., Marsden, J. R., Vanthienen, J., & Zhao, J. L. (2016). Transformational issues of big data and analytics in networked business. *MIS Quarterly*, *40*(4).
2. Bansal, S., Chowell, G., Simonsen, L., Vespignani, A., & Viboud, C. (2016). Big data for infectious disease surveillance and modeling. *The Journal of Infectious Diseases*, *214*(suppl_4), S375–S379.
3. Kaelbling, L. P., Littman, M. L., & Moore, A. W. (1996). Reinforcement learning: A survey. *Journal of Artificial Intelligence Research*, *4*, 237–285.
4. Abouelmehdi, K., Beni-Hessane, A., & Khaloufi, H. (2018). Big healthcare data: Preserving security and privacy. *Journal of Big Data*, *5*(1), 1–18.
5. Seifert, J. W. (2008, August). Data mining and homeland security: An overview. Library of Congress Washington DC Congressional Research Service.
6. Liang, H., Luo, M., Wang, R., Lu, P., Lu, W., & Long, L. (2018). Big data in health care: Applications and challenges. *Data and Information Management*, *2*(3), 175–197.
7. Ginter, P. M., Duncan, W. J., & Swayne, L. E. (2018). *The Strategic Management of Health Care Organizations*. John Wiley & Sons, Canada.
8. Esposito, C., Ciampi, M., & De Pietro, G. (2014). An event-based notification approach for the delivery of patient medical information. *Information Systems*, *39*, 22–44.
9. Wang, Y., Kung, L., & Byrd, T. A. (2018). Big data analytics: Understanding its capabilities and potential benefits for healthcare organizations. *Technological Forecasting and Social Change*, *126*, 3–13.
10. Muthu, B., Sivaparthipan, C. B., Manogaran, G., Sundarasekar, R., Kadry, S., Shanthini, A., & Dasel, A. (2020). IoT-based wearable sensor for disease prediction and symptom analysis in the healthcare sector. *Peer-to-Peer Networking and Applications*, *13*(6), 2123–2134.
11. Davenport, T. H., Cox, D., Lee, H. S., Zachman, J. A., & MacMillan, J. Praise for The Real Work of Data Science.
12. O'Driscoll, A., Daugelaite, J., & Sleator, R. D. (2013). 'Big data, Hadoop and cloud computing in genomics. *Journal of Biomedical Informatics*, *46*(5), 774–781.
13. Balaraman, P., & Kosalram, K. (2013). E-hospital management & hospital information systems-changing trends. *International Journal of Information Engineering & Electronic Business*, *5*, 50–58.
14. Wang, Y., Kung, L., & Byrd, T. A. (2018). Big data analytics: Understanding its capabilities and potential benefits for healthcare organizations. *Technological Forecasting and Social Change*, *126*, 3–13.

15. Langwick, S. A. (2008). Articulate (d) bodies: Traditional medicine in a Tanzanian hospital. *American Ethnologist*, *35*(3), 428–439.
16. Luxton, D. D. (2016). An introduction to artificial intelligence in behavioral and mental health care. In *Artificial Intelligence in Behavioral and Mental Health Care* (pp. 1–26). Academic Press.
17. Ben-Assuli, O., Heart, T., Shlomo, N., & Klempfner, R. (2019). Bringing big data analytics closer to practice: A methodological explanation and demonstration of classification algorithms. *Health Policy and Technology*, *8*(1), 7–13.
18. Geetha, A., & Nasira, G. M. (2014, December). Data mining for meteorological applications: Decision trees for modeling rainfall prediction. In *2014 IEEE International Conference on Computational Intelligence and Computing Research* (pp. 1–4). IEEE.
19. Ali, J., Khan, R., Ahmad, N., Maqsood, I. (2012). Random forests and decision trees. *International Journal of Computer Science Issues(IJCSI).*
20. Lewis, D. D. (1998, April). Naive (Bayes) at forty: The independence assumption in information retrieval. In *European Conference on Machine Learning* (pp. 4–15). Springer, Berlin, Heidelberg.
21. Kramer, O. (2013). K-nearest neighbors. In *Dimensionality Reduction with Unsupervised Nearest Neighbors* (pp. 13–23). Springer, Berlin, Heidelberg.
22. Likas, A., Vlassis, N., & Verbeek, J. J. (2003). The global k-means clustering algorithm. *Pattern Recognition*, *36*(2), 451–461.

12 Predictive Analytics in Healthcare Using Machine Learning Tools and Techniques

Shobana R., A. Ambikapathy, and Kalpana Hazarika
Galgotias College of Engineering and Technology

Amit Kumar Gupta
ABES College of Engineering

CONTENTS

DOI: 10.1201/9781003050827-12

12.1 INTRODUCTION

This chapter introduces you to the field of machine learning (ML) applications in areas of healthcare. Modern technology is driving the global healthcare industry with increasing costs and population. World healthcare now shifts from volume-based models to outcome-based models. The way they operate must employ better access, reduced value, quality treatment, and customer satisfaction. We can meet the objectives by using advanced analytics on healthcare data. ML, an advanced predictive analytics algorithm [1], has been proved to achieve the goals and is widely used by the healthcare industry nowadays. In this chapter, we discuss the importance of ML in the field of healthcare with some case studies on predictions of various diseases.

In short, we discuss the following topics in this chapter:

- Predictive analysis-an overview
- Basis of ML
- Steps in applying the algorithm
- Study of ML algorithm
- Study of ML tools
- Issues and challenges
- Case study – a review

12.2 PREDICTIVE ANALYTICS – AN OVERVIEW

The biggest problem is bridging the gap between operational cost and patient satisfaction. With a growing population, trying one treatment after another for an underlying disease is not predictable. A physician not only earns patient dissatisfaction but also leads to other patients discouraging from the treatment, high cost due to the conduction of successive trial methods. Hence as a solution, Predictive analytics is drawn in the healthcare industry. Predictive analytics is a type of data analytics that uses past and present data to make future predictions. It uses a wide range of possible forecasts by the usage of analytics techniques [2]. The approaches to predictive analytics include regression analytics and ML analytics. ML is widely used over regression analytics due to its uniform efficiency in the prediction of applications, even with a large dataset. These models have applications in inpatient readmission, document classification, and maintenance.

12.3 DEFINE MACHINE LEARNING

Machine learning is a form of data analytics built with the idea of bringing out techniques that learns when the data fed into the ML system undergo training by itself through optimized function based on data and the task.

Due to the increased usage of modern technologies, the popularity of maintaining healthcare data by electronic health records (EHR) has grown against the traditional collection of data, which requires efforts in collecting and preserving complex EHR data [3]. ML also makes specific clinical predictions from the dataset.

ML has many algorithms for data classification and a prediction that helps to achieve speed, accuracy, and medical staff for quality treatment. The few lists of applications of ML are as follows [4]:

- Disease identification
- Robotic surgical tools
- Drug discovery
- Smart health records
- Outbreak prediction
- Improved radiotherapy

We can thus outline ML as a discipline of AI that behaves as a human intelligence simulation within a computer system (Figure 12.1).

ML plays a vital role in all these areas, starting from maintaining patient records to the emergency unit of patient care. The advancement limits patients' waiting time and lessens hospital visits by attending patients remotely and assisting them over the phone.

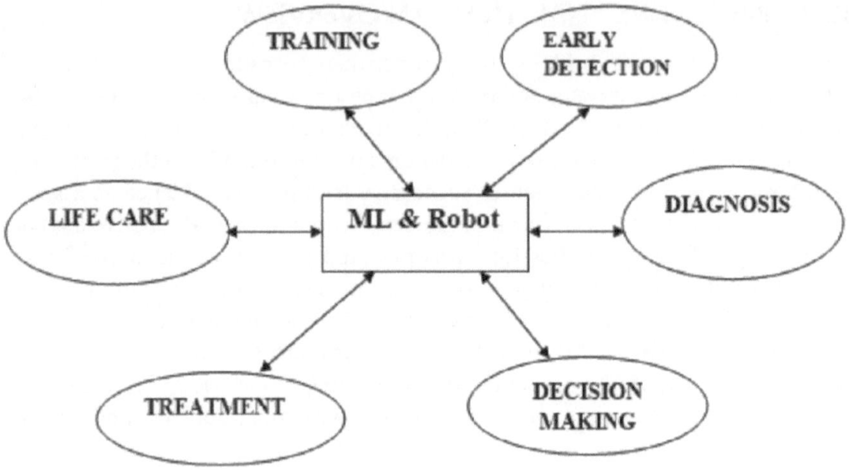

FIGURE 12.1 Machine learning system.

12.3.1 Steps in Applying ML

ML is widely used by individuals and organizations that target improving the overall efficiency of the system. The learning undergoes different tasks before delivering an ideally aided model. The following are the steps used in applying ML[5].

12.3.2 Collecting Data

Data usually collected from sources are in the form of text or SQL or spreadsheet. We can carry out the analysis on a standard electronic platform. We also classify the data of ML as structured, unstructured, and semi-structured data. They are the primary predictor of any ML. The more the data, better will be the output model. The quality and quantity of data decide the accuracy of the model.

12.3.3 Data Preparation

Data preparation involves a lot of human intervention. It requires the selection of data that includes checking of missing data, removal of unrelated data, and repetition of data used. The selected data needs to be preprocessed by formatting, cleaning, and sampling steps. We can transform them for use by enhancing features such as scaling, attribute decomposition, and aggregation.

12.3.4 Training a Model

The ML has a large number of algorithms for data classification and identification. We choose data depending on the application and requirement. They are further added weights and parameters for training.

12.3.5 Performance Evaluation

It is necessary to test the performance of the proposed model. The algorithm's behavior is estimated by how well it responds, learning from present and past events. Prepare the same set of data for training and testing, and assess the model by providing the test dataset.

12.3.6 Improving the Performance

The performance of the model is always measured using Metrics. The weights or parameters of the network are changed to predict a better model to minimize the error. Advanced approaches are required to improve the quality of the model at regular intervals (Figure 12.2).

After all these processes are carried out, and if the observed model is found to perform perfectly, it is then used for its proposed function. The observed model will be used in research-related tasks or for prediction or automated tasks.

12.4 MACHINE LEARNING ALGORITHMS

Algorithms operate on the data and extract the information from data using computational methods. Depending on the nature of learning, we classify ML algorithms into four categories [6] (Figure 12.3).

12.4.1 Supervised Learning

The majority of applications today in ML uses supervised learning. With input variables (X) and output variables(Y), the relationship between x and y can be mapped as $Y=f(X)$. This concept is utilized in supervised learning. Each time a new set of X is taken, Y can be easily obtained using the above function. It is

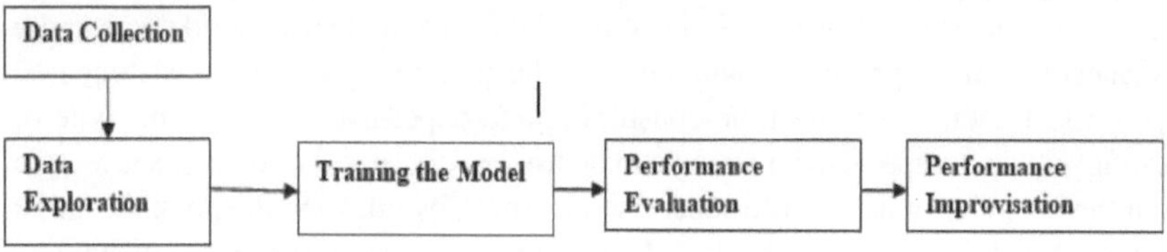

FIGURE 12.2 Steps in applying ML.

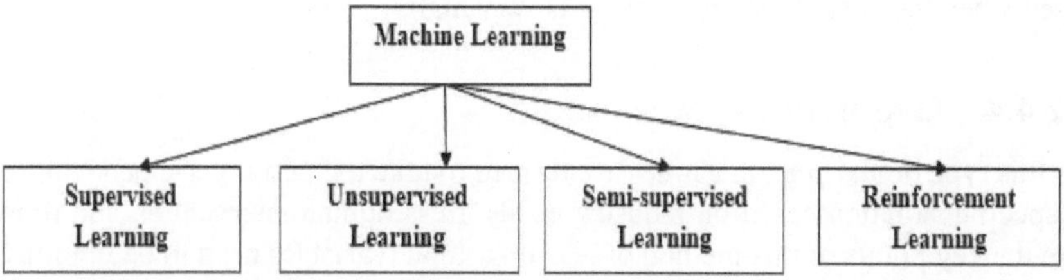

FIGURE 12.3 Machine learning algorithm classification.

called supervised learning because the process of learning an algorithm from a training dataset can be considered as a teacher overseeing the learning process. The algorithm stops learning when the desired level of efficiency is reached.

Supervised learning includes two categories of algorithms, namely classification, and Regression. When the learning is done to predict the output category, it comes under the classification-based Supervised learning such as 'Yes' or 'No.' When the output is a continuous value such as 'money,' 'weight,' and 'number,' then such learning is called Regression-based supervised learning. Generally, this learning is used for constructive models. This learning is useful for prediction, forecasting, and finding a relation between quantitative data.

12.4.2 Unsupervised Learning

When only the input variables (*X*) are available, and the output variable (*Y*) is not available, the relationship between *X* and *Y* cannot be mapped. There is hence no teacher as compared to supervised learning and no correct answers. Such nature of learning is called unsupervised learning. Algorithms are left on their own to interact with the data using trial-and-error methods and produce the output under operational conditions. This learning mainly focuses on modeling the proposed structure in the data to learn more about the behavior of that data.

Unsupervised learning includes two categories of algorithms, namely, Clustering and Association. When the data is unlabeled, clustering deals with finding the structure or pattern among those unlabeled data. Association allows one to establish associations among variables in a large database.

12.4.3 Semi-Supervised Learning

It is a type of learning where a large amount of input data (*X*) and only some amount of output data (*Y*) is available for learning. They come under both the type of supervised and unsupervised learning. Today most of the data in the real world come under this category. The unprocessed data are very cheap and easy to collect as compared to processed data, which are time-consuming and expensive. Through this type of learning, either unsupervised training can be used on these sets of data to learn more about the behavior of data or use supervised learning by guessing the possible output of unprocessed data feeding it back to the network and view the unseen new outputs.

The semi-supervised learning algorithms have their applications in the areas of speech analysis, content classification, and face recognition. This learning is used when the cost of data processing is very high.

12.4.4 Reinforcement Learning

In this type of learning, machines are trained to take the best precise behavior in a specific situation based on industry needs. Less human intervention and time are the key points of this method of learning. Input variables need to be in initial conditions at the time of initialization. Output can be of any type. The model is trained with the available inputs. The user decides rewards and punishments based on the output model. The model continues to learn until the best prediction is achieved. It learns from its experience to achieve maximum efficiency. Robotics application uses this type of algorithm for learning through trial and error for yielding the best rewards. Few glimpses of various learning and their corresponding methods and algorithms used in ML are listed below (Table 12.1).

12.5 MACHINE LEARNING TOOLS

Tools play an important role in ML. Selecting the correct tool is as necessary as applying the best algorithm for the purpose. They help to convert data into usable form and also make work easier, faster, and less time-consuming ML tools can be categorized as follows:

- Platform vs. Library
- GUI vs. CLI vs. API
- Local Tools vs. Remote Tools
- ML using R

12.5.1 Platform vs. Libraries

Platform Tools

- They provide the facility at each step from the beginning until the end of the ML project.
- They provide resources to complete the entire project in ML.
- The interface can be a graphical or command line.
- They can be used for general purpose and in all environments.

Library Tools

- They provide resources to complete a part of the entire project in ML.
- The interface can be application-specific programming on which it runs.
- They are for a specific purpose and specific environment.

The lists of tools under this category are summarized below.

12.5.2 Graphical User Interface vs. Command Line Interface vs. Application Program Interface

Graphical User Interface – GUI

- Programmers or less technical users can use this tool to design the project.
- Interface focuses on a structured process for users.
- The interface also focuses on interactive information displays such as visualization, which involve browser, point, and clicks.

Command-Line Interface – CLI

- This permits non-programmers to use this tool for designing the ML project.

TABLE 12.1
Machine Learning Algorithms

S.No.	Type of Learning	Methods	Algorithms
1.	Supervised Learning	Decision Tree Technique	Classification and Regression Tree (CART)
			Iterative Dichotomiser 3 (ID3)
			C4.5 and C5.0
			Chi-square automatic interaction detection (CHAID)
			Multivariate adaptive regression splines (MARS)
			M5
		Bayesian Methods	Naïve Bayes
			Gaussian Naïve Bayes
			Multinomial Naïve Bayes
			Averaged One-Dependence Estimators (AODE)
			Bayesian Belief Network (BBN)
			Bayesian Network (BN)
		Artificial Neural Networks	Perceptron
			Multilayer Perceptrons (MLP)
			Back-Propagation
			Stochastic Gradient Descent
			Hopfield Network
			Radial Basis Function Network (RBFN)
2.	Unsupervised Learning	Clustering Methods	k-Means
			k-Medians
			Expectation-Maximization (EM)
			Hierarchical Clustering
		Association Methods	Apriori algorithm
			Eclat algorithm

(*Continued*)

TABLE 12.1 (*Continued*)
Machine Learning Algorithms

S.No.	Type of Learning	Methods	Algorithms
3.	Supervised/ Unsupervised Learning	Regression Methods	Ordinary Least Squares Regression (OLSR) Linear Regression Logistic Regression Stepwise Regression Multivariate Adaptive Regression Splines (MARS) Locally Estimated Scatter plot Smoothing (LOESS)
4.	Reinforcement Learning		Deep Q Network (DQN) Deep Deterministic Policy Gradient (DDPG) Q-Learning with Normalized Advantage Functions (NAF) Proximal Policy Optimization (PPO) Trust Region Policy Optimization (TRPO)

- They are focused on small programs or sub-task modes of the ML project.
- ML tasks are framed in terms of input–output requirements.
- This allows recording, scripting programs, and parameters.

Application Program Interface – API
- It permits flexibility in decision-making.
- It allows the user to create their tools for the ML project.
- User can integrate ML into software projects.
- It allows users to combine with library tools, create new tools, and use them.

12.5.3 Local vs. Remote

Local
- Tools can be downloaded, installed, and used locally.
- Memory data and algorithms are tailored.
- Users can have control over their configuration and parameters.

- Fewer algorithms.
- Locally downloaded tools into your system to meet up all needs.

Remote

- Remotely accessible tools to work on its system through a remote server
- Runs on a larger dataset
- Runs across multiple servers and shared memory.
- Less control over run configuration.
- They are also referred to as Machine learning as a Service (MLaaS).

12.5.4 ML Using R

Programmers who want to analyze and carry out projects using numerical techniques and plots prefer R over other tools. The style of coding is easy, and it is open-source. It has a collection of inbuilt functions called packages that can be shared among users. They automatically load when running on the environment and also have many data structures present in them. The features of the dataset can be evaluated and plotted using styles available in the R platform. Strip Chats, QQ Plots, dot Plots, and kernel density plots are some plots available in this platform. All ML projects must include data preparation, data generation, method application, and model construction. The following are some ways for creating predictive models in R [7].

- By creating a model with the basic feature.
- Estimation of the model properties using printing, graphs, and other strategies.
- Outcome analysis for samples using predicted process (Figure 12.4).

A glimpse of various ML tools under each category is discussed in this chapter (Table 12.2).

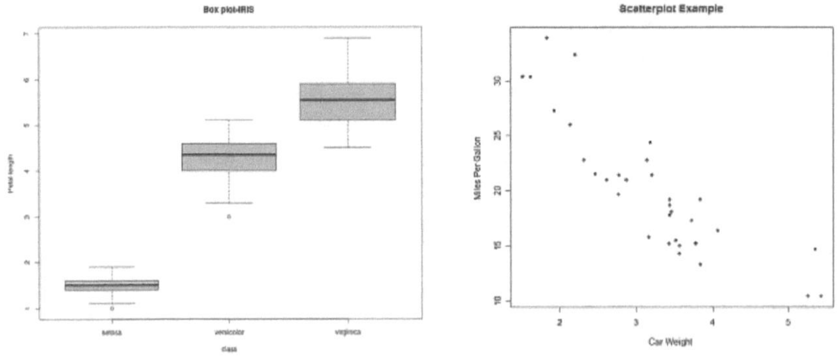

FIGURE 12.4 Plotting examples using R.

TABLE 12.2
Machine Learning Tools

Classifications	Tools	Characteristics
Platforms vs. Libraries	WEKA-Waikato Environment for Knowledge Analysis.	• Enables ML tasks to be done without programming. • Contains data preprocessing and modeling techniques. • Supports Regression, clustering, association rules, and visualization. • Freely available modern computing platform.
	R	• R is the most popular and highly used platform in the field of analytics. • R is a version of Lisp 3⁄4. • It will read and execute R scripts from the R extension file loaded directly or typed. • R allows graphics to be shown on the screen or saved to disk and can also create models that can be queried and modified.
	Scikit-learn	• A Python module, designed on top of SciPy, for ML. • They are used for research in mathematics and science. • The libraries can be used for collaborative 'working bench'/embedded and reused into integrated applications.
	Accord	• The Accord. NET Architecture provides for. NET ML, mathematics, statistics, computer vision, computer auditing, and various methods and techniques related to scientific computing. • Includes libraries that include a more traditional set of functions.

(Continued)

TABLE 12.2 (*Continued*)
Machine Learning Tools

Classifications	Tools	Characteristics
	Java Statistical Analysis Tool	• JSAT is a platform in ML with an inbuilt Java library. • Algorithms are independently implemented using an object-oriented network.
GUI vs. CLI vs. API	Orange	• A suitable method for fast comparisons over estimation techniques. • Sleek mobile software where a variety of steps and estimation techniques can be used to process the data. • Uses GUI widget and connector build.
	Waffles	• Waffles are a series of command-line tools written in the C++ language. • Scriptfriendly, and intended for use in process related to automation. • They automatically tune their parameters without an explicit requirement of any filters and transforms. • The conversion of the process to the native application requires minimal effort.
	Deeplearning4J	• Deep learning is an open-source library for java and widely supports AI algorithms. • Integrated with Hadoop and Spark software. • Has Scala API and uses Python API for network configuration. • DL4J is used on localized GPUs and CPUs.
Local Vs. Remote	Shogun	• An open-source platform is written in C++. • Supports many languages such as python, R, JAVA, and Ruby. • And platforms such as Linux and macOS interfaces. • Easier implementation on all ML algorithms.

(*Continued*)

TABLE 12.2 (*Continued*)
Machine Learning Tools

Classifications	Tools	Characteristics
	To Learn	• An open-source language developed by Google. • Also called 'Batteries Included' allows developers to use ML libraries directly written in Go. • It is built to fix simplicity and customizability than all other platform libraries.
	Google Prediction API	• Includes pattern matching skills and ML capabilities. • API may predict a numerical value after learning from the training data, to represent a new data element. • Examples: To guess what things or items a user likes or how much time he spends on a specific date.
	AWS	• Includes visualization tools and wizards that guide the process of building ML models without studying complex ML algorithms and technology.

12.6 EVALUATION METRICS

One of the main tasks in designing any model is evaluating the performance. There are various metrics to test the performance of the ML algorithm. The metrics tell us how well the machine is adaptive or non-adaptive to the unseen new data. Classification and Regression are widely used problems in ML. A glimpse of a few evaluation metrics in ML are listed below [8].

12.6.1 Confusion Matrix

The output can be of two or more types. The output is usually a matrix. It evaluates the complete performance of the model. A confusion matrix is a table with two classes called 'actual' and 'predicted,' and both the classes can have the same set of combinations, namely, True Positives (TP), True Negatives (TN), False Positives (FP), and False Negatives (FN) (Figure 12.5).

True Positives (TP) – A case where actual and predicted values are true.

True Negatives (TN)–A case where actual and predicted values are false.

False Positives (FP)–A case where the value of actual is false and predicted is true.

False Negatives (FN)–A case where the value of actual is true and predicted is false.

The confusion matrix is usually used in classification-related problems.

12.6.2 Accuracy

It is most commonly used in classification algorithms. It is given as a ratio of the number of correct prediction samples to total samples of predictions. It is calculated using the following formula:

$$\text{Accuracy} = \frac{\text{Number of Correct Predictions}}{\text{Total Number of Prediction Samples}}$$

		Actual 1	**Actual** 0
Predicted	1	True Positives (TP)	False Positives (FP)
Predicted	0		True Negatives (TN)

FIGURE 12.5 Confusion matrix representation.

12.6.3 Precision

There are many scenarios where accuracy cannot be used for performance evaluation. In scenarios like data being imbalanced, it always predicts a high accuracy rate, which makes the model just train continuously without any learning. Precision is used as a metric under such cases.

This metric is given as a ratio of the total samples of True Positives, which are positive to the total number of positively predicted values.

$$\text{Precision} = \frac{\text{True Positives}}{\text{True Positives} + \text{False Positives}}$$

12.6.4 Recall

The recall is an evaluation metric that is given as the fraction of samples correctly predicted to the total number of predicted values.

$$\text{Recall} = \frac{\text{True Positives}}{\text{True Positives} + \text{False Negatives}}$$

12.6.5 F1 Score

F1 score is defined as a metric that gives the weighted average of precision (*P*) and recall (*R*). The F1 score can be either 1 or 0, depending on the nature of the model.

$$\text{F1 score} = 2 * \frac{(P * R)}{(P + R)}$$

12.6.6 The Area under the Curve

It is an evaluation criterion that estimates based on threshold values variation. Ideally, it can be plotted by taking True Positive Rate (Recall) vs. False Positive Rate (Specificity), at various threshold values. Higher the AUC, better will be the performance of the model (Figure 12.6).

12.6.7 Mean Square Error

Cost function and lost function serve the purpose of estimating the performance of the algorithms to train the model accordingly.

The loss function is the difference between the outputs of the model and the actual data value. Least square, logistic loss function, and hinge loss are some of the loss function techniques.

The cost function is defined as the summation of the lost function of every training data point.

Mean square error, MSE, finds the average of squares of error between the predicted and actual values.

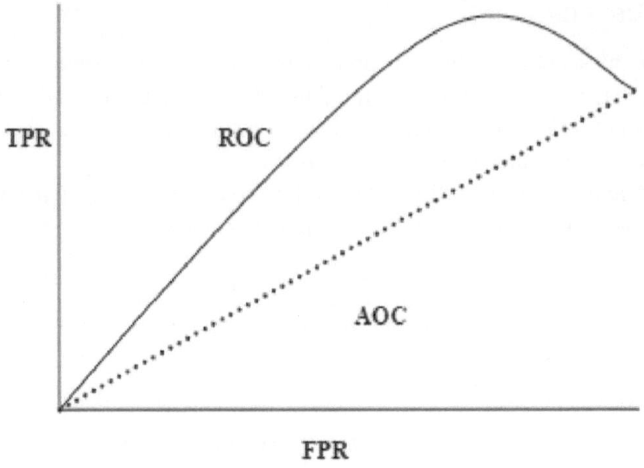

FIGURE 12.6 AOC curve.

$$\text{MSE} = \frac{1}{n}\sum\left(Y - \hat{Y}\right)$$

where

Y – Actual output values

$\hat{Y}$ – Predicted output value

12.6.8 Mean Absolute Error

Mean absolute error (or mean absolute deviation) is a measure of error between the predicted and target values. The absolute value of each error is recorded, and the mean value of all recorded errors is calculated to find MAE.

$$MSE = \frac{1}{n}\left|Y - \hat{Y}\right|$$

where

Y – Actual output values

$\hat{Y}$ – Predicted output value

12.7 MACHINE LEARNING IN VARIOUS FIELDS

AI is everywhere. One of the popular applications of AI is ML. ML allows machines to make decisions on their own from a huge dataset. Industries from the economic sector and telecommunication to education, apply predictive analytics to gain critical information from data and expand their business decisions. Companies such as Google, Accenture, Amazon, Tesla, and Johnson and Johnson use ML technology to enhance their customer service, productivity,

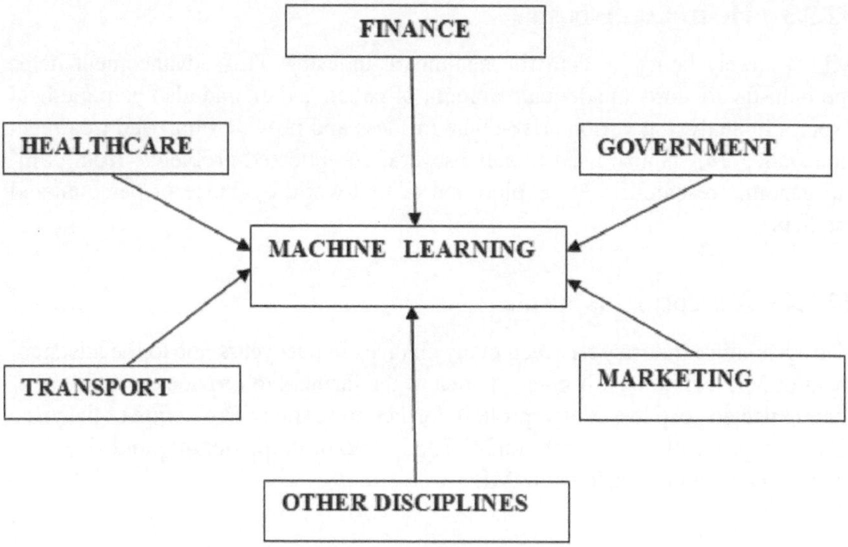

FIGURE 12.7 Machine learning in various fields.

capital works, and new product delivery. For instance, Amazon uses AI for e-commerce. For forecasting people's needs through a feature called style snap, Johnson and Johnson use features of AI for drug discovery. An overview of some of the industries getting benefited by using analytics is discussed in this section[9] (Figure 12.7).

12.7.1 Transportation Industry

The key feature of any predictive analytics data is patterns. A sequence of data called 'historical data' can be obtained. It is used to describe the behavior of data. A key aspect for the transportation industry is the study of data to recognize patterns; through which efficient routes can be designed. Traffic has increased in many places around the world due to the rise in the number of cars on the roads and the slow growth of infrastructures. Many researchers have used ML to build models that control traffic. A prototype is proposed to predict and control traffic at different times.

12.7.2 Economics

In the field of economics and business, ML algorithms are used to have a tight insight into data. The majority of fraudulent online activities are made with the assistance of automated engines. These algorithms work in patterns, which can be extracted from data. A model is now built that takes these extracted data to detect fake patterns and prevent fraud in online transactions.

12.7.3 Healthcare Industry

ML is largely being used in the healthcare industry. This advancement helps the industry to store an adequate amount of patient's data and also help medical experts in analyzing various risk-related illness and provide improved treatment in advance. ML is also used to solve several complicated problems, from clinical genomic research to the exploration of real-world evidence of personalized medicine.

12.7.4 Agriculture Industry

The agriculture industry has seen many changes in past years due to the advancement of ML technology. It gives an idea to the farmers to provide an estimate of their expected cost, losses, and profit in forthcoming times. Prediction of the price of the crop, weather and soil conditions, soil texture, air quality, and livestock management can be made using ML.

12.7.5 Sales and Marketing

All web platforms suggest a few items based on the search history of the customer. ML is used in this field to analyze the buyer's purchase history and provide more recommendations based on their interest in promoting items. Also, it helps in controlling product flow and handles product inventory demand and maintains supply. Microsoft tool called Azure is used by major industries for solving the purpose.

12.8 MACHINE LEARNING PREDICTIONS IN HEALTHCARE

ML algorithms play a significant role in identifying complex patterns in massive data. Medicinal data are complex proteomic and genomic measurements. ML thus finds its application in healthcare due to its handling capacity of complex data. ML is also used in the treatment and identification of various diseases. ML algorithms in medical systems can make better decisions about patient treatment plans by creating an integrated healthcare infrastructure. For the prediction of various disease, ML algorithms provides better insight. Few glimpses of major diseases in which prediction methods are successfully applied are discussed in this section.

12.8.1 Predictions of Stroke

Stroke is a common and frequently occurring disease that affects more than 500 million people worldwide. It costs about $689 billion annually, affecting many countries and families. This chronic disease poses acute events. Stroke is generally caused by a thrombus in a vessel called cerebral infection. Treatment of this disease and prediction is very much important. Immense research on the prediction of stroke is being carried out by data scientists.

The prediction of a stroke at an early stage using fuzzy state machines and PCA is carried out. Maninini et al. also have proposed another prediction of stroke by wearing a mask device to collect data of neurologial patterns and alert medical experts when a deviation from normal pattern occurs using Support Vector Machine (SVM). Both methods have provided an accuracy of 90.5% and 84.5%, respectively [10].

12.8.2 Predictions on Thyroid Disorder

ML methods are used to anticipate thyroid illness. UCI repository datasets are used by support vector machines and decision tree classification-based algorithms to predict thyroid disorder.

Papageorgiou EI and Papandrianos[7] NI have suggested many thyroid diagnostic strategies using a fuzzy map that uses data mining algorithms. This method has provided an accuracy of 89% on analysis.

12.8.3 Prediction of Diabetic Diseases

Diabetes is a chronic illness and a major public health disease worldwide. It evolves due to elevated blood sugar levels. It was also stated to lead to the development of Alzheimer's disease when the levels remain high for a long time. Many techniques have been used to provide quality treatment for diabetes.

Datasets are considered from the UCI repository. Kumari and Chitra have suggested the identification of diabetes disease using SVM that uses ML algorithms [11]. An accuracy of 78% in achieving the target is reported. Sarwar and Sharma have also suggested a Naïve Bayes algorithm based on diabetic detection using real-time values from different sectors of society of India and reported an accuracy of 95% in the identification of disease ahead and quality treatment [12].

12.8.4 Prediction of Bladder Volume

As a result of spinal cord injury or aging or other neurological disorder, specific complications arise in inpatient health conditions. One of them is the failure of bladder storage and urination function. Few researchers proposed a real-time approach by developing DSP for pressure and urine sensing volume by monitoring neural activity created by the bladder wall [13].

Fuzzy is used to categorize the bladder fullness into three levels. The ML algorithms are applied for estimation and early detection of volume and pressure prediction. The mean volume error of estimation of about $5.8\% \pm 3.1\%$ is achieved using the proposed method.

12.8.5 Detection of Breast Cancer

Cancer can be considered as one of the most deadly and least understood afflictions of modern times. A major mortal rate of people dying due to cancer is high in the last few years. AI challenge remains in the detection of Early or 'pre-cancer' symptoms as current methods rely on analysis done from scanned images, hereditary testing information, or multiple techniques that are costly and time-consuming [14]. Many types of research are being conducted on the early detection of breast cancer.

Shajahaan et al. [15] have proposed a breast cancer detection algorithm using Classification and Regression Tree (CART) and a classifier called Naïve Bayes as a model to detect the advancement of cancer by utilizing the dataset from the University of Wisconsin hospitals. An accuracy of 92.42% and 97.42% is observed using the proposed method. Decision Tree algorithms, SVM, and Artificial Neural Network (ANN)-based models were also proposed to predict the reoccurrence of cancer in another research. This research studied the performance of the above three algorithms through their characteristic(sensitivity and accuracy) data. The proposed work showed that SVM-based classification algorithm predicts the recurrence of breast cancer with high precision and less error.

Prediction of various other diseases like tuberculosis (TB) and asthma is also being carried out using ML techniques [16].

12.9 CONCLUSION

At the end of this chapter, it is evident that ML has been transformed from mere learning subject to practically being implemented in many billion-dollar industries. Abundant works on ML have drawn many to research using ML in large applications successfully. This also shows that ML has given healthcare a new tool to work with and led to the discovery of a new path to focus on early predictions of many deadly diseases. A lot of difficulties and limitations in dealing with an enormous amount of noisy, undistributed, unstructured, and dynamic data are being faced. ML's capability in dealing with the above challenges gives it a certain degree of exploitation in the field of analytics. ML in healthcare will enable the field to advance significantly in the future, improving the quality of life of people ahead.

REFERENCES

1. Nithya, B. 2016, "Study on predictive analytics practices in health care system," *IJETTCS*, vol. 5, pp. 98-102.
2. Nithya, B., Ilango, V. 2017, "Predictive Analytics in Health Care Using Machine Learning Tools and Techniques". In *2017 International Conference on Intelligent Computing and Control Systems*, Madurai, Tamilnadu, pp. 492–99.
3. Anderson, J., Parikh, J., Shenfeld, D. K., et al. 2016, "Reverse engineering and evaluation of prediction models for progression to type 2 diabetes: Application of machine learning using electronic health records," *Journal of Diabetes Science and Technology*, vol. 10, pp. 6–18.
4. Tseng, C. J., Lu, C. J., Chang, C. C., Chen, C. D. 2014, "Application of machine learning to predict the recurrence-proneness for cervical cancer," *Neural Computing and Applications*, Springer, May 2014, vol. 24, Issue 6, pp. 1311–1316.

5. Mitchell, T. 1997, *Machine Learning*, McGraw Hill, New York.
6. Beam, A., Kohane, I, 2018. "Big data and machine learning in health care," *JAMA*, vol. 319, Issue 13, pp. 1317–1318.
7. Lantz, B. 2013, *Machine Learning with R*, Packt Publishing Ltd., ISBN 978-1-78216-214-8.
8. Ghassemi, M., Naumann, T., Schulman, P., Beam, A. L., Ranganath, R. Opportunities in machine learning for healthcare, 2018, arXiv preprint arXiv:1806.00388.
9. Jordan, M. I., Mitchell, T. M. 2015. "Machine learning: Trends, perspectives, and prospects," *Science*, vol. 349, p. 255.
10. Kumari, V. A., Chitra, R. 2013. "Classification of diabetes disease using support vector machine," *International Journal of Engineering Research and Applications*, vol. 3, pp. 1797–1801.
11. Martinelli, L., Kopilaš, V., Vidmar, M., Heavin, C., Machado, H., Todorović, Z., Buzas, N., Pot, M., Prainsack, B., Gajović, S. 2020. Face masks during the COVID-19 pandemic: A simple protection tool with many meanings. *Frontiers in Public Health*, 8, p. 606635. doi: 10.3389/fpubh.2020.606635. eCollection 2020.
12. Sarwar, A., Sharma, V. 2012. "Intelligent Naïve Bayes approach to diagnose diabetes type-2. Special Issue," *International Journal of Computer Applications and Challenges in Networking, Intelligence, and Computing Technologies*, vol. 3, pp. 14–16.
13. Mendez, A., Sawan, M., Minagawa, T., Wyndaele, J. J. 2013. "Estimation of bladder volume from afferent neural activity," *IEEE Transactions on Neural Systems and Rehabilitation Engineering*, vol. 25, pp. 704–15
14. Wang, X., Oldani, M. J., Zhao, X., Huang, X., Qian, D. 2014. A review of cancer risk prediction models with genetic variants. *Cancer Informatics*, vol. 13, Issue S2, pp. 19–28. doi: 10.4137/CIN.S13788. PMID: 25288876; PMCID: PMC4179686
15. Shajahaan, S., et al. 2013, "Application of Data Mining techniques to model breast cancer data," *International Journal of Emerging Technology and Advanced Engineering*, vol.3, pp. 362–369.
16. Anderson, J., Parikh, J., Shenfeld, D. K., et al. 2016, "Reverse engineering and evaluation of prediction models for progression to type 2 diabetes: Application of machine learning using electronic health records," *Journal of Diabetes Science and Technology*, vol. 10, pp. 6–18.

13 A Collective Study of Machine Learning (ML) Algorithms and Its Impact on Various Facets of Healthcare

Roshan Lal and Sandhya Tarar
Gautam Buddha University

CONTENTS

DOI: 10.1201/9781003050827-13

13.1 INTRODUCTION TO MACHINE LEARNING

Machine learning (ML) comes under the umbrella term of artificial intelligence that serves the exact purpose of deriving knowledge from information. It provides a system with the capability to learn and improve based on past results, without being told exactly what to do [1]. ML techniques are used to find the patterns within complex data that would be quite hard for professionals to look for. These can be used to predict future events, or help in complex decision-making [2].

Traditionally, most of the problems were solved based on a logic proposed by a human along with data as an input, but ML proposed that data and solutions be given as input to discover the rules behind a problem [3]. To predict the rules behind a problem, the machine would have to try many different approaches, which would give different accuracies, this process is called learning. Hence came the name derivation of the name "Machine Learning".

ML being incontestable is more of an authoritative and dominant aspect of technologies in today's scenario. Its perspective has never been given any prominence or importance [4]. It will turn out to be an outrageous headline for the coming world. The turnout of information to worthy knowledge is presented by ML as a gizmo. In the first half of the century, there has been an outburst of data that was never seen before. Therefore, there is a need to analyze this data and to find patterns associated with it to have a better understanding and grasp of it [5]. Concepts and theories of ML are applied instinctually to perceive the resourceful patterns within the complex forms of data which in terms might be a difficult task to overcome. These future events are detected/predicted by certain hidden patterns and knowledge about a problem for performing and making a variety of complex decisions.

In the present day and age, where there is a lot of technological exposure, most people are unknowingly already making constant interaction with applications of ML in their day-to-day lives. At any hour when we make a Google search, concentrate on a track, or even take a photograph, navigating routes at the backend is ML , persistently studying and enhancing interaction after interaction. It has also been the backend for challenging world advances like cancer detection, growing new tablets, and self-driven vehicles [6]. It works on teaching the machines an ability that is found naturally in living beings, which is the ability to learn from past experiences. It is a subfield of a vast topic known as artificial intelligence (AI) and is widely used in data analytics. The learning of the machine can be in the form of new skills, it can be in the form of recognizing patterns, etc. With the help of ML, the machine need not be programmed beforehand to perform every task, but rather it can be taught to self-learn to do them [7].

The main aim here is to teach the machine to design a program that can take input and give the desired output. It no longer remains the old form of computing where we create the program that takes the input to give output. The algorithms that are used in ML make it possible for the machine to learn directly from the data rather than having to specifically provide the equation needed to solve the inputs [8]. With self-learning, the machine also becomes self-improving; with the increase in data provided to it the more accurate it gets and more efficient becomes its performance.

13.2 MACHINE LEARNING VS. DATA MINING

ML while often confused with data mining is quite different. Even though both the methods traverse, study. and analyze data to get a pattern in them, they use the information differently [9]. Data mining provides the learned information to humans so that they can interpret these patterns and make use of them as required, whereas ML uses the information to give inferences on the basis for future inputs belonging to the same area; thus, it creates a self-working program that can then be modified to use as is required [10].

13.3 ESSENTIAL KEYWORDS

a. **Dataset**: A collection of data that contains the essential features for solving a problem.
b. **Features**: Essential chunks of data that make the problem understandable for us. They are stored in the algorithms of ML.
c. **Model**: It is the mathematical representation of model parameters for each input parameter for prediction or the class and action for the regression, classification, and reinforcement techniques.

13.4 STEP-BY-STEP MECHANISM OF ML

a. **Data collection**: Collection of data for the algorithm.
b. **Data preparation**: The collected data is cleaned, edited, and prepared in a usable or required format, removing the factors that are not needed and essentially keeping the major factors that are dominantly required for the predictions. Dimension reduction may also take place depending on the dataset.
c. **Training**: The training phase, also called the fitting stage, is where the ML algorithm learns from the broadcasted data that was passed from the previous stage.

d. **Evaluation**: Testing is done to assess the performance and efficiency in terms of error of the applied ML algorithm on the dataset.
e. **Tuning**: The original performance is maximized in this stage.

13.5 HISTORY OF MACHINE LEARNING

Essentially, we can say that a mathematical representation can be used to create the relation between the representations of the phenomenon. The foremost engineer Ada Lovelace states that machines are capable enough to understand the world without the assistance of a human.

It took two centuries, some core objectives are quite critical in the concepts of ML. Irrespective of the clause, the obtained knowledge can be graph plotted as data points. then comes the concept and algorithms of ML with the authentic piece of knowledge [11].

13.6 PROBABILITY THEORY

Thomas Bayes, the mathematician he originates the ideas of propagating probability theory into ML algorithms. The world is filled with probability. All our actions are uncertain; they come up with different possibilities. Bayesian probability refers to computing the uncertainty of an event on which ML is based.

As a result of this, we consider our probabilities based on the information available about an event, rather than counting the numbers of repeated trials [12]. Let's consider an example for this, on predicting a Cricket Match, instead of counting the number of matches won by India against Australia, the Bayesian considers to-the-point information of them like their current forms, starting teams.

The core advantage of taking the Bayesian approach is that the rare possibilities can be based on their relevant features and reasoning [13].

13.7 MACHINE LEARNING APPROACHES

There are quite a few approaches that can be taken into consideration in ML. The supervised approach and the unsupervised approach are quite well-known approaches. They are mostly used in the field due to their nature, which is discussed later. ML is at the beginners' level and more complex in nature but is used to generate remarkable outcomes.

The "No free lunch theorem" is the most interesting and popular in ML. It states that there cannot be only one algorithm that can work well for all the functions. Every function which we try to solve has its way to predict/solve the ideology. In light of this, there are a lump-sum of Algorithm and approaches that help in solving such problems with their methodologies [14]. Certain new methodologies of ML and Artificial Intelligence are coming up to best suit these problems.

Figure 13.1 describes the broad categories in which ML can be subdivided. They branch into the supervised learning approach and unsupervised approach. Supervised learning can be further divided into two categories – classification and regression. Unsupervised learning can be further classified into clustering models.

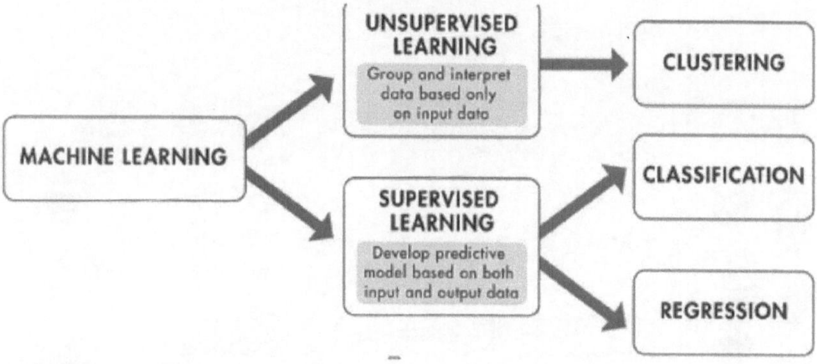

FIGURE 13.1 Machine learning approaches.

ML works according to the algorithms it is based on; there are three different broad types of algorithms a ML model could be following these are as follows.

13.7.1 Supervised Machine Learning

This algorithm focuses on building models in times of uncertainty by making predictions based on evidence [25]. The algorithm works with input data that is labeled and also whose output is known and labeled. It takes these to derive the working that gave the output for the following input and then uses this knowledge to provide a prediction of output for the new data whose output was not previously known. The main point of focus is that it works with labeled data.

An example could be figuring out if the given equipment will be useful or not based on the labeled data points and the output for them. The algorithm gets better over time by comparing the predicted output with the given output to compare the error, after which it makes modifications for the betterment of the predicted result [15].

Supervised learning is often useful where historical data can give insight into future possibilities. Predicting fraud credit card transactions is one of the many uses. It uses classification techniques and regression techniques for developing predictive models.

a. **Classification techniques**: the technique predicts discrete responses. Example: can tell apart spam from genuine email, etc.
b. **Regression techniques**: The technique is used for giving and predicting continuous responses.

For example, it can predict fluctuations in the demand for power, etc.

Figure 13.2 describes the concepts of classification and regression. Supervised learning explains the concepts of classification and regression with the help of an example.

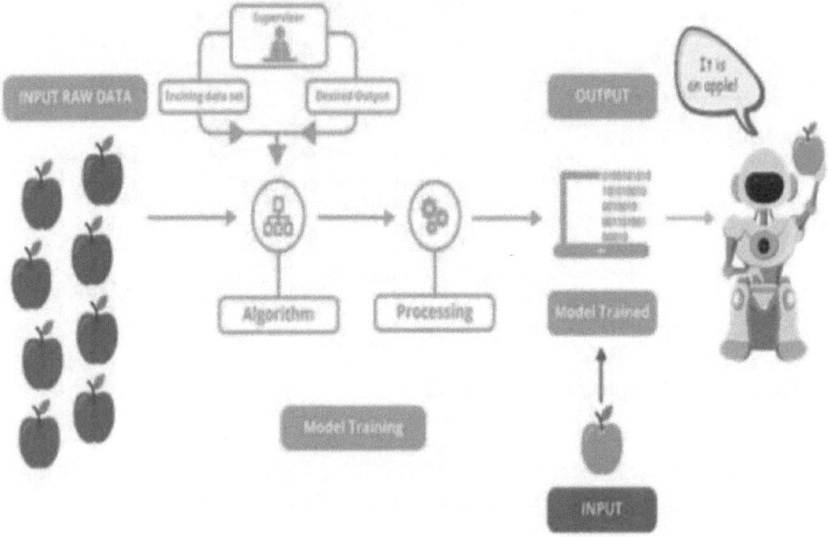

FIGURE 13.2 Supervised learning.

13.7.2 Unsupervised Machine Learning

Unsupervised learning is used when we are faced with data that has no labels. The system does not know the "output or right answer" in advance. The algorithm has to derive meaning from the input, i.e. the algorithm has to figure out what the input data signifies. It finds these hidden patterns or the intrinsic structures within the set of data [16]. The inference it draws is based on its patterns rather than the labeled output provided. The kind of data which works best for this algorithm is transactional data.

An example could be to feed a large number of customer databases and then segregating those customers into groups such that each group will require a similar type of action being taken over them for maximum success in productive marketing campaigns.

The techniques that are most used could be self-organizing or nearest-neighbor mapping, it could be clustering or decomposition.

Clustering: This unsupervised learning technique is the most common among all. It is used for data analysis that focuses on exploration by finding patterns and grouping similar patterns. Figure 13.3 describes the concept of clustering.

13.7.3 Reinforced Machine Learning

The reinforcement learning algorithm is most sought after in the fields of navigation, gaming, and robotics. It works based on trial and error to discover the best path to follow to gain maximum rewards [17].

There are three primary components of reinforced ML – the learner, the environment, and actions. The aim is to choose the action that will yield the maximum

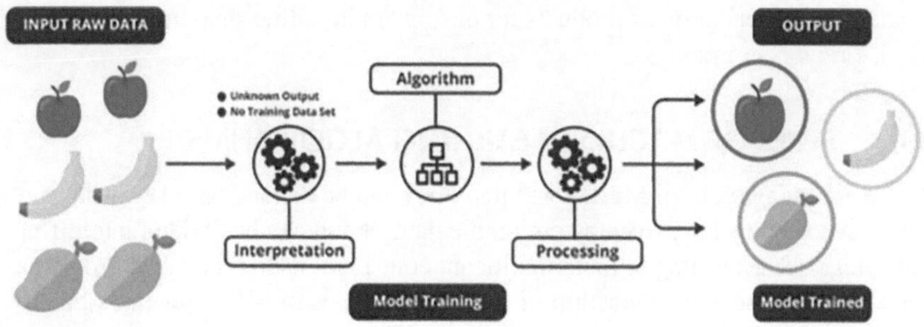

FIGURE 13.3 Unsupervised learning.

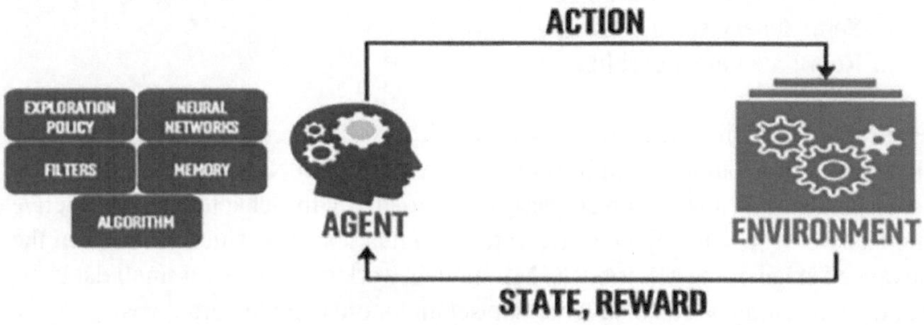

FIGURE 13.4 Reinforcement learning.

reward in a constrained time. The best way to make the agent learn well is to award it over good policy so it eventually follows the best policy [18]. Figure 13.4 describes the concept of reinforcement learning with the help of an example, highlighting the reward points of the agent.

13.8 MACHINE LEARNING APPLICATIONS

ML is a very important field and according to researchers we are in the golden era of artificial intelligence techniques; there are various learning methods of ML methodologies. The ML technique is used to solve many real-world complex problems which cannot be solved with traditional approaches. The following are some real-world applications of ML techniques:

a. Customer segmentation
b. Speech recognition
c. Speech synthesis
d. Stock market analysis and forecasting
e. Error detection and prevention
f. Sentiment analysis
g. Emotion analysis

h. Recommendation of products for customers in online shopping
i. Fraud prevention

13.9 TYPES OF MACHINE LEARNING ALGORITHMS

There are many types of ML algorithms. They can be defined based on what specific purpose they have to carry out on the dataset and can be divided into different categories according to their significant contributions to real-world problems. So we can define each algorithm of the various types of ML algorithms one by one in detail.

a. Supervised algorithm
b. Unsupervised algorithm
c. Semi-supervised algorithm
d. Reinforcement algorithm

Figure 13.5 explains the different ML algorithms that are used. There are mainly three types of popular ML algorithms namely supervised, unsupervised, and reinforcement algorithms, which are described broadly in this chapter. More effective models are deep learning and deep reinforcement learning which are used in the advanced stage of the ML models. ML models work effectively on small datasets, but deep learning works on a large dataset and more complex problems.

13.9.1 Supervised Algorithm

Supervised learning is the process of learning in which a function maps an input to an output based on a data-driven set. Algorithms that fall under this category analyze the training data after which an inferred function produces that can further be used to map the training and testing data with label data. Supervised learning algorithms are easiest to understand and very simple to implement [19]. The supervised learning algorithm is very similar to teaching a child with the use of flashcards. For example, we assign the dataset role of a teacher whose aim

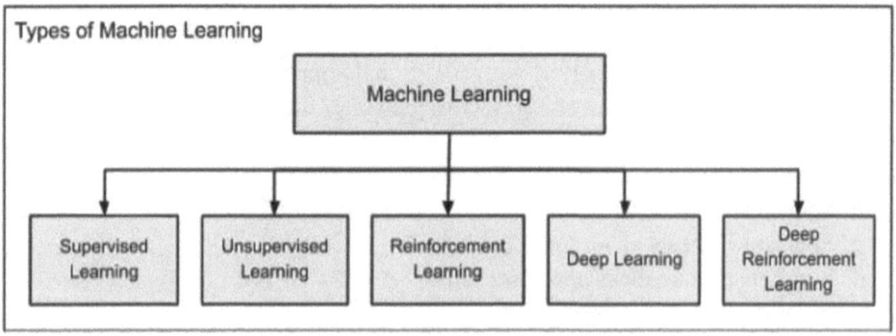

FIGURE 13.5 Types of machine learning algorithms.

is to train the child, which is the machine. After the machine gets trained by the dataset, it will be capable of predicting or deciding based on a new dataset given to the machine [20]. Some of the supervised learning algorithms have supervised learning goals that aim to carry out learning by mapping input over output.

Let us consider the following example: let the weather forecast be the input and the output be visitors on the beach. The aim is to get a relation between how weather plays an effect on the number of visitors on the beach. This will be done with the help of supervised learning because there are labels, which leads us to say that the labeled data acquired in this example is the set of past outcomes of input and output which declares how the model should work that is being "supervised". Here in this beach example, the input can be the weather forecast temperature and the algorithm of ML will predict the number of visitors. While training, the requirement of generalization is to be maximized so the relation of original "general" is defined by supervised learning [21].

Now there are two main terms associated with generalization: they are overfitting, which is when the model is over-trained and under-fitting. There's a bit of drawback in this type of learning that is supervised learning which is when you provide supervision it creates a bias to the learning. BIAS is a typical assumption created by the model itself to make the target function accurately. But the model we create needs to be unbiased, reliable, and similar to what it is shown to us. In addition to this, the supervised learning requires a huge amount of data before learning. The toughest and most costly task is to obtain reliable labeled data hence a keyword is given to data called NEW OIL.

To understand better let's consider the previous example again. Here the resultant output from a supervised ML model is the group from a finite set {low, average, high} for the number of visitors present on the beach:

Considering the input to be:

$$\{\text{temperature} = 27\} \rightarrow \text{Model} \rightarrow \text{Output } \left\{\text{visitors} = \textbf{high}\right\}$$

When such kind of case is present, it's said to be deciding to classify the input, hence being called Classification.

Whereas on the other hand, the output can also be a scalar quantity (a number):

$$\text{Input } \{\text{temperature} = \textbf{27}\} \rightarrow \text{Model} \rightarrow \text{Output } \left\{\text{visitors} = \textbf{350}\right\}$$

Then such a case is called Regression.

Figure 13.6 describes the concepts of classification and regression. Figure 13.6a explains the classification of the ML models, and Figure 13.6b explains the concepts of regression of the ML models.

Classification

Classification refers to the grouping of similar kinds of data points into distant parts to classify them. But the question arises: how to create these algorithms?

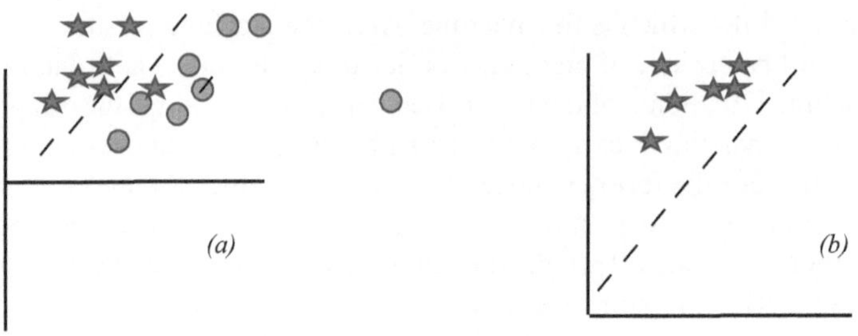

FIGURE 13.6 (a) Classification and (b) Regression.

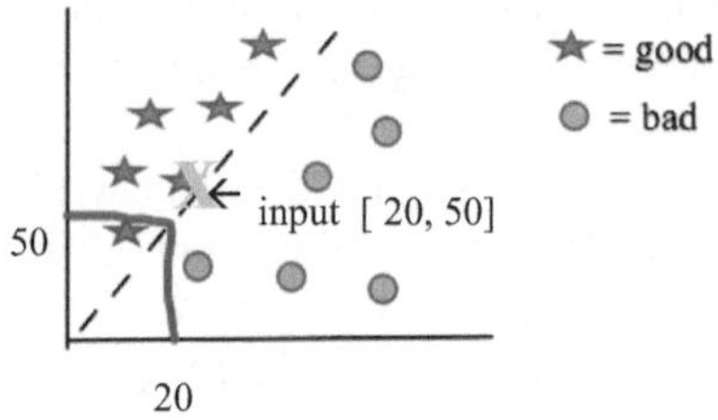

FIGURE 13.7 Classification model.

The answer to this question is, there are several ways to discover algorithms focusing majorly on the data and the replies for the discovery of the algorithms that linearly distinguish data points [22].

Linear distinguishing or linear reparability is a key concept in ML which the help of which we can conceptualize whether distinguished data can be bifurcated by a line or not. Classification heads ahead to allocate the optimum way to differentiate the data with a line [23].

Decision boundaries are the lines drawn to distinguish different categories. Decision **surface is the area used to define a class**. If a data point falls within the boundaries on the decision surface, it will be assigned that certain class.

Figure 13.7 explains the concept of the classification model of ML. This classification model divides similar types of data points into different groups. This is explained with the help of good and bad data points.

Regression

Regression being the other form of supervised learning leads us to tell us there is a difference between classification and regression. In regression, the output is a scalar quantity, i.e. a number is obtained instead of a class being obtained as the output. Therefore, Regression is more powerful than classification while predicting number-based questions.

Figure 13.8 explains the concept of regression of the ML model which is used to compute the trends of the dataset and presents the trends of the present and

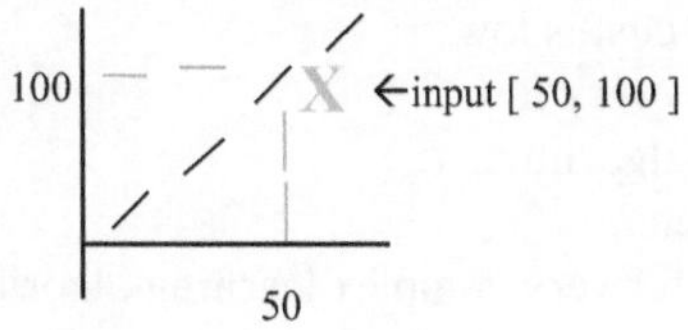

FIGURE 13.8 Regression model.

future based on the past dataset of the problems. To find the future trends of the market, regression models are the best models of the ML models.

Receiving Emails that are spam is a type of classification and analyzing the patterns in the stock market and other assets for the predictive buy/sell or profit of all these financial trading is done by Regression.

Classification as well as Regression both can be more complex tasks to be performed, for example, speech and audio tasks, image classifications, chatbots, and many others [24].

Some algorithms that come under supervised algorithms are as follows:

There are various algorithms of supervised learning which are described broadly in this section one by one.

a. Simple linear regression
b. SVM algorithm
c. Random forest
d. Decision tree
e. Logistic regression
f. KNN()
g. Naïve Bayes

13.9.1.1 Simple Linear Regression Algorithm

Simple linear regression is great learning for the data analytical process. It works with two variables out of which one is the dependent variable and the other one is an independent variable and finds the relationship between the two as a linear function. The following model function is considered and worked upon:

$$y = b_0 + b_1 x \tag{13.1}$$

where b_0 is the slope and b_1 is the y-intercept.

Advantages of Simple linear regression

a. The speed of the machine is fast.
b. The machine does not require complex calculations.
c. The machine runs very fast when the amount of data is huge.
d. Data is labeled in this type.
e. It is simple and easy to apply and understand.

f. The computational cost is low.
g. Space complexity is very low.
h. It is a high latency algorithm.
i. It is a good interpreter.
j. It provides ground for very complex learning algorithms.

Disadvantages of Simple linear regression

a. The machine does not work on non-linear data.
b. First, it has to be checked whether the data is linear or not linear. If it is simple linear, then it may be implemented.
c. The machine requires training dataset and test dataset.
d. The machine assumes data is normally distributed in real time.
e. Machine oversimplifies many real-world problems.

13.9.1.2 Random Forest Algorithm

A random forest algorithm is a supervised learning process that can be used for both regressions as well as classification [25]. It is used for classification tasks. As we know that a forest is made up of trees, and it is a very robust forest. Random forest algorithm creates a decision tree on sample data and then gets the prediction from each tree and lastly selects the best solution. It is an ensemble method that is better than a single decision tree. Random forest algorithm reduces the overfitting as compared to a single decision tree [26].

Advantages of Random forest algorithm

a. Random forest algorithm retains good accuracy.
b. Random forest algorithm does not scale data.
c. Random forest algorithm has very little variance.
d. Random forest algorithm reduces the overfitting.
e. Random forest algorithm retains good accuracy.
f. Random forest algorithm works very well on large-scale data.
g. Random forest algorithms possess high accuracy.
h. Random forest algorithms are more flexible rather than decision trees.

Disadvantages of Random forest algorithm

a. Random forest algorithm is very complex.
b. A random forest algorithm has time-consuming process than a decision tree.
c. Random forest algorithm has very hard construction.
d. Random forest algorithms require more computational resources.
e. Random forest algorithm takes more time in prediction with another algorithm.

13.9.1.3 Logistic Regression Algorithm

In logistic regression, the concept of the probability of odds is used, which lies between 1 and 0. Some real-life examples exist in situations where we have a student who can either pass or fail, a lottery that can either be won or lost, a person who is potentially sick or healthy. The cost function in the linear equation is taken as the mean squared error (MSE) [27]. If the same is used in a logistic regression algorithm, with the help of gradient descent one can find the global minima of the function that will converge if the function is convex in nature.

Advantages of Logistic regression

a. Implementation is very easy.
b. Efficiency increases once the data is trained.
c. It can easily overfit in high-dimensional datasets.
d. It performs well when the datasets are linear in nature.
e. It is highly interpretable.

Disadvantages of Logistic regression

a. Prediction of discrete functions can be done using this type of regression.
b. The dependent variable is bounded and restricted to the discrete set of numbers.
c. In logistic regression, if the number of features is more than the number of observations, it is not used because it may lead to induce overfitting in the model.
d. It is the linear assumption between the dependent variables and the independent variables. In the real world, the data is rarely linearly differentiable.

13.9.1.4 KNN (K-Nearest Neighbors)

KNN is a simple algorithm of ML used to solve classification problems. KNN means K-nearest neighbors where K is the number of neighbors in the KNN algorithm that does not require learning during the training period [28]. KNN takes datasets during real-time predictions. KNN is a very fast algorithm rather than others like SVM, linear and logistic regression, etc. KNN algorithm requires no training before predictions. So adding new data cannot impact the accuracy of KNN. KNN algorithm is very easy to implement requires only two parameters the value of K and the distance function.

Advantages of KNN algorithm

a. KNN algorithm has no training period of datasets so it is called lazy learner or instance-based learning.
b. KNN is very faster algorithm rather than others.
c. Adding new data may not impact the accuracy.
d. KNN is easy to implement due to only two parameters.
e. KNN is a very effective algorithm on small dataset rather than others.

Disadvantages of KNN algorithm

a. KNN algorithm does not work very well on a large dataset.
b. KNN algorithm has low performance due to the cost of the distance between the new point.

c. KNN does give optimum results with high-dimensional datasets.
d. KNN requires feature scaling before being applied to any data.
e. KNN is sensitive to noisy data, missing values, and outliers.
f. KNN needs to manually remove outliers.

13.9.1.5 Decision Tree Algorithm

As the name suggests, this algorithm works with the help of trees, which are flowchart-like structures with leaf nodes. In the tree, every leaf node depicts a label class, whereas the nodes inside (internal nodes) are tested on the attributes or features. A decision tree is an algorithm of ML which is used to solve both types of problems like regression and classification [29]. Decision tree leads to overfitting the data which is overcome by a random forest algorithm. The decision tree algorithm's output can be interpreted by humans in a very simple way.

Advantages of Decision Tree algorithm

a. It can be used in both regression and classification.
b. It is very easy to implement and simple to understand.
c. It can handle variables of both continuous nature and categorical nature.
d. It can handle missing values automatically.
e. It is robust to outliers.
f. It takes less training period as compared to random forest.
g. It uses if-else statements which are very easy to understand.
h. It can handle non-linear parameters efficiently.

Disadvantages of Decision Tree algorithm

a. It leads to overfitting data.
b. It leads to wrong predictions.
c. It has a very high variance due to the overfitting of the dataset.
d. It is very Unstable due to adding a new point.
e. It has a little bit of data noise that can make it unstable which leads to wrong predictions.
f. It is not used with large datasets; if the data size is large, then one single tree can grow complex and leads to overfitting.

13.9.1.6 Support Vector Machine (SVM)

This ML algorithm can be used for both classification and regression; however, it is used mostly to classify data. SVM was introduced in the 1960s and was further refined in the 1990s [30]. This algorithm aims at plotting data as a refined point in an n-dimensional plane such that the coordinate of the point is equal to the value of the features. SVM transforms the non-linear data into linear data and is used to find the best and optimal results of the problems [18].

Advantages of Support Vector Machine (SVM)

a. It has a good generalization with the help of which we can prevent overfitting from taking place.
b. SVM efficiently handles the non-linear data.
c. It follows and uses the Kernel method.
d. Classification and regression problems can both be solved using SVM.
e. SVM has more model stability.

Disadvantages of Support Vector Machine

a. SVM has difficulty choosing an appropriate kernel function.
b. SVM requires more extensive memory.
c. SVM has an algorithmic complexity.
d. SVM requires feature scaling.
e. SVM takes a long time for training data.
f. SVM is more difficult to understand.
g. SVM is interpreted by a human.

13.9.1.7 Naïve Bayes

A Naïve Bayes classifier is based on the probabilistic ordering of ML models. The heart of the classifier is the theory of Bayes:

$$P(A/B) = P(B/A)\ P(A)/P(B) \tag{13.2}$$

Using the hypothesis from Bayes, we can find the probability of an occurrence of A after B occurred. The proof here is *B* and the hypothesis is *A* [21]. It is meant to be free of metrics/highlights. It is the proximity between one specific variable and the other. It is either referred to as being faithful or naïve.

Advantages of Naïve Bayes

a. Naïve Bayes gives better performance than other models if the predictors are independent of each other.
b. As compared to other models, the Naïve Bayes theorem requires a smaller amount of training data; therefore, training period is less.
c. Naïve Bayes can handle multicategory tasks.
d. It is suitable for incremental training of data.
e. It is easy to implement.
f. It is relatively easy to understand, is relatively simple, and is used for the classification of text.
g. It explains the results easily.

Disadvantages of Naïve Bayes

a. In real-life problems, it is very rare to have predictors that are independent of each other.

b. In this algorithm, there is a need to calculate the prior probability.
c. It is sensitive to the form of input data.

13.9.2 Unsupervised Learning Algorithm

The unsupervised learning algorithm is a major category of ML algorithms that uses previously undetected patterns in a dataset that has no pre-existing labels, taking into account a minimum of human supervision [31]. In contrast to the supervised learning algorithm, it uses human-labeled data. It is self-organization that allows the modeling of probability. Unsupervised learning algorithm gains through the observation, also finding structures in the dataset simultaneously. Once the machine has been given a dataset, then it automatically starts finding the relations and in the dataset with the help of clustering analysis.

Some of the followings examples of unsupervised algorithm are as follows:

a. K-Means algorithm
b. Apriori algorithm
c. Clustering algorithm

13.9.2.1 K-Means Algorithm

K-means algorithm is an iterative method that creates subgroups by predefined K distinct clusters. Each point belongs to one group without any overlapping. With the help of centroids, the data points in the clusters are clubbed which are similar and closer to the other intra-cluster points simultaneously keeping the different clusters as far away from other clusters as possible [32]. The algorithm works as follows:

a. The number of clusters (K) that will be used is specified.
b. Initial centroids are selected by randomly picking K data points from the dataset.
c. The Sum of mean distance is computed between the centroids and all data points.
d. Each point is put under the closest calculated centroid and hence the closest cluster.
e. Centroid is updated by assigning the new value as the average of the data points present in the cluster.
f. Steps (c) to (e) are repeated until there is no change in the assignment of the clusters of the data points. Hence, it is an iterative algorithm.

Advantages of K-Means

a. It is easy to implement.
b. It scales to larger datasets.
c. There will be a definite convergence.
d. It easily adapts to new examples.

Disadvantages of K-Means

a. The value of K is difficult to predict and is often chosen manually.
b. There may be different final results with different initial partitions.
c. It does not work effectively with clusters of different densities and sizes.
d. Centroids are affected by outliers and therefore should be considered to be removed or clipped before making subgroups.

13.9.2.2 Apriori Algorithm

This unsupervised learning algorithm is often used for data mining and applies the downward closure property. According to this property, in a subset of data items, the frequent itemsets occurring in the dataset are frequent; however, the converse may or may not hold [33]. It identifies individual items that occur frequently in the transactional databases. Two parameters are used in this algorithm, namely being "support" and "confidence". The prior is associated with the frequency of the item's occurrence whereas confidence refers to the conditional probability [34]. The main steps of the algorithm are:

1. The candidate set is generated, i.e. support of each item in the database is calculated.
2. The candidate set is updated by eliminating or removing the items that have lower support than the given threshold.
3. Frequent itemsets are joined and steps 1 and 2 are repeated until no new itemsets can be joined in the final set.

Advantages of Apriori algorithm

a. It is easy to understand.
b. It is easy to implement.
c. The results are easy to communicate to the end-user as they are intuitive in nature.
d. Various scans are created for candidate sets.
e. It can easily be parallelized.

Disadvantages of Apriori algorithm

a. It is slow in comparison and the bottleneck is candidate generation in this algorithm.
b. It requires a large number of database scans.
c. It assumes the transaction database is a memory resident.
d. Performance time is more.

13.9.2.3 Clustering Algorithm

Clustering is a data analysis technique. It aims at identifying and creating subgroups of the given data and based on the given attributes of the sample space [35]. The differentiation and division of the data points are based on the similarity between the points.

Advantages of Clustering

a. It is relatively easy to implement.
b. It is easy to understand.

Disadvantages of Clustering

a. In this algorithm, it is difficult to detect if the dataset is homogeneous or contains clusters.
b. Variables cannot be ranked by their contribution to the heterogeneity of the dataset.
c. Special patterns (zone of mixing, noise, outliers, etc.) cannot be detected.
d. Inability to estimate a correct number of clusters.

13.9.3 Semi-Supervised Learning Algorithm

The semi-supervised learning algorithm is a process of combining a supervised approach and an unsupervised approach [36]. A semi-supervised learning algorithm is used when less amount of labeled data is used with a greater amount of unlabelled data while training a dataset. It is simple to understand and easy to implement on the given dataset.

Advantages of Semi-Supervised Learning

a. It is helpful when labeled data is hard to get or expensive.
b. It can be used in NLP.
c. It has a lot of medical applications.
d. Unlabelled data is cheaper than labeled data.

Disadvantages of Semi-Supervised Learning

a. It makes a lot of assumptions like the smoothness, clusters, manifolds, etc.
b. Some techniques require a very specific setup.

13.9.4 Reinforcement Learning Algorithm

The reinforcement learning algorithm is an algorithm in which we use the ability of an agent to interact with the environment, which leads us to find the best outcome. It is based on the concept that is followed is the hit-and-trial approach. The agent is rewarded with a point for a correct answer, i.e. plus point, or penalized with a point for a wrong answer, i.e. minus point. On the grounds of the reward points gained by the machine, it retrains itself. Once it has been trained, it becomes ready to predict new data that is given to it. An example of the reinforcement learning algorithm is

a. Markov Decision Process
b. Q-Learning
c. Chess Game

13.9.4.1 Characteristics of Reinforcement Learning algorithm

a. Determination of the subsequent data that will be received will due to the agent's actions.
b. There is no concept of a supervisor, only a real number or reward signal exists.
c. Feedback is always delayed.
d. Time plays an important role.
e. Sequential decision-making in terms of reward points.

13.9.4.2 Reinforcement Learning Algorithm Applications

There are various applications of reinforcement learning algorithm

a. Airspace control system
b. Robot motion control system
c. Data processing
d. Planning for business strategy
e. Robotics for industrial automation
f. Reward points for gaming

Advantages of Reinforcement Learning algorithm

a. It can be used to solve complex problems.
b. It gives long-term results.
c. Errors can be corrected during the training process.
d. If there is an absence of a dataset, it can learn from experience.

Disadvantages of Reinforcement Learning algorithm

a. This type of learning can be non-stationary.
b. It may lead to an overload of states.
c. Realistic environments may have partial observability.
d. Parameters can be the affect the speed of learning.
e. Reward design should be involved.

13.10 SMART HEALTHCARE

Health is an integral aspect of life. Traditional health systems can be divided into three layers, i.e. the range layer, the transfer layer, and the investigation layer. The Body Area Network (BAN) sensor layer [37] collects detection data by design, sends it to a center or base station (BS) via an intelligence device or advanced mobile telephone. The door or the BS sends the doors to an inquiry process (for example, a cloud server farm). AI and input mining calculations on the cloud server farm extract and verify the information. Ultimately, the program acquires the well-being and health treatment reference metrics of consumers [38]. People nowadays work like machines to get more cash, which is why they are neglecting their welfare. The food they eat changes in this context, and they change their way of life. There is constant stress on the weight of cash production. This results in very young conditions such as blood pressure, diabetes, and cholesterol [39]. This leads to improvements in health. All these factors lead to a lack of care that raises the risk of heart failure. The heart is the most vital organ of the human body, and it affects the other essential bodies in the body. The purpose of the program proposed is the creation, along with the background of the patient, of a computerized system for the follow-up of complete clinical subtlety. Clinicians will use this program to seek patient advice [40]. The aim is to help persistent prices, stock management, and the age of clear insights in various data structures at clinics [41].

Medical judgments are mostly based on physicians, instinct, and knowledge instead of the rich data in the database. (1). Unwanted biases (2). Clinical mistakes and (3). The insurance costs for the patients were massive, impacting the standard of care. The proposal is to define these issues through the integration of emotional network options and PC-based knowledge. The system for the mining of information can create a comprehensive situation to improve the quality of clinical choices. The support for the quality of clinical structure means effective patient care and strong medical control. Poor clinical choices will lead to obviously unsatisfactory catastrophic outcomes. There should not just be emergency clinics [42] for the expense of clinical trials. These findings can be obtained using ample computer-based knowledge and emotionally supportive networks. To locate hidden information for a production model, the social security sector gathers considerable knowledge on public assets which, surprisingly, are not "mined". Related instances and relations are not yet discovered. Such structures as the decision tree and the naïve Bayes will lead to a cure [43].

In 2009, IBM introduced a global initiative to encourage the achievement of global progress by using smarter systems, one of which was "smart healthcare". It is an intelligent under-structure in which information is transmitted with the help of devices such as sensors, and communicated further through the Internet of Things (IoT). This transmitted information is then processed with the help of cloud computing. Smart healthcare uses new wearable and mobile technology to connect users to healthcare institutions thereby managing and keeping a record of their health status as well [44].

Smart Healthcare being essentially important for today's world as Acute myocardial infarction is being one of the most dangerous disorders facing patients. The key to the management of cardiovascular disease is to analyze a broad variety of datasets, compare them, and my knowledge for prediction, diagnosis, management, and treatment of chronic conditions such as cardiac diseases [45]. Big data analysis is well known in the business world for its use in broad datasets monitoring, contrast, and management, and the prediction, avoidance, management, and treatment of cardiovascular diseases can be effectively implemented. Here in this report, we'll be using smart clothing techniques along with some ML algorithms to predict heart attacks. Heartbeats can be easily monitored through our wrist, the wrist of the neck, within the elbow, behind the knee and in the lower leg, and joints; all these places can be easily be monitored by smart clothing like shirt, t-shirts or lower pants which will be equipped with nanochips to monitor your heart rate and provide you time to time updates on your handsets.

Sometimes there may present abnormalities or ruptures deep beneath the skin in the human body. For diagnosis of such conditions, diagnostic equipment such as MRI, CT, and PET is required. There are some diseases such as heart attacks and epilepsy that can be predicted before they have occurred. But the increase in population has put a lot of strain and pressure on the modern healthcare systems [46], increasing the demand for the resources required for healthcare procedures and diagnosis [47]. IoT provides us with a way to reduce the pressure on the healthcare facilities, by being a potential solution to curb this pressure parallelly maintaining the standard and quality of healthcare to its optimum level [48].

Smart healthcare creates a sphere that promotes interaction between the user, the healthcare field party, the services the user requires, and a real-time diagnosis of the user. Truly, smart healthcare is the higher stage and upgrade of the traditional healthcare system. It is a multidimensional sector that includes disease monitoring, diagnosis, treatment, and health-related decision-making, simultaneously ensuring that the patients get the services they need [49]. There are multiple connections and participants connected with a single disease case including patients, doctors, hospital facility, and research facilities.

Patients can use wearable mobile devices to keep in check and monitor their levels at all times, seek medical assistance through virtual means followed by clinical decision support that guides the patient on how to improve their health and diagnosis [50].

Doctors work through integrated platforms like Laboratory Information Management System, Electronic Medical Record, Picture Archiving and Communication Systems, etc. Optimum precision on surgeries can be achieved by combining the power of man and technology, namely the surgical robots supported by reality technology.

Healthcare facilities or hospitals manage the supply chain and decision material using radio-frequency identification technology to collect information of the patient and to assist in the diction making of the diagnosis.

13.11 ROLE OF IoT IN SMART HEALTHCARE SYSTEM

Today, with the rise in population, the world has seen a subsequent need for healthcare to tackle the rise in diseases. The healthcare industry is a dynamic and ever-changing sector that has been incorporating new technologies to improve the quality of services for the patients in terms of efficiency, management cost, and speeding diagnostic treatments and procedures [52]. Healthcare that is provided in hospitals and other medical environments can be stressful, especially for the age group mainly consisting of children and senior citizens. The existing healthcare systems following the traditional approach are still being used in many parts of the world but they are unable to accommodate all the needs efficiently. Irrespective of having technologies and quality infrastructure, medical services may not be affordable nor approachable to everyone. Taking this to the next level, a relatively new and upcoming paradigm – smart healthcare, is being introduced in many medical facilities. Unlike traditional healthcare facilities, smart

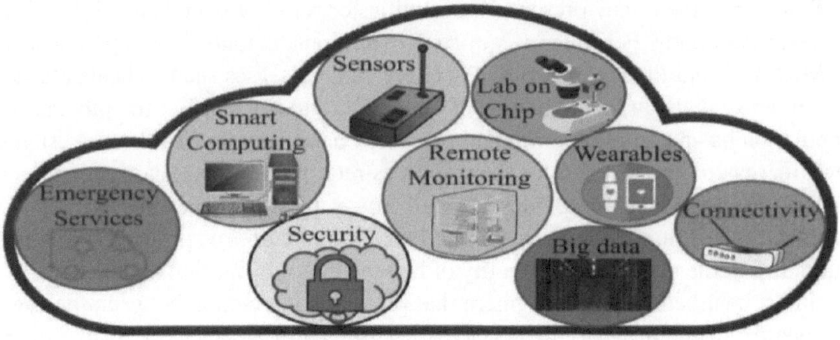

FIGURE 13.9 The IoT in smart healthcare.

healthcare focuses on empowering and educating the patients to self-manage in emergencies [53]. It gives the user a better quality of service and experience, going hand in hand with the best optimization of the available resources to their maximum potential. Although the first take of smart health was with the association of the smart cities, the paradigm has been generalized from smart cities to any smart domain like smart hospitals, smart infrastructure, smart homes, etc. Smart healthcare brings the best out of the capabilities of sensing, mobile devices, and network infrastructure [54]. It supports remote monitoring of patients as the medical assistance and services can be extended by practitioners without having any geographical barriers and lowers the cost of the treatment.

Figure 13.9 explains the concept of IoT in smart healthcare with the help of this diagram.

13.12 SMART HEALTHCARE SYSTEM AND INTERNET OF THINGS (IoT)

IoT had taken over many aspects of our life, and its applications are implemented in many fields such as healthcare, agriculture, transportation, etc. Due to IoT, smart healthcare can efficiently be multidisciplinary with many big data, cloud computing, ML, VLSI, Artificial Intelligence, embedded systems, and many more. It can be implemented at many levels, its complexity and cost varying based on the functionality, required precision of the setup, and its application.

Smart Healthcare can be broadly classified based on application, the IoT network connectivity technologies, end-users, medical equipment, services, and system management. Devices in the setup should be capable of communicating with one another to enter the IoT landscape [55]. The application decides the setup and precision for effective connectivity. In some scenarios, the transmission of a couple of bytes may be is required intermittently or in a heavy stream of data. Some use cases might require low latency with high robustness because of the mission-critical nature of the application, while others may focus on a lower QoS.

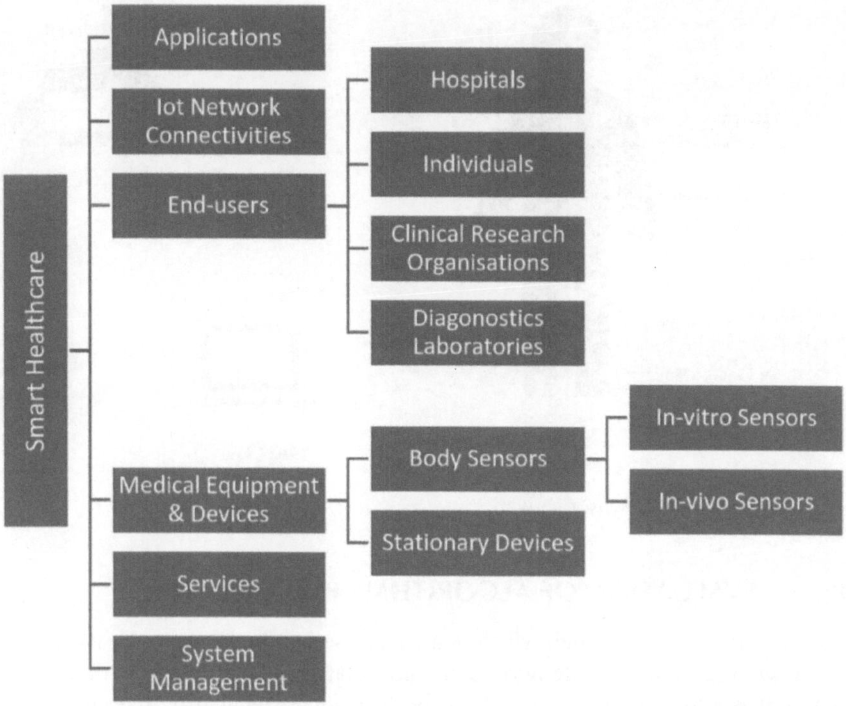

FIGURE 13.10 Smart healthcare classification.

Figure 13.10 explains healthcare system applications can be expanded using the IoT network connectivity technologies. Several devices can be integrated through wireless technology through which IoT can be implemented in monitoring remote healthcare systems. In healthcare scenarios, the hospital network is maintained with the help of ground cables and Wi-Fi for consistent internet connectivity and heavy data traffic support.

Medical equipment and devices that can be implemented in smart healthcare can be divided into two categories: body sensors and stationary devices. Body sensors or a network of body sensors which is known as Body Sensor Network (BSN), and it is an extensively used technology used in the healthcare system which functions on IoT. It is an integration of the combination of low-power wireless node sensors with the help of which the human body can be monitored and gathering critical information can be done in unreceptive environments [56]. Many body sensors together make a BSN. IoT can be implemented in a large variety of medical setup and equipment to personalize the patient systems. It plays a significant role in many applications implemented in a smart healthcare system, ranging from monitoring daily physical activities to managing chronic diseases at the end of the spectrum [57]. Figure 13.11 explains the concept of information gathering using IoT with the help of sensors.

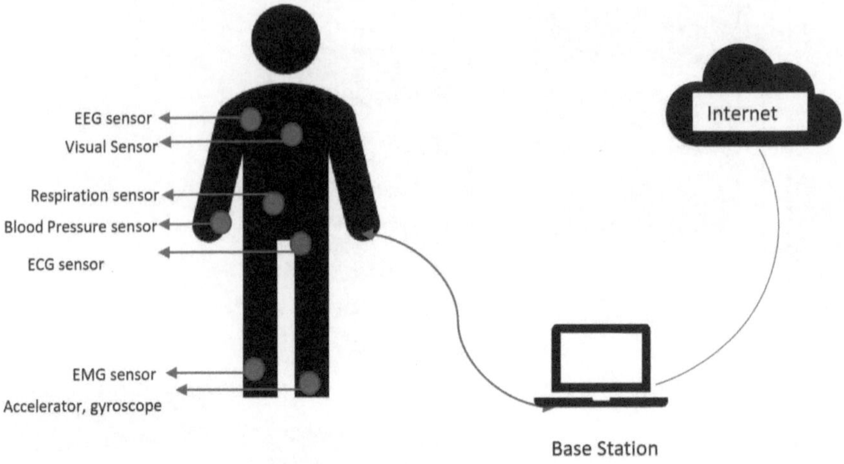

FIGURE 13.11 Body sensor network.

13.13 EVALUATION OF ALGORITHMS PERFORMANCE

On evaluating of ML model, which is a supervised and unsupervised ML algorithm, we take into consideration confusion matrix or commonly called error matrix that allows visualization of the performance executed by the algorithm [58]. The confusion matrix not only gives the errors in the dataset but also gives an understanding of the type of errors (Types I and II).

In a binary classifier confusion matrix, two classes are predicted: "positive" and "negative". In the prediction of heart attack, "positive" will mean they are predicted to have a heart attack, and "negative" would mean they are not predicted to have a heart attack. There are four labels in the confusion matrix:

i. True Positive (TP)
ii. False Positive (FP; Type I)
iii. True Negative (TN)
iv. False Negative (FN; Type II)

In Type I error, the result is a false positive, i.e. it is a False true. In Type II error, the result of the datapoint is a false negative, i.e. it is a False false.

Figure 13.12 depicts the true and false with the help of positive and negative values.

There are various parameters associated with the confusion matrix such as sensitivity, specificity, FPR, accuracy, precision, Recall, and $F1$-score to assess the performance of the ML models that have been applied, which are calculated by the following 1 to 7 formula.

$$\text{TPR (sensitivity)} = \frac{|\text{TP}|}{|\text{TP}| + |\text{FN}|} \tag{13.3}$$

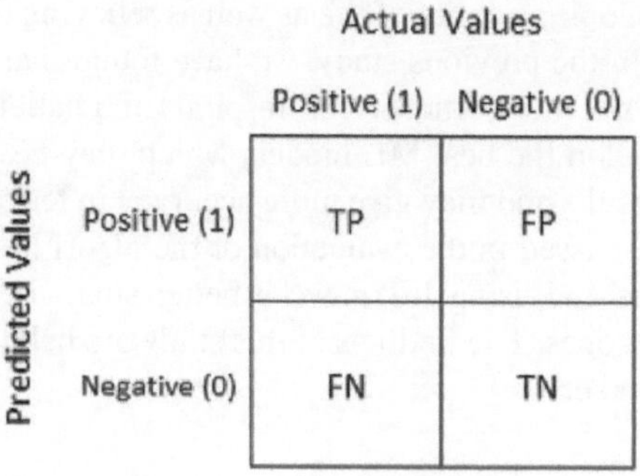

FIGURE 13.12 Confusion matrix.

$$\text{TPR}\left(\text{specificity}\right) = \frac{|\text{TN}|}{|\text{FP}| + |\text{TN}|} \tag{13.4}$$

$$\text{FPR} = \frac{|\text{FP}|}{|\text{FP}| + |\text{TN}|} \tag{13.5}$$

$$\text{Accuracy} = \frac{|\text{TP}| + |\text{TN}|}{|\text{TP}| + |\text{TN}| + |\text{FP}| + |\text{FN}|} \tag{13.6}$$

$$\text{Precision} = \frac{|\text{TP}|}{|\text{TN}| + |\text{FP}|} \tag{13.7}$$

$$\text{Recall} = \frac{|\text{TP}|}{|\text{TP}| + |\text{FN}|} \tag{13.8}$$

$$F\text{-Measure} = 2 \cdot \frac{\text{precision} \cdot \text{recall}}{\text{precision} + \text{recall}} \tag{13.9}$$

13.14 DISCUSSION

In the case of ML models, almost all the algorithms either do not provide high accuracy or sometimes provide misleading results which can be a future threat in smart healthcare research. This chapter focuses on the significant research areas of smart healthcare analytics to understand the data used and the kind of ML algorithms applied to these datasets to generate fruitful results. We also compare the results of these investigations to find insights from healthcare datasets and techniques. We also show a better prediction analysis after the comparative study in terms of metric computational performance like accuracy, recall, precision, and $F1$-score. The results of smart healthcare analytics using ML algorithms and

IoT aim to improve people's quality of life as well as relieving the issue of medical facilities shortages. In the previous study, we have found that the existing smart healthcare system is not more efficient for hospitals and patients. So, using these findings, we can develop the best ML models which may be used for achieving the best prediction results and may give more accuracy in terms of computational metrics which are discussed in the evaluation of the algorithm. Then we can say that proposed models and using IoT may be better smart healthcare system as compared to previous ones. The findings of this study are helpful for health sector analysts and policymakers.

13.15 CONCLUSION AND FUTURE SCOPE

This chapter introduced clever and powerful coronary illness forecast techniques utilizing information mining. We went through an efficient methodology for the extraction of designs from the coronary illness information distribution centers to get a proficient forecast of coronary illness. The clinical conclusion is considered a critical yet mind-boggling task that should be done correctly and effectively. The robotization of the same would be exceptionally helpful. Information mining has the potential to create an information-rich condition that can help to essentially improve the nature of clinical choices. This framework understood the discernment of information and asset and tackled the issues of rigid system asset sending. In particular, the framework initially checked and broke down the physical soundness of the shrewd dress clients considering the edge subjective figuring design and significant advancements. Health knowledge is rising rapidly in the coming years [59]. Models of public service benefits are also changing; the present state of social security offers significant uses and opportunities for adoption as key new variables. While benefits are not necessary and powerless, social insurance providers have to efficiently collect vast quantities of information from transparent tools, processes, and programs, and potentially lose a lot of dollars in wages and rewards. Large-scale information analysis will enhance care, save lives, and minimize costs by defining affiliations and bringing examples and patterns into the information.

REFERENCES

1. Milanko, S., & Jain, S. (2020). LeftRight: Quantifying strength training performance using a wearable sensor. *Smart Health*, 100115. doi:10.1016/j.smhl.2020.100115
2. Rastegari, E., & Ali, H. (2020). A bag-of-words feature engineering approach for assessing health conditions using accelerometer data. *Smart Health*, 100116. doi:10.1016/j.smhl.2020.100116
3. Tu, L., Hao, T., Bi, C., & Xing, G. (2019). BreathCoach: A smart in-home breathing training system with bio-feedback via VR Game. *Smart Health*, 100090. doi:10.1016/j.smhl.2019.100090
4. Nigam, S. (2019). Transforming healthcare through "Big Data".
5. Shen, Z., & Spruit, M. (2019). A systematic review of open source clinical software on GitHub for improving software reuse in smart healthcare. *Appl. Sci.* 9, 150.

6. Zhuhadar, L.P., & Thrasher, E. (2019). Data Analytics and its advantages for addressing the complexity of healthcare: A simulated Zika case study example. *Appl. Sci.* 9, 2208.
7. Saint-Pierre, C., Prieto, F., Herskovic, V., & Sepúlveda, M. (2019). Relationship between continuity of care in the multidisciplinary treatment of patients with diabetes and their clinical results. *Appl. Sci.* 9, 268.
8. Chui, K.T., & Lytras, M.D. (2019). A novel MOGA-SVM multinomial classification for organ inflammation detection. *Appl. Sci.* 9, 2284.
9. Chiu, P.-Y., Tang, H., Wei, C.-Y., Zhang, C., Hung, G.-U., Zhou, W. (2019). NMD-12: A new machine-learning derived screening instrument to detect mild cognitive impairment and dementia. *PLoS One.* 14, e0213430.
10. Ünver, H.M., Kökver, Y., Duman, E., Erdem, O.A. (2019). Statistical edge detection and circular hough transform for optic disk localization. *Appl. Sci.* 9, 350.
11. Polat, H., & Mehr, H.D. (2019). Classification of pulmonary CT images by using hybrid 3D-deep convolutional neural network architecture. *Appl. Sci.*, 9, 940.
12. Chang, W., Liu, Y., Xiao, Y., Xu, X., Zhou, S., Lu, X., & Cheng, Y. (2019). Probability analysis of hypertension-related symptoms based on XGBoost and clustering algorithm. *Appl. Sci.* 9, 1215.
13. Rew, J., Choi, Y.-H., Kim, H., & Hwang, E. (2019). Skin aging estimation scheme based on lifestyle and dermoscopy image analysis. *Appl. Sci.* 9, 1228.
14. Wang, J., Wang, C., & Zhang, W. (2018). Data analysis and forecasting of tuberculosis prevalence rates for smart healthcare based on a novel combination model. *Appl. Sci.* 8, 1693.
15. Lee, S., & Kim, C. (2018). Estimation of association between healthcare system efficiency and policy factors for public health. *Appl. Sci.* 8, 2674.
16. Sundaravadivel, P., Kougianos, E., Mohanty, S.P., & Ganapathiraju, M.K. (2018). Everything you wanted to know about smart health care: Evaluating the different technologies and components of the Internet of Things for better health. *IEEE Consum. Electron. Mag.* 7(1), 18–28.
17. Lytras, M.D., & Visvizi, A. (2018). Who uses smart city services and what to make of it: Toward interdisciplinary smart cities research. *Sustainability.* 10, 1998.
18. Chui, K.T., Lytras, M.D., & Visvizi, A. (2018). Energy sustainability in smart cities: Artificial intelligence, smart monitoring, and optimization of energy consumption. *Energies.* 11, 2869.
19. Lytras, M.D., Visvizi, A., Daniela, L., Sarirete, A., & De Pablos, P.O. (2018). Social networks research for sustainable smart education. *Sustainability.* 10, 2974.
20. Yan, J., Liu, J., & Tseng, F.-M. (2018). An evaluation system based on the self-organizing system framework of smart cities: A case study of smart transportation systems in China. *Technol. Forecast. Soc. Chang.*
21. Bansal, D., Chhikara, R., Khanna, K., & Gupta, P. (2018). Comparative analysis of various machine learning algorithms for detecting dementia. *Procedia Comput. Sci.* 132, 1497–1502.
22. Clarke, N., Barrick, T., & Garrard, P. (2018). Predicting dementia from written texts using feature extraction and machine learning. *Alzheimer. Dement.* 14, P538.
23. Kotsilieris, T., Pintelas, E., Livieris, I.E., & Pintelas, P. (2018). Reviewing machine learning techniques for predicting anxiety disorders; Technical Report TR01-18; University of Patras, Patras, Greece.
24. Ijaz, M.F., Alfian, G., Syafrudin, M., & Rhee, J. (2018). Hybrid prediction model for Type 2 diabetes and hypertension using DBSCAN-based outlier detection, synthetic minority over sampling technique (SMOTE), and random forest. *Appl. Sci.* 8, 1325.

25. Mun, S., Park, J., Dritschilo, A., Collins, S., Suy, S., Choi, I., & Rho, M. (2018). The prostate clinical outlook (PCO) classifier application for predicting biochemical recurrences in patients treated by stereotactic body radiation therapy (SBRT). *Appl. Sci.* 8, 1620.
26. Chen, M., Zhang, Y., Qiu, M., Guizani, N., & Hao, Y. (2018). SPHA: Smart personal health advisor based on deep analytics. *IEEE Commun. Mag.* 56(3), 164–169.
27. Menger, V., Scheepers, F., & Spruit, M. (2018). Comparing deep learning and classical machine learning approaches for predicting inpatient violence incidents. *Clin. Text. Appl. Sci.* 8, 981.
28. Harshitha, S., Narramneni, P., & Raghavee, N.S. (2018). Body sensor using Internet of Things (IoT). *ARPN J. Eng. Appl. Sci.* 13, 2916–2922.
29. Chui, K.T., Alhalabi, W., Pang, S.S.H., Pablos, P., Liu, R.W., & Zhao, M. (2017). Disease diagnosis in smart healthcare: Innovation, technologies and applications. *Sustainability.* 9, 2309.
30. François, C. (2017). Deep learning with Python.
31. Baker, S.B., Xiang, W., & Atkinson, I. (2017). Internet of Things for smart healthcare: Technologies, challenges, and opportunities. *IEEE Access.* 5, 26521–26544.
32. Abinaya, R., Varsha, V., & Hariharan, K. (2017). An intelligent street light system based on piezoelectric sensor networks. *4th International Conference on Electronics and Communication Systems (ICECS)*, pp. 138–142. IEEE, Coimbatore.
33. Kumar, S.P., Samson, V.R.R., Sai, U.B., Rao, P.M., & Eswar, K.K. (2017). Smart health monitoring system of patient through IoT. *International Conference on IoT in Social, Mobile, Analytics and Cloud (I-SMAC)*, pp. 551–556. IEEE, Palladam.
34. Liu, Y., & Choi, K.S. (2017). Using machine learning to diagnose bacterial sepsis in the critically ill patients. *Proceedings of International Conference on Smart Health*, pp. 223–233. Springer, Hong Kong.
35. Chai, Y., He, L., Mei, Q., Liu, H., & Xu, L. (2017). Deep learning through two-branch convolutional neuron network for glaucoma diagnosis. *Proceedings of International Conference on Smart Health*, pp. 191–201, Springer, Cham.
36. Zhang, J., Luo, Y., Jiang, Z., & Tang, X. (2017). Regression analysis and prediction of mini-mental state examination score in Alzheimer's disease using multi-granularity whole-brain segmentations. *Proceedings of International Conference on Smart Health*, pp. 202–213. Springer, Cham.
37. Viegas, R., Salgado, C.M., Curto, S., Carvalho, J.P., Vieira, S.M., & Finkelstein, S.N. (2017). Daily prediction of ICU readmissions using feature engineering and ensemble fuzzy modeling. *Expert Syst. Appl.* 79, 244–253.
38. Gope, P., & Hwang, T. (2016). BSN-care: A secure IoT based modern healthcare system using body sensor network. *IEEE Sen. J.* 6(5), 1368–1376.
39. Ullah, K., Shah, M.A., & Zhang, S. (2016). Effective ways to use Internet of Things in the field of medical and smart health care. *International Conference on Intelligent Systems Engineering (ICISE)*, pp. 372–379, Islamabad.
40. Mohanty, S.P., Choppali, U., & Kougianos, E. (2016). Everything you wanted to know about smart cities: The Internet of Things is the backbone. *IEEE Consum. Electron. Mag.* 5(3), 60–70.
41. Scheffler, R., Cometto, G., Tulenko, K., Bruckner, T., Liu, J., Keuffel, E.L., Preker, A., Stilwell, B., Brasileiro, J., & Campbell, J. (2016). *Health Workforce Requirements for Universal Health Coverage and the Sustainable Development Goals.* World Health Organization: Geneva.
42. Dong, Y., Wang, Q., Zhang, Q., & Yang, J. (2016). Classification of cataract fundus image based on retinal vascular information. *Proceedings of International Conference on Smart Health*, pp. 166–173, Springer, Wiesbaden.

43. Agrawal, S., & Agrawal, J. (2015). Survey on anomaly detection using data mining techniques. *Procedia Comput. Sci.* 60, 708–713.
44. Perrier, E. (2015). *Positive Disruption: Healthcare, Ageing & Participation in the Age of Technology.* The Mckell Institute: Sydney.
45. Zhu, N., Diethe, T., Camplani, M., Tao, L., Burrows, A., Twomey, N., Kaleshi, D., Mirmehdi, M., Flache, P., & Craddock, I. (2015). Bridging E-health and the Internet of Things: The sphere project. *IEEE Intell. Syst.* 30(4), 39–46.
46. Solanas, A., Patsakis, C., Conti, M., Vlachos, I., Ramos, V., Falcone, F., Postolache, O., Pérez-Martínez, P.A., Di Pietro, R., Perrea, D.N., & Martinez-Balleste, A. (2014). Smarthealth: A context-aware health paradigm within smart cities. *IEEE Commun. Mag.* 52(8), 74–81.
47. Pointer, S.C. (2014). *Australian Institute of Health and Welfare.* Australian Institute of Health and Welfare: Canberra.
48. Quinlan, J.R. (2014). *C4.5: Programs for Machine Learning.* Elsevier: Amsterdam.
49. Niewolny, D. (2013). How the Internet of Things is revolutionizing healthcare, freescale semiconductor white paper. http://freescale.com/healthcare.
50. Banaee, H., Ahmed, M.U., & Loutfi, A. (2013). Data mining for wearable sensors in health monitoring systems: A review of recent trends and challenges. *Sensors.* 13(-12), 17472–17500.
51. Fortino, G., Giannantonio, R., Gravina, R., Kuryloski, P., & Jafari, R. (2012). Enabling effective programming and flexible management of efficient body sensor network applications. *IEEE Trans. Hum. Mach. Syst.* 43(1), 115–133.
52. Hinselmann, G., Rosenbaum, L., Jahn, A., Fechner, N., Ostermann, C., & Zell, A. (2011). Large-Scale learning of structure–activity relationships using a linear support vector machine and problem-specific metrics. *J. Chem. Inf. Model.* 51(2), 203–213.
53. Visa, S., Ramsay, B., Ralescu, A.L., & Van Der Knaap, E. (2011). Confusion matrix-based feature selection. *MAICS.* 710, 120–127.
54. Szepesvári, C. (2010). Algorithms for reinforcement learning. *Synth. Lect. Artif. Intell. Mach. Learn.* 4(1), 1–103.
55. Zhu, X., & Goldberg, A.B. (2009). Introduction to semi-supervised learning. *Synth. Lect. Artif. Intell. Mach. Learn.* 3(1), 1–130.
56. Watson, P. (2008). Naive Bayes classification using 2D pharmacophore feature triplet vectors. *J. Chem. Inf. Model.* 48(1), 166–178.
57. Iyengar, S., Bonda, F.T., Gravina, R., Guerrieri, A., Fortino, G., & Sangiovanni-Vincentelli, A. (2008). A framework for creating healthcare monitoring applications using wireless body sensor networks. *ICST 3rd International Conference on Body Area Networks*, Tempe, AZ.
58. Leung, K.M. (2007). *Naive Bayesian Classifier* (pp. 123–156). Polytechnic University Department of Computer Science/Finance and Risk Engineering: New York.
59. Konovalov, D.A., Coomans, D., Deconinck, E., & Vander Heyden, Y. (2007). Benchmarking of QSAR models for blood-brain barrier permeation. *J. Chem. Inf. Model.* 47(4), 1648–1656.

Index